Transition Metal Hydrides

Transition Metal Hydrides

Robert Bau, EDITOR

University of Southern California

A symposium sponsored
by the ACS Division
of Inorganic Chemistry
at the 2nd Joint Conference
of the Chemical Institute
of Canada and the American
Chemical Society, Montreal,
May 30–June 2, 1977.

ADVANCES IN CHEMISTRY SERIES **167**

AMERICAN CHEMICAL SOCIETY

WASHINGTON, D. C.   1978

249672

546.6

17 APR 1980

TRA.

Library of Congress CIP Data
Transition metal hydrides.
    (Advances in chemistry series: 167 ISSN 0065-2393)

    Includes bibliographies and index.

    1. Transition metal hydrides—Congresses.
    I. Bau, Robert. II. American Chemical Society. Division of Inorganic Chemistry. III. Chemical Institute of Canada. IV. American Chemical Society. V. Title.

QD1.A355 no. 167 [QD172.T6] 540'.8s [546'.6]
                                                78-7327
ISBN 0-8412-0390-3 ASCMC8 167 1–411 1978

Advances in Chemistry Series

Robert F. Gould, *Editor*

Advisory Board

FOREWORD

ADVANCES IN CHEMISTRY SERIES was founded in 1949 by the American Chemical Society as an outlet for symposia and collections of data in special areas of topical interest that could not be accommodated in the Society's journals. It provides a medium for symposia that would otherwise be fragmented, their papers distributed among several journals or not published at all. Papers are reviewed critically according to ACS editorial standards and receive the careful attention and processing characteristic of ACS publications. Volumes in the ADVANCES IN CHEMISTRY SERIES maintain the integrity of the symposia on which they are based; however, verbatim reproductions of previously published papers are not accepted. Papers may include reports of research as well as reviews since symposia may embrace both types of presentation.

CONTENTS

PREFACE

R esearch in transition metal hydride chemistry is currently being conducted at a very vigorous pace. Covalent metal hydride complexes have been implicated as intermediates in homogeneous catalytic reactions, ternary metal hydrides are being actively investigated as potential hydrogen storage devices, and cluster hydride complexes have been synthesized whose cores resemble little fragments of metals with hydrogen atoms attached. Crystallographic methods are extending the limits of accuracy by which hydrogen positions can be determined, and new forms of metal–hydrogen bonding are being discovered. It was against such a backdrop that a symposium on Transition Metal Hydrides was organized at the Joint American Chemical Society/Chemical Institute of Canada Meeting in Montreal earlier last year (May 30–June 2, 1977). This book represents the proceedings of that symposium.

This volume summarizes recent results of some of the leading investigators in transition metal hydride research. Readers interested in more extensive background material are urged to consult some of the many excellent books on the subject, such as "Transition Metal Hydrides" edited by E. L. Muetterties (Marcel Dekker, Inc., New York, 1971), which covers covalent metal hydride complexes, and "Metal Hydrides" edited by W. M. Mueller, J. P. Blackledge, and G. G. Libowitz (Academic, New York, 1968), which comprehensively covers work in binary and ternary metal hydrides. Also available in the covalent metal hydride area are excellent reviews by Ginsberg [*Transition Metal Chemistry* (*1965*) **1**, 112], and Kaesz and Saillant [*Chemical Reviews* (*1972*) **72**, 231]. In this book we have not tried to be comprehensive; rather, our purpose is to update recent developments in both major areas of metal hydride research.

I wish to thank the Inorganic Division of the American Chemical Society for providing partial financial support for holding the Symposium, and above all, I would like to thank the contributing authors, without whose help this book would obviously not have come into existence.

University of Southern California ROBERT BAU
Los Angeles, California
December, 1977

Relationships between Carbonyl Hydride Clusters and Interstitial Hydrides

P. CHINI, G. LONGONI, S. MARTINENGO, and A. CERIOTTI

Istituto di Chimica Generale dell'Università e Centro del CNR, Via G. Venezian 21, 20133 Milano, Italy

Carbonyl hydride clusters based on small isolated polyhedra have not been found to contain interstitial hydride, whereas hydrogen atoms have been found to occupy partially the interstitial positions in clusters based on multihole polyhedra, such as $[Rh_{13}(CO)_{24}H_{5-n}]^{n-}$ (n = 2,3,4) and $[Ni_{12}(CO)_{21}H_{4-n}]^{n-}$ (n = 2,3). This behavior parallels that of simple metallic interstitial hydrides and suggests a strong competition between metal–metal and metal–hydrogen bonds into the hole. For hydrogen, this competition is particularly severe because of the limiting conditions imposed by its orbital's s character, although interstitial hydrides are expected to be exceptionally stable in the presence of a high number of μ and μ_3 ligands.

Potentially, four metal atoms give rise to the simplest closed polyhedron, the tetrahedron, that could accommodate a hydrogen atom in an interstitial position, and nearly 40 different examples of tetranuclear carbonyl hydride clusters are known (1, 2), as shown in Table I. However, steric crowding between the carbonyl groups will prevent the formation of a tetrahedron in tetranuclear clusters containing 16 and 15 carbonyl groups (1), and open structures have been found by x-ray analysis for $[Re_4(CO)_{15}H_4]^{2-}$ (3), $[Re_4(CO)_{16}]^{2-}$ (4), and $ReOs_3(CO)_{15}H$ (5). These open structures are also in agreement with the excess deviation from the magic number of 60 valence electrons.

With 13 carbonyls, the metal atoms can adopt the usual tetrahedral arrangement although considerable steric crowding occurs with the smallest metal atoms, as shown in the short contacts present in the dianion $[Fe_4(CO)_{13}]^{2-}$ (6). (The shortest van der Waals contacts between the carbon atoms have been found experimentally to depend strongly on the relative inclination of the carbonyls: parallel carbonyls present a minimum distance of 3.0–3.1 Å, while this distance decreases to 2.5–2.6 Å at the relative angle of 90°–110°; this effect is clearly related to the expected oval shape of the ligand (1).) Cluster opening occurs by

0-8412-0390-3/78/33-167-001/$05.00/0

Table I. The Tetranuclear Carbonyl Hydrides (*1, 2*)

$[Re_4(CO)_{16}(OCH_3)H_4]^{3-}$	$(68)^a$
$Re_2Ru_2(CO)_{16}H_2$	(64)
$MnOs_3(CO)_{16}H$	(64)
$ReOs_3(CO)_{16}H$	(64)
$[Re_4(CO)_{15}H_4]^{2-}$	$(64$ —◁$; 1t + 3\mu)$
$ReOs_3(CO)_{15}H$	$(62; ◇ ;1t)$
$[Re_4(CO)_{13}H_4]^{2-}$	$(60;T;4\mu)$
$MnOs_3(CO)_{13}H_3$	(60)
$ReOs_3(CO)_{13}H_3$	(60)
$Fe_4(CO)_{13}H_2$	(60)
$Ru_4(CO)_{13}H_2$	$(60;T;2\mu)$
$Os_4(CO)_{13}H_2$	(60)
$FeRu_3(CO)_{13}H_2$	$(60;T;2\mu)$
$FeOs_3(CO)_{13}H_2$	(60)
$FeRu_2Os(CO)_{13}H_2$	(60)
$FeRuOs_2(CO)_{13}H_2$	(60)
$[Fe_4(CO)_{13}H]^-$	$(62; ◇)$
$[Ru_4(CO)_{12}(RC_2R')H]^+$	(60)
$[Re_4(CO)_{12}H_6]^{2-}$	$(60;T;6\mu)$
$Re_4(CO)_{12}H_4$	$(56;T;4\mu_3)^b$
$Ru_4(CO)_{12}H_4$	(60)
$Ru_4(CO)_{12-n}L_nH_4$	$(60;-;4\mu)$
$Os_4(CO)_{12}H_4$	(60)
$FeRu_3(CO)_{12}H_4$	(60)
$FeOs_3(CO)_{12}H_4$	(60)
$[Ru_4(CO)_{12}H_3]^-$	(60)
$Co_2Os_2(CO)_{12}H_2$	(60)
$[Ir_4(CO)_{12}H_2]^{2+}$	(60)
$[Ir_4(CO)_8L_4H_2]^{2+}$	(60)
$FeCo_3(CO)_{12}H$	(60)
$FeCo_3(CO)_{12-n}L_nH$	$(60;T;1\mu_3)^b$
$RuCo_3(CO)_{12}H$	(60)
$OsCo_3(CO)_{12}H$	(60)
$[Fe_3Ni(CO)_{12}H]^-$	(60)
$Ir_4(CO)_{11}H_2$	(60)
$[Ir_4(CO)_{11}H]^-$	$(60;T;1t)$
$[Ir_4(CO)_{10}H_2]^{2-}$	$(60;T;2t)$
$Os_2Pt_2(CO)_8L_2H_2$	(58)
$Co_4(\eta\text{-}C_5H_5)_4H_4$	$(60;T;4\mu_3)$
$Ni_4(\eta\text{-}C_5H_5)_4H_3$	$(63;T;3\mu_3)^b$

^{a} Number of valence electrons; cluster structure from x-ray, T = tetrahedron; number and type of H bonds.
^{b} Neutron diffraction data.

protonating $[Fe_4(CO)_{13}]^{2-}$ to give the monoanion $[Fe_4(CO)_{13}H]^-$, which adopts a butterfly structure (7) (Figure 1). This particular reaction indicates that there is no space for the incoming proton either on the surface or in the interior of the cluster, and it disproves previous claims that the hydrogen was interstitial (8).

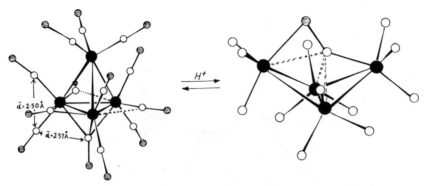

Figure 1.   Structural change associated with $[Fe_4(CO)_{13}]^{2-}$ dianion protonation

Similarly, the much-debated case of $FeCo_3(CO)_{12}H$ has been settled by a low temperature x-ray diffraction study (9) of its derivative, $FeCo_3-(CO)_9[P(OMe)_3]_3H$, and has been confirmed recently by neutron diffraction (10). This study shows that the hydrogen is under the basal plane.

Whereas edge- and face-bridging hydrogen atoms are associated mainly with metal–metal orbitals already present in the tetrahedral skeleton (*see* last section) and therefore have limited steric requirements, terminal hydrides occupy a full coordination position. It is not surprising then that the only examples of tetrahedral species with terminal hydrides contain 11 and 10 carbonyls, $[Ir_4(CO)_{11}H]^-$ and $[Ir_4(CO)_{10}H_2]^{2-}$. (The assignment for $[Ir_4(CO)_{11}H]^-$ is based on the structure determination of the analogous $[Ir_4(CO)_{11}Br]^-$ and on the similarity of the ir spectra of the two species (11).)

Penta- and hexanuclear carbonyl hydrides are much less common. Only

Table II.   Penta-, Hexa-, and Heptanuclear Carbonyl Hydrides

Carbonyl Hydride	Structure from X-ray	Number and Type of H Bonds	Ref.
$Os_5(CO)_{15}H_2$	trig. bipyr.	2μ	12
$[Os_5(CO)_{15}H]^-$	trig. bipyr.	μ	12
$Ru_6(CO)_{18}H_2$	octah.	$2\mu_3$	13
$Os_6(CO)_{18}H_2$	capped square pyramid	$\mu + \mu_3(?)$	14
$[Ru_6(CO)_{18}H]^-$	octah.	formyl type $)\tau = -6.5)^a$	15
$[Os_6(CO)_{18}H]^-$	octah.	μ_3	14
$[Co_6(CO)_{15}H]^-$	octah.(?)	$O\text{-}\text{-}\text{-}H\text{-}\text{-}\text{-}O$ $(\tau = -13.2)^b$	16, 17
$[Rh_6(CO)_{15}H]^-$	octah.^{c}	terminal	16, 18
$Os_7(CO)_{19}(C)H_2$	capped trig. prism	$\mu + \text{term.}(?)$	19

^{a} Although an interstitial position has been assigned to the $[Ru_6(CO)_{18}H]^-$ anion hydride (15), the experimental data are more compatible with a formyl situation.

^{b} Hydrogen bonding to oxygen atoms in the $[Co_6(CO)_{15}H]^-$ anion is suspected because of extremely low field position of the signal (21).

^{c} Based on the structure of the analogous $[Rh_6(CO)_{15}I]^-$ anol ^1H NMR data.

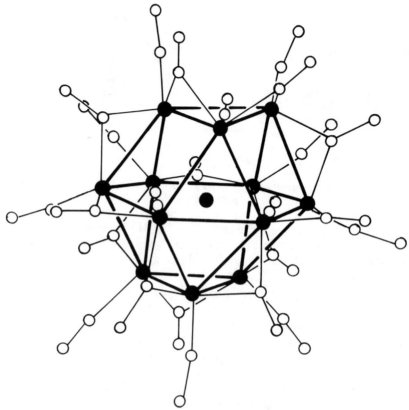

Figure 2.   The hcp structure of the anions $[Rh_{13}(CO)_{24}H_{5-n}]^{n-}$ (n = 2,3,4)

nine different species (Table II) have been reported.   A new situation where the hydride is associated with one or more carbon atoms to give a formyl situation (20) or with hydrogen bonding to the oxygen atoms (21) is recognized readily from the particular low field position of the ^{1}H NMR absorptions.   In these cases, deprotonation is facile in Lewis basic solvents and results in loss of the ^{1}H NMR signal because of exchange with the solvent.   Sometimes the ir spectrum in solution can correspond to the deprotonated species.   This situation is probably more common than the tables indicate.

In the penta-, hexa-, and heptanuclear carbonyl hydride clusters, terminal hydrides are observed only in the less crowded species, and again no evidence for the presence of interstitial hydride is found.

Our inability to observe interstitial hydride in simple, isolated carbonyl hydride polyhedra is in contrast with the situation in larger multihole polyhedra. In the twinned cube-octahedron of $[Rh_{13}(CO)_{24}H_{5-n}]^{n-}$ (n = 2,3,4) (22) in Figure 2, ^{1}H NMR spectra show conclusively the interstitial nature of the hydride (23).   In the distorted icosahedron of $[Ni_{12}(CO)_{21}H_{4-n}]^{n-}$ (n = 2,3), the same conclusion has been reached more directly from neutron diffraction data (*24*). These five hydrides have a ^{1}H NMR signal in the usual high field region (28–39 τ), supporting our previous assignment of the low field signals reported in Table II. In all these cases, the occupation of holes leads to a significant increase in the corresponding metal–metal distances, parallel to the general trend observed in the simpler μ and μ_3 hydrides (*25, 26*).

Features of Simple Metallic "Interstitial" Hydrides

The examples in Table III, show that the hydrogen atoms occupy tetrahedral holes at the beginning of the transition series. As we move along the transition series, we observe the interstitial hydride shift toward octahedral holes and the hydrides of the heavier elements become progressively unstable. Palladium is exceptional since it is the only heavy element of group VIII that gives a simple hydride. Hydride formation is accompanied in most cases by a change in metallic lattice type and in all cases by a considerable increase in metal–metal distances.

The shift from tetrahedral to octahedral interstitial position is accompanied by a considerable increase in the hydride atom's apparent dimension, which can indicate that the late transition metals are more electron rich and more prone to give up a partial negative charge in favor of the electronegative interstitial atom.

Table III. Structural Features of Representative Interstitial Hydrides (*27, 28*)[a]

Ti	TiD$_{1.97}$	Cr	CrH	Ni	NiH$_{0.8}$
hcp	ccp	bcc	hcp	ccp	ccp
2.93[b]	3.14[b]	2.49[b]	2.71[b]	2.49[b]	2.64[b]
	rH = 0.35 tet.		rH = 0.55 oct.		rH = 0.55 oct.
	$\Delta H_f^0 = -29.6$				$\Delta H_f^0 = -2.1$
	kcal mol^{-1}				
Hf	HfD$_{1.63}$			Pd	PdH$_{0.7}$
hcp	ccp			ccp	ccp
3.16[b]	3.31[b]			2.75[b]	2.89[b]
	rH = 0.37 tet.				rH = 0.60 oct.
	$\Delta H_f^0 = -33$				$\Delta H_f^0 = -10$
La	LaH$_2$[c]	LaH$_{2.92}$[c]	Ce	CeH$_2$	CeH$_{2.80}$
hcp × 2	ccp	ccp	hcp	ccp	ccp
3.742[b]	4.019[b]	3.973[b]	3.936[b]	3.958[b]	3.928[b]
	rH = 0.51 tet.	tet. + oct.		rH = 0.46 tet.	tet. + oct.

increase in R
$\longrightarrow$
~10^6 ohm-cm

increase in R
$\longrightarrow$
~10^4 ohm-cm

[a] H positions based on neutron diffraction data. In cp structures: 2 tetrah. holes ($r + R = 1.225 R$) and 1 octah. hole ($r + R = 1.414 R$).
[b] Metal–metal distances, Å.
[c] H positions based on NMR data.

In all these cases, hydride formation corresponds to partial occupation of the available holes, reminiscent of multihole polyhedra behavior. Occupation of all available holes would require a limiting stoichiometry MH_3, and would correspond to occupation of the unique hole in isolated polyhedra. This situation is known for some rare earth hydrides (*see* Table III). Significantly, transformation of the metallic dihydride to the trihydride occurs with a decrease in apparent metal–metal distances and with a large increase in resistivity. These observations indicate a salt-like character and the disappearance of metal–metal bonds (27).

Finally, the heavier group VIII transition metals' reluctance to form stable interstitial hydrides could be related to the higher values of the metal–metal interactions (1), as discussed in the following section.

Competition between Metal–Metal and Metal–Hydrogen Bonds

We have shown that: A) interstitial hydride formation is observed only with partial occupation of the available holes, B) occupation of the interstitial position in isolated polyhedra is not observed, and C) occupation of all the holes in a close-packed lattice cancels metal–metal interactions. Therefore, it seems that interstitial hydrogen can be tolerated only in a fraction of the total number of holes, and with the weakening of metal–metal interactions. This behavior indicates strong competition between metal–metal and metal–hydrogen bonds, which is unique for hydrogen because interstitial carbon can stabilize some unusual arrangements in carbonyl carbide clusters (29, 30).

Table IV. Symmetry Relation .

	s	A_{1g}	
Six metal	p_z	A_{1g}	
atoms of	$d_z{}^2$	A_{1g}	
the octa-	$d_{x^2} - y_2$		A_{2g}
hedron	d_{xy}		
	$d_{xz} + d_{yz}$		
	$p_x + p_y$		
	54 MO =	$3A_{1g}$ +	A_{2g}
Topological	center	A_{1g}	
correspond-	8 faces	A_{1g}	
dence	12 edges	A_{1g}	
		$3A_{1g}$ +	
	s	A_{1g}	
Atom in	$p_x + p_y + p_z$		
the hole	$d_z{}^2 + d_{x^2-y^2}$		
	$d_{xy} + d_{xz} + d_{yz}$		
	9 AO =	A_{1g}	

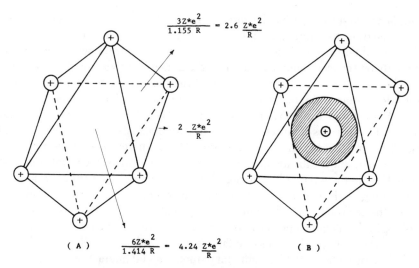

$$\frac{3Z*e^2}{1.155\,R} = 2.6\,\frac{Z*e^2}{R}$$

$$2\,\frac{Z*e^2}{R}$$

(A) $\quad\dfrac{6Z*e^2}{1.414\,R} = 4.24\,\dfrac{Z*e^2}{R}\quad$ (B)

Figure 3. Representation of the metallic field in an octahedral hole:
(A) void and (B) filled

In a simple triangulated polyhedron, an octahedron for instance, electronic density distribution among the center, the triangular faces, and the edges is expected to be related to the number and values of the Coulombic fields, Figure 3A. Introduction of an extra atom into the hole would give a new potential well (Figure 3B) mainly because the triangular and edge fields would change into

ships in an Octahedral Hole (O_h)

$$
\begin{array}{lllllllll}
 & & E_g & & & & T_{1u} & & \\
 & & E_g & & & & T_{1u} & & \\
 & & E_g & & & & T_{1u} & & \\
 & & E_g & & & & & & T_{2u} \\
A_{2u} & & E_u & & & T_{2g} & & & \\
 & & & T_{1g} & T_{2g} & T_{1u} & T_{2u} & & \\
 & & & T_{1g} & T_{2g} & T_{1u} & T_{2u} & &
\end{array}
$$

$+ \quad A_{2u} + 4E_g + E_u + 2T_{1g} + 3T_{2g} + 5T_{1u} + 3T_{2u}$

$$
\begin{array}{llll}
A_{2u} & & T_{2g} & T_{1u} \\
 & E_g & T_{2g} & T_{1u} & T_{2u} \\
A_{2u} + E_g & + & 2T_{2g} + 2T_{1u} + T_{2u}
\end{array}
$$

$$
\begin{array}{lll}
 & & T_{1u} \\
E_g & & \\
 & T_{2g} & \\
+ \quad E_g & + & T_{2g} + T_{1u}
\end{array}
$$

tetrahedral and triangular fields displaced towards the octahedron interior. At the same time, the original central field will increase because of the addition of the central positive nucleus, but its behavior in the central region will depend strongly on the electronic population of the added atom's subshell. The result will be a more homogeneous and localized field in a new internal region where the added atom will now interact with most of the metal–metal bonds.

As shown in Table IV, only 21 of the 54 MOs derived from the six metal atoms at the corners of an octahedron bear a topological correspondence (31) to the fields previously considered in Figure 3A and therefore are expected to be most responsible for the metal–metal bonds system. An extra central atom will introduce, in the intermediate region sketched in Figure 3B, further limitations depending on the symmetry of the available orbitals. Hydrogen is unique because the $1s$ orbital can only match the A_{1g} combinations; all the other MOs involved in the metallic bonds will become nonbonding in the region around the octahedron center, and their electrostatic repulsion with the A_{1g} electronic density will decrease substantially the amount of metal–metal interactions. This result can be proved by considering the symmetry products involved in the electronic interactions (32).

The insulator effect on metal–metal bonding for the interstitial hydrogen atom in an octahedral hole can be extended readily to other geometries. Table V shows that a similar effect is expected in a tetrahedral hole.

The same effect could be involved in the increase of the metal–metal distances observed in the presence of μ and μ_3 hydrogen atoms (0.05–0.40 Å) (25, 26) because in C_{2v} and C_{3v} symmetries, only the A_1 combinations are hydrogen-allowed. In these cases, however, geometric considerations indicate only partial disturbance of the metal–metal interactions.

Table V.   Symmetry Relationships in a Tetrahedral Hole (T_d)

		A_1			T_2
	s	A_1			T_2
Four metal atoms	p_z	A_1			T_2
of the	d_{z^2}	A_1			T_2
tetrahedron	$d_{x^2-y^2} + d_{xy}$		E	T_1	T_2
	$p_x + p_y$		E	T_1	T_2
	$d_{xz} + d_{yz}$		E	T_1	T_2
	36 MO =	$3A_1$	$+\quad 3E$	$+\quad 3T_1$	$+\quad 6T_2$
	center	A_1			
Topological	4 faces	A_1			T_2
correspondence	6 edges	A_1	E		T_2
		$3A_1$	$+\quad E$	$+$	$2T_2$
	s	A_1			
Atom in the	$d_{x^2-y^2} + d_{z^2}$		E		
hole	$d_{xy} + d_{xz} + d_{yz}$				T_2
	$p_x + p_y + p_z$				T_2
	9 AO =	A_1	$+\quad E$	$+$	$2T_2$

The only authentic example of an interstitial hydride in an isolated polyhedron is the noncarbonyl cluster $Nb_6I_{11}H$ (33), better described by the formula $[Nb_6(\mu_3\text{-}I)_8H](\mu\text{-}I)_{6/2}$. In the isomorphous Nb_6I_{11}, eight iodides are face bridging over the eight octahedral faces, and each apex is occupied by an iodide common to two octahedra (34, 35). In this compound, the eight μ_3 iodides are expected to counteract the interstitial hydrogen atom since these face bridging iodides will distort the main metal–metal field outside the octahedron and make it much less sensitive to the interstitial hydride. This situation is reminiscent, for instance, of the contemporary presence of both a bridging carbonyl and a bridging hydride between the same metal atoms. In this sense, other exceptionally stable interstitial hydrides are expected in the presence of many μ_3 or μ ligands.

Literature Cited

1. Chini, P., Heaton, B. T., "The Tetranuclear Carbonyl Clusters," *Top. Curr. Chem.* (1977) **71**, 1.
2. Kaesz, H. D., "Hydrido Transition-Metal Cluster Complexes," *Chem. Br.* (1973) **9**, 344.
3. Albano, V. G., Ciani, G., Freni, M., Romiti, P., *J. Organomet. Chem.* (1975) **96**, 259.
4. Bau, B., Fontal, B., Kaesz, H. D., Churchill, M. R., *J. Am. Chem. Soc.* (1967) **89**, 6374.
5. Churchill, M. R., ADV. CHEM. SER. (1978) **167**, 36.
6. Doedens, J., Dahl, L. F., *J. Am. Chem. Soc.* (1966) **88**, 4847.
7. Manassero, M., Sansoni, M., Longoni, G., *J. Chem. Soc., Chem. Commun.* (1976) 919.
8. Farmery, K., Kilner, M., Greatrex, R., Greenwood, N. N., *J. Chem. Soc. A* (1969) 2339.
9. Huie, B. T., Knobler, G. B., Kaesz, H. D., *J. Chem. Soc., Chem. Commun.* (1975) 684.
10. Koetzle, T. F., McMullan, R. K., Bau, R., Teller, R. G., Tipton, D. L., Wilson, R. D., ADV. CHEM. SER. (1978) **167**, 61.
11. Giordano, G., Canziani, F., Martinengo, S., Albano, V. G., Ciani, G., Manassero, M., Chini, P., unpublished data.
12. Eady, C. R., Guy, J. J., Johnson, B. F. G., Lewis, J., Malatesta, M. C., Sheldrick, M., *J. Chem. Soc., Chem. Commun.* (1976) 807.
13. Churchill, M. R., Wormald, J., *J. Am. Chem. Soc.* (1971) **93**, 5670.
14. McPartlin, M., Eady, C. R., Johnson, B. F. G., Lewis, J., *J. Chem. Soc., Chem. Commun.* (1976) 883.
15. Eady, C. R., Johnson, B. F. G., Lewis, J., Malatesta, M. C., Machin, P., McPartlin, M., *J. Chem. Soc., Chem. Commun.* (1976) 945.
16. Longoni, G., Martinengo, S., Chini, P., unpublished data.
17. Chini, P., *J. Chem. Soc., Chem. Commun.* (1967) 29.
18. Chini, P., Martinengo, S., Giordano, G., *Gazz. Chim. Ital.* (1972) **102**, 330.
19. Eady, C. R., Johnson, B. F. G., Lewis, J., *J. Chem. Soc., Dalton Trans.* (1977) 838.
20. Casey, C. P., Neumann, S. M., *J. Am. Chem. Soc.* (1976) **98**, 5395.
21. Mann, B. E., personal communication.
22. Albano, V. G., Ceriotti, A., Chini, P., Ciani, G., Martinengo, S., Anker, W. M., *J. Chem. Soc., Chem. Commun.* (1975) 859.
23. Martinengo, S., Heaton, B. T., Goodfellow, R. J., Chini, P., *J. Chem. Soc., Chem. Commun.* (1977) 39.
24. Dahl, L. F., Broach, R. W., Longoni, G., Chini, P., Schultz, A. J., Williams, J. M., ADV. CHEM. SER. (1978) **167**, 93.

25. Love, R. A., Chin, H. B., Koetzle, T. F., Kirtley, S. W., Whittlesey, B. R., Bau, R., *J. Am. Chem. Soc.* (1976) **98**, 4491.
26. Churchill, M. R., De Boer, B. G., Rottella, F. J., *Inorg. Chem.* (1976) **15**, 1843.
27. Wells, A. F., "Structural Inorganic Chemistry," 4th ed., Clarendon, Oxford, 1975.
28. Mackay, K. M., "Hydrides" in *Comprehensive Inorganic Chemistry,*" Vol. 1, pg. 23, Pergamon, London, 1973.
29. Chini, P., Longoni, G., Albano, V. G., *Adv. Organomet. Chem.* (1976) **14**, 285.
30. Albano, V. G., Chini, P., Ciani, G., Sansoni, M., Strumolo, D., Heaton, B. T., Martinengo, S., *J. Am. Chem. Soc.* (1976) **98**, 5027.
31. Kettle, S. F. A., *Theor. Chim. Acta* (1965) **3**, 211.
32. Fantucci, P., personal communication.
33. Simon, A., *Z. Anorg. Allg. Chem.* (1967) **355**, 311.
34. Bateman, L. R., Blount, J. F., Dahl, L. F., *J. Am. Chem. Soc.* (1966) **88**, 1082.
35. Simon, A., Schnering, H. G., Schafer, H., *Z. Anorg. Allg. Chem.* (1967) **355**, 295.

RECEIVED July 19, 1977.

Recent Neutron Diffraction Studies of Metal–Hydrogen–Metal Bonds

JEFFREY L. PETERSEN—Department of Chemistry, West Virginia University, Morgantown, WV 26506

LAWRENCE F. DAHL—Department of Chemistry, University of Wisconsin, Madison, WI 53706

JACK M. WILLIAMS—Chemistry Division, Argonne National Laboratory, Argonne, IL 60439

Single-crystal neutron diffraction studies were performed for $Mo_2(\eta^5\text{-}C_5H_5)_2(CO)_4(\mu_2\text{-}H)(\mu_2\text{-}P(CH_3)_2)$ and the $[Et_4N]^+$ and $[(Ph_3P)_2N]^+$ salts of the $[Cr_2(CO)_{10}(\mu_2\text{-}H)]^-$ monoanion to obtain detailed structural information on the M–H–M bond. Results for the molybdenum complex provided the first evidence that a bridging hydrogen atom can be located symmetrically within a bent M–H–M bond even without crystallographically imposed symmetry. Neutron diffraction studies of the $[Et_4N]^+$ salt surprisingly show that the bridging hydrogen atom is disordered between two symmetry-related sites and displaced ca. 0.3 Å from the crystallographic center of symmetry; the Cr–H–Cr bond angle is 158.9(6)°. The large, anisotropic thermal ellipsoid found for the bridging hydrogen atom in the $[(Ph_3P)_2N]^+$ salt prevented an unambiguous interpretation of the Cr–H–Cr geometry at room temperature.

As a result of the recognized role of transition metal hydrides as reactive intermediates or catalysts in a broad spectrum of chemical reactions such as hydroformylation, olefin isomerization, and hydrogenation, transition metal hydride chemistry has developed rapidly in the past decade (1). Despite the increased interest in this area, detailed structural information about the nature of hydrogen bonding to transition metals has been rather limited. This paucity of information primarily arises since, until recently, x-ray diffraction has been used mainly to determine hydrogen positions either indirectly from stereochemical considerations of the ligand disposition about the metals or directly from weak peaks of electron density in difference Fourier maps. The inherent limi-

0-8412-0390-3/78/33-167-011/$05.00/0

tations of x-ray diffraction methods prevent us from giving a precise representation of the hydrogen bonding in these systems.

To extend our fundamental understanding of the stereochemical aspects of metal–hydrogen interactions, a collaborative effort to examine the structures of a carefully selected number of metal hydrides via single-crystal neutron diffraction methods has been initiated at Argonne National Laboratory. Since the neutron-scattering cross section for the hydrogen atom is of the same order of magnitude as those for the heavier transition metals, neutron diffraction provides the opportunity to resolve not only the nuclear position but also to obtain meaningful thermal parameters for each hydrogen atom in complexes containing metal–hydrogen bonds.

In particular, we have been interested in metal hydrides in which the hydrogen atom occupies a bridging position between at least two metal atoms. To date, two hydride systems containing metal–hydrogen–metal bonds have been studied by neutron diffraction at Argonne. $Mo_2(\eta^5\text{-}C_5H_5)_2(CO)_4(\mu_2\text{-}H)(\mu_2\text{-}P(CH_3)_2)$ and the $[Cr_2(CO)_{10}(\mu_2\text{-}H)]^-$ monoanion were selected because they represent two possible geometries for the bridging hydrogen atom in a metal–hydrogen–metal bond. From earlier x-ray work, the Mo–H–Mo bond in the former complex (2) was assumed to be bent whereas the Cr–H–Cr bond in the latter (3, 4) was presumed to be the first example of a linearly protonated metal–metal bond. Suitable crystals of both materials were obtained from saturated solutions by slow solvent evaporation, and the necessary neutron diffraction data were measured with the automated neutron diffractometer located at the CP-5 reactor at Argonne National Laboratory. Computer analysis of the data was carried out using the IBM 370/195 system at Argonne.

Neutron Diffraction Study of $Mo_2(\eta^5\text{-}C_5H_5)_2(CO)_4(\mu_2\text{-}H)(\mu_2\text{-}P(CH_3)_2)$

A single-crystal neutron diffraction investigation of $Mo_2(\eta^5\text{-}C_5H_5)_2(CO)_4$-$(\mu_2\text{-}H)(\mu_2\text{-}P(CH_3)_2)$ was performed to determine precisely the location and thermal motion for the bridging hydrogen atom. Previous x-ray work by Doedens and Dahl (2) provided the first evidence for the existence of symmetric three-center, electron pair metal–hydrogen–metal bond in this particular hydride complex. Assuming that the bridging hydrogen atom occupies a distinct coordination site, they concluded that because of the equivalent electronic environment of the two molybdenum atoms, the hydrogen atom was associated equally with both molybdenum atoms in a bent Mo–H–Mo bond. Although the two molybdenum atoms are chemically equivalent, the fact that they are not crystallographically related avoids the intrinsic uncertainty problem that is encountered when the hydrogen atom is constrained by crystallographically imposed site symmetry. This problem has been discussed previously by Ibers (5, 6, 7) for several different hydrogen-bonded systems and makes it difficult to distinguish between a truly symmetric single-well potential as opposed to a statistically symmetric (or time-averaged) double-well potential for a crystallo-

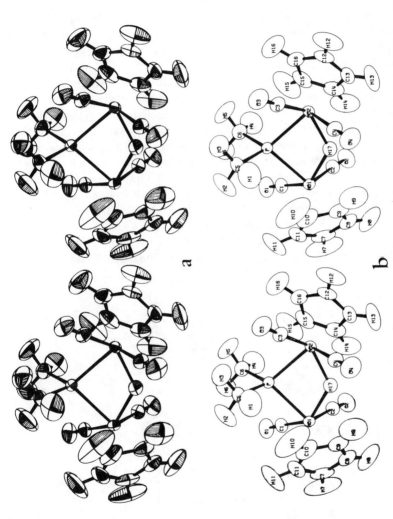

Figure 1. Stereoscopic drawings of the molecular configuration of $Mo_2(\eta^5\text{-}C_5H_5)_2(CO)_4(\mu_2\text{-}H)(\mu_2\text{-}P(CH_3)_2)$ showing: (a) thermal ellipsoids of nuclear motion for all atoms scaled to enclosed 50% probability; (b) the atom labeling. The entire molecule possesses a pseudotwofold axis passing through the bridging hydrogen and phosphorus atoms.

graphically symmetric hydrogen atom. However, under more favorable conditions such as those that exist for the bridging hydrogen atom in $Mo_2(\eta^5\text{-}C_5H_5)_2(CO)_4(\mu_2\text{-}H)(\mu_2\text{-}P(CH_3)_2)$, a neutron diffraction study often gives the opportunity to differentiate between a symmetric single-well and a statistically symmetric double-well hydrogen bond in the absence of any imposed site symmetry. Since the x-ray study (2) shows no indication of any crystal order–disorder phenomenon associated with the nonhydrogen atoms, the molecular structure is well behaved. Consequently, one expects that the neutron-determined hydrogen position and thermal parameters should provide a clear picture of the nature of the Mo–H–Mo bond.

The molecular configuration for $Mo_2(\eta^5\text{-}C_5H_5)_2(CO)_4(\mu_2\text{-}H)(\mu_2\text{-}P(CH_3)_2)$, as determined by our neutron diffraction study (8), is illustrated stereographically in Figure 1. The determined position of the bridging hydrogen atom, H(17), demonstrates that the hydrogen atom occupies a distinct coordination site. The molecular configuration possesses a pseudotwofold axis that passes through the bridging phosphorus and hydrogen atoms that link together the two $Mo(\eta^5\text{-}C_5H_5)(CO)_2$ moieties. The molecular packing of the four molecules in a nonstandard $C\bar{1}$ triclinic cell, that is shown stereographically in Figure 2, indicates the absence of any abnormal intermolecular interactions which possibly could affect the environment of the bridging hydrogen atom.

A least-squares plane through the four atoms in the $Mo_2(\mu_2\text{-}H)(\mu_2\text{-}P)$ core, as depicted in Figure 3, shows that they are coplanar within 0.02 Å. The bridging

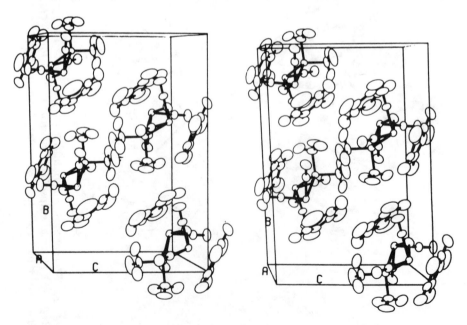

Figure 2. Stereoscopic view of the four $Mo_2(\eta^5\text{-}C_5H_5)_2(CO)_4$-$(\mu_2\text{-}H)(\mu_2\text{-}P(CH_3)_2)$ molecules in the triclinic unit cell of symmetry $C\bar{1}$

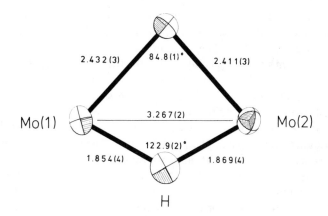

*Figure 3.   The $Mo_2(\mu_2\text{-}H)(\mu_2\text{-}P)$ plane with appropriate
internuclear separations and bond angles and their
corresponding standard deviations*

hydrogen atom within the bent Mo–H–Mo system of angle 122.9(2)° is essentially equidistant from the two molybdenum atoms with Mo(1)–H(17) and Mo(2)–H(17) distances of 1.851(4) and 1.869(4) Å, respectively. The small and statistically borderline difference of 0.019(7) Å (i.e., 2.7σ) between the two Mo–H distances can be attributed to a slight asymmetry in the potential energy surface associated with the bridging hydrogen nucleus. This asymmetry might result from the inseparable electronic and steric effects imposed by the small asymmetry of the bridging phosphorus atom, as manifested by Mo(1)–P and Mo(2)–P bond distances of 2.432(3) and 2.411(3) Å, respectively, that differ by 0.021(4) Å (i.e., 5.3 σ). In general, one should expect that the asymmetry observed for a M–H–M bond will strongly depend on the asymmetric effects dictated by the metal atoms and other ligands in the molecule. The slightly longer Mo(1)–Mo(2) distance of 3.267(2) Å, when compared with the corresponding value of 3.235(1) Å for unprotonated $Mo_2(\eta^5\text{-}C_5H_5)_2(CO)_6$, is consistent with the findings from other structural studies (9, 10, 11, 12, 13) which have shown that bridging hydrogen atoms involving analogous bent M–H–M systems produce comparably longer metal–metal bond lengths. The markedly smaller C_5H_5 (centroid)-Mo–H angle of 110.5° (av) compared with the C_5H_5(centroid)–Mo–CO angles of 119.8° (av) and 122.5° (av) and the C_5H_5(centroid)–Mo–P angle of 124.9° (av) is consistent with a previously discussed observation by Frenz and Ibers (14), who have pointed out that ligands adjacent to a hydride ligand are displaced toward the hydrogen atom.

By examining the size, shape, and orientation of the thermal ellipsoid associated with the bridging hydrogen atom, we concluded that the hydrogen atom in the bent Mo–H–Mo system is described preferably as an effectively symmetric atom oscillating around a single equilibrium point rather than being randomly distributed between two equilibrium positions in the crystal lattice. The thermal motion of the bridging hydrogen atom compares in magnitude with that for the molybdenum and phosphorus atoms in the $Mo_2(\mu_2\text{-}H)(\mu_2\text{-}P)$ core and indicates a relatively rigid structure. The low thermal motion supports the premise of a single equilibrium site since a composite of the thermal motion around two reasonably separated equilibrium sites would produce a much larger and elongated thermal ellipsoid. Information regarding the shape and orientation of the thermal ellipsoid of the bridging hydrogen atom can be extracted by comparing the root-mean-square components and directions of the thermal displacement for the bridging hydrogen and phosphorus atoms. The thermal ellipsoids of both the hydrogen and phosphorus atoms are positioned symmetrically with respect to the $Mo_2(\mu_2\text{-}H)(\mu_2\text{-}P)$ fragment, with the largest principal axis component essentially perpendicular to the $Mo_2(\mu_2\text{-}H)(\mu_2\text{-}P)$ plane (viz., within $2(3)°$ for the hydrogen atom and $16(4)°$ for the phosphorus atom). Within the plane, the thermal ellipsoids of the four atoms are essentially isotropic within experimental error. Despite the greater mass and additional coordination to two methyl groups, the estimated isotropic thermal displacement component of 0.156 Å for the phosphorus atom within the $Mo_2(\mu_2\text{-}H)(\mu_2\text{-}P)$ plane is only slightly smaller than the corresponding value of 0.192 Å for the hydrogen atom. If the hydrogen atom were disordered between two equilibrium sites in a symmetric double-minimum potential well, the rms thermal displacement components should indicate a high degree of thermal anisotropy in the plane. Moreover, the maximum thermal displacement for the hydrogen and phosphorus atoms of 0.270(5) Å and 0.185(4) Å, respectively, (perpendicular to the plane) are much greater than the estimated isotropic thermal cross sections in the plane. Consequently, a detailed analysis of the thermal ellipsoids for the bridging hydrogen and phosphorus atoms in $Mo(\eta^5\text{-}C_5H_5)_2(CO)_4(\mu_2\text{-}H)(\mu_2\text{-}P(CH_3)_2)$ provides definitive evidence that the bridging hydrogen atom resides in a symmetric, single-minimum potential well and is vibrating with its primary direction of thermal motion normal to the $Mo_2(\mu_2\text{-}H)(\mu_2\text{-}P)$ plane.

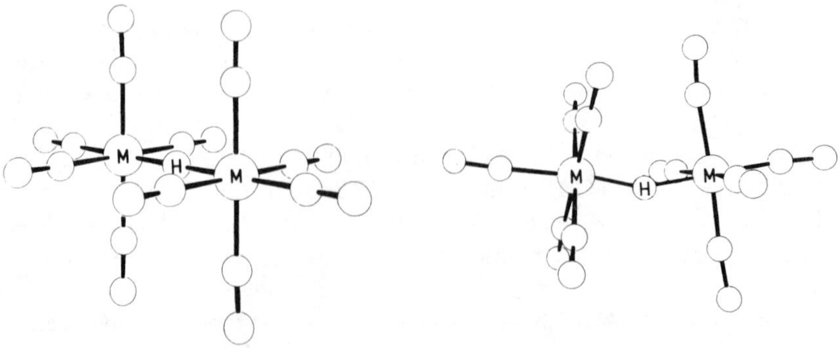

Figure 4. The two possible geometries that have been observed for the $[M_2(CO)_{10}(\mu_2\text{-}H)]^-$ *monoanion—the linear-eclipsed* (left) *and bent-staggered* (right) *configurations*

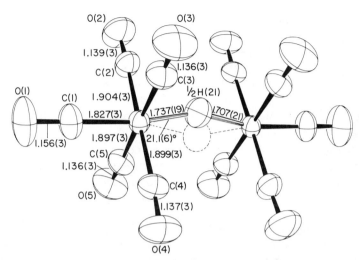

Figure 5. Architecture of the $[Cr_2(CO)_{10}(\mu_2\text{-}H)]^-$ *monoanion of crystallographic* C_i-*1 symmetry showing the approximate* D_{4h} *geometry of the metal carbonyl framework and the two centrosymmetrically related (half-weighted) sites of the bridging hydrogen atom in the bent Cr–H–Cr molecular fragment. Internuclear distances and bond angles are given with their estimated standard deviations.*

Neutron Diffraction Studies of the $[Cr_2(CO)_{10}(\mu_2\text{-}H)]^-$ Monoanion in the Tetraethylammonium and Bis(triphenylphosphine)iminium Salts

The outcome of several x-ray diffraction studies (*3, 4, 15*) of the $[M_2(CO)_{10}(\mu_2\text{-}H)]^-$ monoanion (M = Cr, Mo, W) has indicated that the overall metal carbonyl framework can adopt either a bent–staggered or linear–eclipsed configuration as shown in Figure 4. In particular, from an earlier x-ray diffraction study, (*3, 4*), the Cr–H–Cr bond in the tetraethylammonium salt of $[Cr_2(CO)_{10}(\mu_2\text{-}H)]^-$ was presumed to be the first example of a linearly protonated metal–metal bond. Although the intrinsic limitations of the x-ray photographic data did not allow direct determination of the bridging hydrogen position in the monoanion, its proposed location along the metal–metal axis was based on the idealized D_{4h} geometry of the nonhydrogen atoms, the crystallographic C_i symmetry imposed on the monoanion, and the assumption that the metal-coordinated hydrogen atom occupies a regular octahedral metal coordination site. Within these considerations, the hydrogen atom could lie on the crystallographic center of symmetry equidistant from the symmetry-related metal atoms or could be statistically averaged because of a random distribution over equivalent sites displaced from the center of symmetry along the metal–metal axis. Neutron diffraction was used to investigate the presumed linearity of the Cr–H–Cr linkage and to determine whether the bridging hydrogen atom resides in a symmetric single-well or symmetric double-well potential.

The neutron diffraction-determined molecular configuration (16) of the $[Cr_2(CO)_{10}(\mu_2\text{-}H)]^-$ monoanion for the tetraethylammonium salt is shown in Figure 5. Rather surprisingly, at room temperature the Cr–H–Cr bond is non-linear, with the bridging hydrogen atom randomly disordered between two centrosymmetrically related sites. This bent, disordered structure results in a significant displacement of the two half-weighted hydrogen positions by ca. 0.3 Å from the crystallographic center of symmetry and thereby produces a nonlinear Cr–H–Cr bond angle of 158.9(6)°. A symmetric disposition of the bridging hydrogen atom around the two chromium atoms is indicated from the experimentally equivalent Cr–H internuclear separations of 1.707(21) and 1.737(19) Å. When compared with the other four carbonyl groups (with average Cr–C and C–O internuclear separations of 1.898(3) and 1.137(1) Å, respectively), the carbonyl group trans to the bridging hydrogen atom reflects considerable π-backbonding with the chromium atom, as indicated by the significantly shorter Cr–C(1) distance of 1.827(3) Å and longer C(1)–O(1) separation of 1.156(3) Å. Although the differences between the axial vs. equatorial metal–carbon and C–O distances are more pronounced in this case, similar results (10, 11, 17–21) have been observed for octahedrally coordinated metal carbonyls of the type M(CO)$_5$X, where X is a poorer electron acceptor than the CO ligand.

Another view of the $[Cr_2(CO)_{10}(\mu_2\text{-}H)]^-$ monoanion is illustrated stereo-graphically in Figure 6. This view, approximately along the metal–metal axis, reveals that the two coplanar, bent Cr–H–Cr moieties are staggered with respect to the eclipsed array of equatorial carbonyl ligands. The plane of each of the two coplanar Cr–H–Cr fragments is oriented at angles of 44.2° and 45.9°, with respect to the two perpendicular mean planes that pass through two chromium atoms and two axial and four equatorial carbonyl ligands. As was found previously (8) for the bridging hydrogen atom in $Mo_2(\eta^5\text{-}C_5H_5)_2(CO)_4(\mu_2\text{-}H)(\mu_2\text{-}P(CH_3)_2)$, the maximum rms component of thermal displacement (viz., $\mu(3) = 0.441(11)$ Å) for the bridging hydrogen atom in the $[Cr_2(CO)_{10}(\mu_2\text{-}H)]^-$ monoanion also is directed normal to the metal–hydrogen–metal plane (viz., 85(4)°). Although the hydrogen disorder does not sufficiently alter the overall pseudo D_{4h} geometry of the metal carbonyl framework, an examination of the sizes, shapes, and orientations of the nuclear thermal ellipsoids of the axial carbonyl atoms reveals that the thermal anisotropy observed for the nonhydrogen structure most likely results from a crystal disorder involving the near superposition of two slightly bent, identical $[Cr_2(CO)_{10}(\mu_2\text{-}H)]^-$ structures of idealized C_{2v} geometry. In contrast to the direction of $\mu(3)$ for the bridging hydrogen atom, the maximum rms components of thermal displacement for the axial carbon and oxygen atoms (viz., $\mu(3) = 0.287(3)$ Å and $\mu(3) = 0.391(4)$ Å, respectively) are essentially directed parallel to the Cr–H–Cr plane. (The acute angles between the directions of $\mu(3)$ for the bridging hydrogen atom and $\mu(3)$ for the axial carbon and oxygen atoms are 86(4)° and 87(4)°, respectively.) Although a corresponding twofold disorder

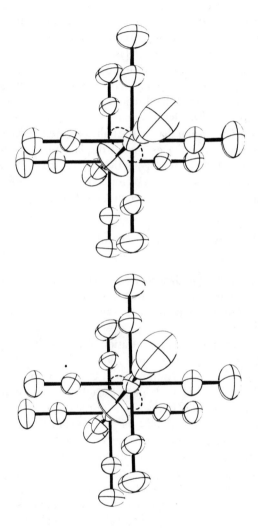

Figure 6. Stereoscopic drawing of the $[Cr_2(CO)_{10}(\mu_2\text{-}H)]^-$ monoanion as viewed approximately along the Cr–Cr axis. The Cr–H–Cr plane is staggered with respect to the eclipsed array of equatorial carbonyls. The thermal anisotropy of the bridging hydrogen atom is primarily directed normal of the Cr–H–Cr plane.

of the single independent, axial carbonyl group is not resolved, the observed thermal anisotropy may be explained as a composite of two half-carbon and two half-oxygen atoms, each displaced ca. 0.08 and 0.15 Å from their respective neutron-determined positions. These estimated displacements were calculated by assuming that the two half atoms for the axial carbon and oxygen atoms can be represented with an isotropic thermal displacement of 0.20 and 0.24 Å, respectively, which are approximately equal to the mean value of the thermal displacements normal to the corresponding maximum displacement, $\mu(3)$. An idealized representation of this composite model is illustrated in Figure 7 for the atoms contained in the Cr–H–Cr plane. Since the axial O–C–Cr bond angle is nearly 180° (i.e., 179.0(2)°), the overlap region in the bond (determined from the extrapolated intersection points of the directed vectors from the two sets of opposite, half-weighted axial carbonyls) is estimated to be nearer to the centrosymmetric midpoint of the Cr–Cr line than to the half-weighted hydrogen atom positions. This result arises from the estimated displacements of the half-weighted carbon and oxygen atoms being smaller than the ca. 0.3 Å displacement of the half-weighted, bridging hydrogen atom from the crystallographic center of symmetry and is in accord with the three-center, electron pair Cr–H–Cr in-

teraction being a closed-type $M \overset{\overset{\textstyle H}{|}}{\diagdown} M$ bond. The results provide

further support that the bridging hydrogen atoms in bent M–H–M fragments are unlike terminal hydrogen atoms since they do not occupy regular metal coordination sites. For example, Bau and co-workers (22, 23, 24) have shown from their neutron diffraction studies of $W_2(CO)_9(NO)(\mu_2\text{-H})$ and $W_2(CO)_8$-$(NO)(P(OCH_3)_3)(\mu_2\text{-H})$ that predicting the location of the bridging hydrogen atom based on an idealized octahedral geometry leads to an overestimation of the W–H–W bond angle and an underestimation of the W–H bond lengths.

The observation of a bent Cr–H–Cr bond in the tetraethylammonium salt without an accompanying substantial deformation of the linear architecture of the nonhydrogen atoms in the $[Cr_2(CO)_{10}(\mu_2\text{-H})]^-$ monoanion reflects the inherent flexibility of the bond. The deformability of the $[M_2(CO)_{10}(\mu_2\text{-H})]^-$ monoanion species to adopt an appreciably bent, staggered carbonyl structure was first reported by Bau and co-workers (23) from neutron diffraction studies of two crystalline modifications of the electronically equivalent, neutral $W_2(CO)_9(NO)(\mu_2\text{-H})$ molecule. Subsequent x-ray diffraction studies (15) of the analogous $[W_2(CO)_{10}(\mu_2\text{-H})]^-$ monoanion found that the nonhydrogen backbone can have either an appreciably bent structure for the bis(triphenylphosphine)-iminium salt or a linear structure for the tetraethylammonium salt, with the W–W separation 0.11 Å less in the bent form. Crystal packing forces probably were responsible (15) for the different molecular configurations of the monoanion in the two lattices. In solution, however, all known salts of the $[W_2(CO)_{10}(\mu_2\text{-H})]^-$ monoanion exhibit the same three-band carbonyl ir absorption spectrum characteristic of the eclipsed carbonyl geometry which suggests an inherent stability of the eclipsed structure for the $[M_2(CO)_{10}(\mu_2\text{-H})]^-$ species in solution.

For the $[M_2(CO)_{10}(\mu_2\text{-H})]^-$ monoanion, a qualitative understanding of the bonding can be rationalized in terms of a three-center molecular orbital representation involving two σ-type metal hybrid orbitals and the hydrogen $1s$ atomic orbital. For a linear, nonhydrogen backbone with the metal hybrids of primarily $d_z{}^2$ and p_z character directed along the M–M internuclear vector, a small perpendicular displacement of the bridging hydrogen atom would be expected to destabilize the a_1 bonding molecular orbital (MO) because of the subsequent reduction of the orbital overlap between the metal and hydrogen centers. However, as the M–H–M fragment moves from a linear to an appreciably bent configuration, an accompanying decrease in the metal–metal separation should provide better metal–metal overlap. This produces a net energy stabilization

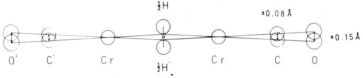

Figure 7. Idealized model of the thermal anisotropy of the atoms located in the Cr–H–Cr plane based on the representation of the nuclear thermal ellipsoids of axial carbonyl atoms as composites of two half-carbon and half-oxygen atoms, each displaced from their neutron-determined positions by ca. 0.08 and 0.15 Å, respectively. The estimated overlap region in the bond lies near the centrosymmetric midpoint of the Cr–Cr line in accord with a closed-type,

$$\underset{Cr\text{———}Cr}{\overset{\overset{\textstyle H}{|}}{}} \quad bond.$$

of the a_1 bonding MO relative to that for the linear Cr–H–Cr bond. A subtle balance of electronic effects must be present in the $[Cr_2(CO)_{10}(\mu_2\text{-H})]^-$ anion since the Cr–H–Cr bond can bend without an accompanying twisting distortion of the entire linear, eclipsed carbonyl structure toward an appreciably bent, staggered configuration.

In the solid state, crystal packing forces should be considered since they can appreciably affect the crystal structure. For example, as previously mentioned, the replacement of the tetraethylammonium cation by the bis(triphenylphosphine)iminium cation for the $[W_2(CO)_{10}(\mu_2\text{-H})]^-$ species dramatically changes the solid-state structure of the monoanion. For further insight into the possible influence of packing forces on the nature of the Cr–H–Cr bond in the $[Cr_2(CO)_{10}(\mu_2\text{-H})]^-$ monoanion, a neutron diffraction study of the corresponding bis(triphenylphosphine)iminium salt, that has been shown by x-ray diffraction methods (25) to possess a linear D_{4h}, nonhydrogen backbone for the monoanion (essentially identical to that in the tetraethylammonium salt), has been performed. A complete set of neutron data was collected out to $2\theta = 70°$ at $T = 22° \pm 2°C$. Because of the high percentage of hydrogen in the unit cell and the relatively high overall molecular thermal motion, reflections measured at higher 2θ values were generally quite weak. A total of 2357 independent reflections with $F_0^2 \geq 0$ was used to refine the structure with a final data-to-parameter ratio of 5.7:1. Our analysis of the neutron data for the bis(triphenylphosphine)iminium salt is illustrated in Figure 8, which depicts two different views of the monoanion, and in Figure 9, which represents a stereographic drawing of the packing in the unit cell. In contrast to the twofold, disordered structure found for the bridging hydrogen atom of the tetraethylammonium salt, only one position corresponding to the bridging hydrogen atom was located in this case, with the maximum nuclear density being at the crystallographic center of symmetry rather than being displaced from it. As shown in these two orientations of the $[Cr_2(CO)_{10}(\mu_2\text{-}H)]^-$ monoanion, the large, highly anisotropic, thermal ellipsoid of the bridging hy-

(a)

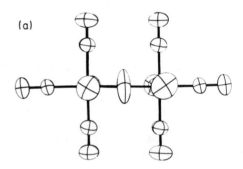

(b)

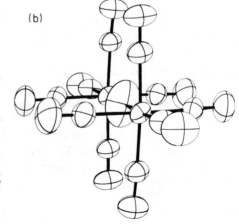

Figure 8.   The molecular structure of the $[Cr_2(CO)_{10}(\mu_2\text{-}H)]^-$ anion for the bis(triphenylphosphine)-iminium salt showing: (a) a view normal to the Cr–Cr axis; (b) a view looking down the Cr–Cr axis. The Cr–Cr internuclear separation is 3.349(13) Å. The thermal ellipsoids of nuclear motion for all atoms are scaled to enclosed 50% probability.

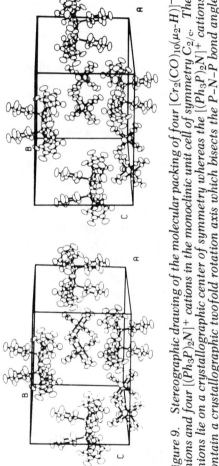

Figure 9. Stereographic drawing of the molecular packing of four $[Cr_2(CO)_{10}(\mu_2\text{-}H)]^-$ anions and four $[(Ph_3P)_2N]^+$ cations in the monoclinic unit cell of symmetry $C_{2/c}$. The anions lie on a crystallographic center of symmetry whereas the $[(Ph_3P)_2N]^+$ cations contain a crystallographic twofold rotation axis which bisects the P–N–P bond angle of 154.8(4)°.

drogen atom indicates considerable vibrational motion normal to the Cr–Cr axis. Although the Cr–H–Cr bond appears to be linear, the large amplitudes of the two rms components of thermal displacement normal to the Cr–Cr vector (i.e., $\mu(2) = 0.422(38)$ Å and $\mu(3) = 0.532(25)$ Å) do not rule out the strong possibility of a radially disordered, bridging hydrogen atom. Therefore, within the limitations of our room-temperature neutron data, one cannot differentiate a truly linear bond from a composite of slightly bent structures. The axial carbonyl ligands for the bis(triphenylphosphine)iminium salt do not clearly indicate a disordered structure, as in tetraethylammonium salt. However, the smaller axial C–O bond distances of 1.140(7) Å in the $[(Ph_3P)_2N]^+$ salt do reflect a small degree of thermal shortening. Attempts to produce a slightly bent Cr–H–Cr structure by disordering the hydrogen position off the crystallographic center of symmetry were not successful. Although these differences are not sufficient to substantiate a linear structure for the Cr–H–Cr bond in the $[(Ph_3P)_2N]^+$ salt, they suggest that crystal packing forces are sufficiently different in the two crystalline environments to affect the solid state structure of the monoanion.

To investigate the general applicability of laser Raman spectroscopy to structural studies of transition metal hydrides, Shriver and co-workers (26) performed low-temperature (10°K) studies on deuterated salts of the $[M_2(CO)_{10}(\mu_2\text{-}H)]^-$ monoanion. Their work indicated a dependence between the frequency of the asymmetric symmetric mode for the M–D–M group and the M–D–M bond angle. For an appreciably bent bond, such as that found in the $[(Ph_3P)_2N]^+$ salt of $[W_2(CO)_{10}(\mu_2\text{-}D)]^-$, a significant decrease in the asymmetric stretching frequency is observed. However, for more nearly linear structures, such as those found for the $[Cr_2(CO)_{10}(\mu_2\text{-}D)]^-$ monoanion, a force-field analysis shows that the vibrational frequency is quite insensitive to small bond angle deformations. The observed vibrational frequencies measured at 10°K for the $[Et_4N]^+$ and $[(Ph_3P)_2N]^+$ salts of 1274 and 1270 cm^{-1}, respectively, indicate that the two salts possess a similar Cr–D–Cr structure at this temperature.

From our neutron diffraction of the $[Cr_2(CO)_{10}(\mu_2\text{-}H)]^-$ monoanion, we see that although the bridging hydrogen atom is symmetrically disposed between the two chromium atoms, the crystalline environment influences the solid state structure of the Cr–H–Cr bond. For the tetraethylammonium salt, a disordered, slightly bent Cr–H–Cr structure has been resolved while for the bis(triphenylphosphine)iminium salt, the highly anisotropic nature of the thermal motion of the bridging hydrogen atom prohibits an unambiguous interpretation of the Cr–H–Cr geometry at room temperature. In either case, however, the potential energy surface associated with the hydrogen nucleus appears to be sufficiently shallow to facilitate an easy bending of the Cr–H–Cr fragment without appreciably perturbing the eclipsed arrangement of the carbonyl ligands. To better understand the nature of the metal–hydrogen–metal bond in this intriguing series of complexes, further theoretical and experimental studies based upon our structural findings are in progress.

Acknowledgment

This work was performed under the auspices of the Division of Basic Energy Sciences of the Department of Energy. We are pleased to acknowledge partial support of the initial stages of this research by a grant to L. F. D. from Research Corporation. The support of the National Science Foundation under Grants CHE-76-07409 and GP-19175X is gratefully acknowledged by J. M. W. and L. F. D. We also express our thanks to Jacques Roziere (Université des Sciences et Techniques du Languedoc, Cedex, France), Robert R. Stewart, Jr. (Miami University), and Paul L. Johnson (Argonne) for their assistance. Computer time for refining the bis(triphenylphosphine)iminium salt was provided to J. L. P. by the West Virginia Network for Educational Telecomputing.

Literature Cited

1. Kaesz, H. D., Saillant, R. B., *Chem. Rev.* (1972) **72**, 231.
2. Doedens, R. J., Dahl, L. F., *J. Am. Chem. Soc.* (1965) **37**, 2576.
3. Handy, L. B., Treichel, P. M., Dahl, L. F., Hayter, R. G., *J. Am. Chem. Soc.* (1966) **88**, 366.
4. Handy, L. B., Ruff, J. K., Dahl, L. F., *J. Am. Chem. Soc.* (1970) **92**, 7312.
5. McGaw, B. L., Ibers, J. A., *J. Chem. Phys.* (1963) **39**, 2677.
6. Ibers, J. A., *J. Chem. Phys.* (1964) **41**, 25.
7. Doedens, R. J., Robinson, W. T., Ibers, J. A., *J. Am. Chem. Soc.* (1967) **89**, 4323.
8. Petersen, J. L., Dahl, L. F., Williams, J. M., *J. Am. Chem. Soc.* (1974) **96**, 6610.
9. Kaesz, H. D., Fellmann, W., Wilkes, G. R., Dahl, L. F., *J. Am. Chem. Soc.* (1964) **87**, 2753.
10. Kaesz, H. D., Bau, R., Churchill, M. R., *J. Am. Chem. Soc.* (1967) **89**, 2775.
11. Churchill, M. R., Bau, R., *Inorg. Chem.* (1967) **6**, 2086.
12. Churchill, M. R., Bezman, S. A., Osborn, J. A., Wormald, J., *Inorg. Chem.* (1972) **11**, 1818.
13. Bau, R., Kirtley, S. W., Sorrell, T. N., Winarko, S., *J. Am. Chem. Soc.* (1974) **96**, 988, and references cited therein.
14. Frenz, B. A., Ibers, J. A., "Transition Metal Hydrides," E. L. Muetterties, Ed., pp. 33–74, Marcel Dekker, New York, 1971.
15. Wilson, R. D., Graham, S. A., Bau, R., *J. Organomet. Chem.* (1975) **91**, C49.
16. Roziere, J., Williams, J. M., Stewart, R. P., Jr., Petersen, J. L., Dahl, L. F., *J. Am. Chem. Soc.* (1977) **99**, 4497.
17. Dahl, L. F., Rundle, R. E., *Acta Crystallogr.* (1963) **16**, 419.
18. Bailey, M. F., Dahl, L. F., *Inorg. Chem.* (1965) **4**, 1140.
19. La Placa, S. J., Hamilton, W. C., Ibers, J. A., *Inorg. Chem.* (1964) **3**, 1491.
20. Ayron, P. A., Ellison, R. D., Levy, H. A., *Acta Crystallogr.* (1967) **23**, 1079.
21. Schubert, E. H., Sheline, R. K., *Z. Naturforsch., Teil B* (1965) **20**, 1306.
22. Andrews, M., Tipton, D. L., Kirtley, S. W., Bau, R., *J. Chem. Soc., Chem. Commun.* (1973) 181.
23. Olson, J. P., Koetzle, T. F., Kirtley, S. W., Andrews, M., Tipton, D. L., Bau, R., *J. Am. Chem. Soc.* (1974) **96**, 6621.
24. Love, R. A., Chin, H. B., Koetzle, T. F., Kirtley, S. W., Whittlesey, B. R., Bau, R., *J. Am. Chem. Soc.* (1976) **98**, 4491.
25. Petersen, J. L., O'Connor, J. P., Ruff, J. K., Dahl, L. F., unpublished data, 1975.
26. Shriver, D., Cooper, C. B., III, Onaka, S., *ADV. CHEM. SER.* (1978) **167**, 000.

RECEIVED July 19, 1977.

3

Location of Terminal Hydride Ligands in Transition Metal Hydrides

JAMES A. IBERS

Department of Chemistry, Northwestern University, Evanston, IL 60201

The location of the positions of terminal hydride ligands in transition metal complexes using x-ray diffraction techniques is examined by reference to some recent structure determinations. These include studies of the complexes $RuHX(CO)(PPh_3)_2$, where $X = TolNNNTol$ and $TolNCHNTol$, $RuH_2(N_2B_{10}H_8SMe_2)(PPh_3)_3 \cdot 3C_6H_6$, $OsH(CSSMe)(CO)_2(PPh_3)_2 \cdot \frac{1}{2}C_6H_6$, $[PtH(PhHNNCMe_2)(PPh_3)_2][BF_4]$, and $[Pt((tert\text{-}Bu)_2P(CH_2)_3P(tert\text{-}Bu)_2)]_2$.

Confusion over the stereochemical role of the hydride ligand was largely cleared up some 12 years ago (*see* for example Ref. *1*). The location from x-ray diffraction data of the hydride position in $RhH(CO)(PPh_3)_3$ (*2, 3*) was a crucial step in this process since this was the first example of the location of a hydride ligand in the presence of other, bulkier ligands on a transition metal. The Rh–H bond length of 1.60(12)Å, a value that remains reasonable today, was determined by Fourier methods based on room-temperature diffraction data whose intensities had been estimated visually. The notion that the hydride ligand is "buried" in the metal orbitals, though seriously undermined by this structure determination, was finally put to rest with the determination of the Mn–H distance of 1.602(16)Å by using neutron diffraction techniques (*4*). Since then the determination of the position of a terminal hydride ligand in a transition metal hydride complex by x-ray diffraction methods has become rather routine. In favorable cases, the position of the hydride ligand can be determined by least-squares procedures to an apparent accuracy of about ± 0.05Å from x-ray intensity data collected by diffractometer methods at room temperature. But not all cases are favorable. The purpose of this present chapter is to present some recent experiences from our laboratory on the determination of hydride positions in various transition metal hydrides (*5–10*). In so doing, we hope to point to some of the potential problems involved. Since this book is devoted to transition metal hydrides, the reasons for our studying the specific compounds discussed below will be given with utmost brevity. But we emphasize that in these days of the

0-8412-0390-3/78/33-167-026/$05.00/0

© American Chemical Society

ubiquitous hydride ligand, the fact that the compounds studied happened to be hydrides was incidental to our chemical interests.

Experimental Procedures

The experimental procedures followed in collecting the diffraction data were standard in this laboratory (*see* for example Ref. *11*). Diffraction data were obtained at room temperature on a Picker FACS-I diffractometer using either filtered copper radiation or monochromatized molybdenum radiation. Efforts were made to collect as large data sets as possible. All data sets were corrected for absorption effects. The compounds to be discussed have molecular weights in the range 880–1070 amu, primarily because they contain PPh$_3$ or related phosphine groups. All least-squares refinements of these structures were carried out by full-matrix methods. Where there were no complicating features, the hydride positions were included as part of the refinement, with their positional and isotropic thermal parameters being varied. Phenyl groups were refined as rigid groups by using techniques first developed for RhH(CO)(PPh$_3$)$_3$ (*3*). Abbreviations used in this chapter are: Ph = phenyl, Tol = *p*-tolyl, Me = methyl, Et = ethyl, *tert*-Bu = *tert*-butyl, and Cy = cyclohexyl. Unless otherwise stated, these group atoms were constrained to vibrate isotropically. Table I lists selected information on the various compounds to be discussed, including some details on data collection and on M–H bond lengths.

Discussion

RuHX(CO)(PPh$_3$)$_2$. The compounds X = TolNNNTol (triazenido) and TolNCHNTol (amidinato) are prepared as shown in Reactions 1 (*12*) and 2 (*6, 13*). These compounds were studied because of our interest in the diverse bonding patterns of the triazenido and isoelectronic amidinato ligands.

RuH$_2$(CO)(PPh$_3$)$_2$ + TolNNHNTol →

$$\text{RuH(TolNNNTol)(CO)(PPh}_3)_2 + \text{H}_2 \quad (1)$$

RuH$_2$(CO)(PPh$_3$)$_2$ + TolN=C=NTol →

$$\text{RuH(TolNCHNTol)(CO)(PPh}_3)_2 \quad (2)$$

The inner coordination spheres of these two complexes are presented in Figure 1. While the Ru–H distance is well defined in the amidinato, it is not in the triazenido complex. Yet reference to Table I indicates that the triazenido structure is "better" if one uses the unreliable criterion that the lower the R index, the better the structure. In this instance, the low R index in the triazenido complex results from an elaboration of the usual group refinement model (*3*), allowing for anisotropic motion of the group atoms. This elaboration introduces a large number of additional variables and provides us with an opportunity to lower the R index! More importantly, we established that the other features of the structure were virtually unaffected by this elaboration. We conclude that with problems of this type, such an elaboration probably is not justified by the expense involved. Why can't the hydride position be located accurately in the

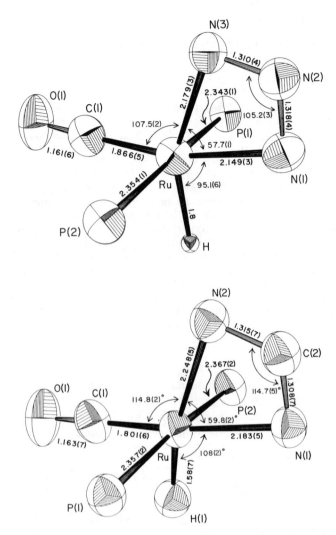

Figure 1. (top) *The inner coordination sphere of RuH-
(TolNNNTol)(CO)(PPh₃)₂;* (bottom) *the inner coordination
sphere of the isoelectronic RuH(TolNCHNTol)(CO)-
(PPh₃)₂.*

triazenido while it was refined isotropically in the amidinato complex? In this
instance the reason is clear: in the triazenido structure there is a CO–H disorder
with about 80% of the CO at the position shown in Figure 1 (top). The overlap
of the 20% CO with the 80% H is sufficient to obscure the position of the hydride
ligand. This disorder is not imposed crystallographically. On the other hand,
the amidinato structure, which crystallizes in a different space group, is perfectly
ordered.

In the triazenido complex, there is a slight trans labilizing effect of the hydride over the carbonyl, as reflected in the trans Ru–N distances. This effect is much more pronounced in the amidinato complex because of the absence of disorder and perhaps because the four-membered ring has opened up so that atom N(2) is more nearly trans to the hydride ligand.

These two structures illustrate a fundamental and disturbing point about diffraction experiments. It is not until the late stages of refinement, after considerable time and money has been spent on the experiment, that one sometimes discovers his inability to define accurately a salient feature of the structure—in this instance the hydride position in the triazenido complex. There is no way from the formula, space group, or films to have predicted this; nor are there any usefully consistent methods that enable one to predict, especially in common low-symmetry space groups, when disorder will occur.

$RuH_2(N_2B_{10}H_8SMe_2)(PPh_3)_3 \cdot 3C_6H_6$. The synthesis of this compound is:

$$RuH_2(N_2)(PPh_3)_3 + 1,10\text{-}N_2B_{10}H_8SMe_2 \ (14) \xrightarrow[\text{1 hr under } N_2]{\text{THF}}$$
$$(\nu_{NN} = 2240 \ cm^{-1})$$

$$RuH_2(N_2B_{10}H_8SMe_2)(PPh_3)_3 \ (15)$$
$$(\nu_{NN} = 2060 \ cm^{-1}) \tag{3}$$

The compound is of interest to us because of our studies of the varying bonding modes of the N_2R species. From the very high value of the NN stretching frequency, we believed (and confirmed) that the complex represents the first example of the linear attachment of an RN_2 group to a transition metal. The structural study (Table I) was straightforward, and the inner coordination sphere is shown in Figure 2. The R index is somewhat higher than expected, probably because of some residual electron density in the region of the SMe_2 group. Nevertheless, the two hydride positions refined successfully. The resultant Ru–H distances, insignificantly different from one another, are as expected. The trans-labilizing effect of the hydride ligand relative to a triarylphosphine group is apparent.

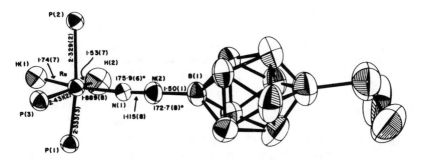

Figure 2.   The inner coordination sphere of $RuH_2(N_2B_{10}H_8SMe_2)$-
$(PPh_3)_3$

Table I.   Some Experimental Details

Compound	Space Group	Formula Units/ Cell
$RuH(TolNNNTol)(CO)(PPh_3)_2$	$C_i^1 - P\bar{1}$	2
$RuH(TolNCHNTol)(CO)(PPh_3)_2$	$C_{2h}^6 - \underline{C}2/c$	8
$RuH_2(N_2B_{10}H_8SMe_2)(PPh_3)_3 \cdot 3C_6H_6$	$C_i^1 - P\bar{1}$	2
$OsH(CSSMe)(CO)_2(PPh_3)_2 \cdot \frac{1}{2}C_6H_6$	$C_i^1 - P\bar{1}$	2
$[PtH(PhHNNCMe_2)(PPh_3)_2][BF_4]$	$C_{2h}^5 - P2_1/c$	4
$[Pt((tert\text{-}Bu)_2P(CH_2)_3P(tert\text{-}Bu)_2)]_2$	$C_{2h}^6 - C2/c$	4^b

$^a R(F_0) = \Sigma||F_0| - |F_c||/\Sigma|F_0|$.

$OsH(CSSMe)(CO)_2(PPh_3)_2 \cdot \frac{1}{2}C_6H_6$. This compound was synthesized by the following route (16):

$$Os(C_2H_4)(CO)_2(PPh_3)_2 + CS_2 \longrightarrow Os(CS_2)(CO)_2(PPh_3)_2$$

(4)

The compound is interesting to us because of the novelty of the CSSMe ligand and the possibility of varying modes of attachment of this ligand to transition metals.

The structural study was completely straightforward (see Table I) and the inner coordination sphere is shown in Figure 3. Note, in particular, the successful refinement of the hydride position (and isotropic thermal parameter) in the presence of a third-row element. As judged by the agreement among what are expected to be chemically equivalent distances within the complex, the assigned standard deviations are reasonable; hence, the Os–H distance has been determined to ±0.06 Å. The complex itself contains 50 nonhydrogen atoms and 36 hydrogen atoms in addition to the hydrogen atom of the hydride ligand. As a result, the compound would present a formidable neutron diffraction study. Such a study might not lead to a much better determination of the hydride position. In our opinion, a neutron diffraction study is not justified on a transition metal hydride containing only a terminal hydride if this hydride position has been reasonably well defined from an x-ray study. It is wiser to apply the limited neutron diffraction resources to the study of structures involving bridging hydride ligands where small positional changes may affect our interpretation of the bonding. Note the trans-labilizing effect of the hydride ligand relative to the CSSMe group.

on Selected Transition Metal Hydrides

No. of Obser-vations	$\left(\dfrac{\sin\theta}{2}\right)_{max}$ (Å^{-1})	No, of Vari-ables	$R(F_0)^a$	M–H (Å)	Ref.
5847	0.562	436	0.038	?	5
5805	0.576	192	0.063	1.58(7)	6
6544	0.479	324	0.072	1.74(7)	7
				1.53(7)	
7005	0.627	182	0.036	1.64(6)	8
5635	0.559	214	0.046	?	9
5350	0.700	204	0.030	No H	10

[b] The dimer has crystallographically imposed C_2 symmetry.

[PtH(PhHNNCMe₂)(PPh₃)₂][BF₄]. The synthesis of this compound is (*9*):

$$trans\text{-PtHCl}(PPh_3)_2 + H_2NNHPh + NaBF_4 \xrightarrow{CH_2Cl_2} PtH(H_2NNHPh)(PPh_3)_2{}^+$$

$$\downarrow acetone$$

$$trans\text{-PtHCl}(PPh_3)_2 + AgBF_4 + PhHNNCMe_2 \xrightarrow{THF} PtH(PhHNNCMe_2)(PPh_3)_2{}^+$$

Because there are at least three ways in which the hydrazone ligand could attach itself to a transition metal and because hydrazone complexes of transition metals barely have been investigated, this type of compound is interesting to us.

As contrasted with the studies discussed above, we were unable to locate and refine the hydride position in this compound (*see* Table I). The inner coordination sphere (Figure 4) shows the hydride ligand drawn at an assumed position. The reasons for our failure are not clear, but we are led to the sobering conclusion that success cannot always be guaranteed in studies of this kind. One might ask if intrinsic differences in the M–H bond might affect our ability to locate the hydride ligand. Obviously, as one goes in a formal sense from M^+–H^- to M^-–H^+, that is, if the hydride were to lose its electron, the hydride ligand would become transparent to x-rays. In principle, then, our inability to locate hydride positions in certain complexes might correlate with such a hypothetical electron transfer. But spectroscopic differences also should manifest themselves. The spectroscopic properties of the Pt–H bond in this hydrazone complex are normal. Thus, $\nu_{Pt-H} = 2220$ cm^{-1} and $\tau_{Pt-H} = 27.36$ ppm. (Compare these values with those of 2119 cm^{-1} and 23.37 ppm in the amidinato complex above where the hydride position was located without difficulty.) We believe it to be far more likely that failure to locate the hydride position in this hydrazone complex results from undetected errors in our data or in our model and not from an intrinsic property of the Pt–H bond.

[Pt((*tert*-Bu)₂P(CH₂)₃P(*tert*-Bu)₂)]₂. We end this brief summary of recent structural studies in our laboratory by discussing a compound that apparently

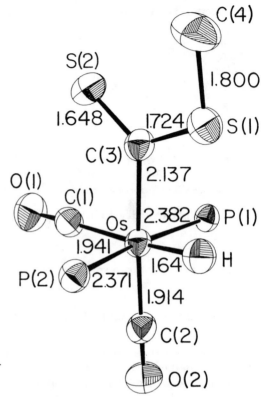

Figure 3. The inner coordination sphere of OsH-(CSSMe)(CO)₂(PPh₃)₂

is not a hydride. We attempted the preparation of the new complexes M(PP), where PP represents a bidentate phosphorus ligand constrained by geometry to present cis phosphorus atoms to the metal M. The resultant bent M(PP) complex should exhibit high reactivity and unusual chemistry. In the course of this work, the dimeric compound [Pt((*tert*-Bu)₂P(CH₂)₃P(*tert*-Bu₂)]₂ precipitated and was characterized (*10*). The preparative scheme is:

$$PtCl_2(diphos) \xrightarrow[THF]{Na/Hg} cis\text{-}PtH_2(diphos)$$

$$-H_2 \Updownarrow +H_2 \qquad (6)$$

$$\tfrac{1}{2}[Pt(diphos)]_2 \rightleftharpoons \{Pt(diphos)\}$$

$$diphos = (tert\text{-}Bu)_2P(CH_2)_3P(tert\text{-}Bu)_2$$

The inner coordination sphere of the dimer is shown in Figure 5. By the normal rules of electron counting, the rare-gas configuration is obeyed at platinum if a Pt═Pt double bond is invoked. Yet the Pt–Pt distance of 2.765(1)Å is a normal or slightly long, single bond. One thus wonders if the compound might be a

$$\tfrac{1}{2}[\text{Pt(diphos)}]_2 + D_2 \rightarrow \text{PtD}_2(\text{diphos})$$

$$\text{PtHD (diphos)}$$

$$\text{PtH}_2 \text{ (diphos)}$$

(7)

$$[\text{Pt(diphos)}]_2 + \text{CHCl}_3 \xrightarrow{\text{toluene}} \text{PtCl}_2 \text{ (diphos)} + \text{CH}_2\text{Cl}_2$$

40%   92%

(8)

hydride. Some chemistry of this dimer (Reactions 7 and 8) suggests that it could be a hydride, presumably containing Pt–H–Pt bridges.

As we noted above, there are instances where one fails to locate the hydride position in a known transition metal hydride. Consequently, failure to locate residual electron density in positions thought to be reasonable for hydride ligands hardly can be taken as strong evidence against a given compound being a hydride. The classic diagnostic tool for detecting hydrides has been the high-field shift of the hydride proton in the NMR spectrum. The present dimer shows no such shift; nor is there ir evidence for the presence of hydride bridges. However, Haymore (17) and Stone (*see* Chapter 8) (18) have demonstrated that the proton signal in the NMR spectrum is not always found at high field but rather may be found at low field. Stone's example, in fact, is the compound $[\text{Pt}(\mu\text{-H})(\text{SiEt}_3)\text{-}(\text{PCy}_3)]_2$, a compound closely related to the present one. No x-ray evidence for

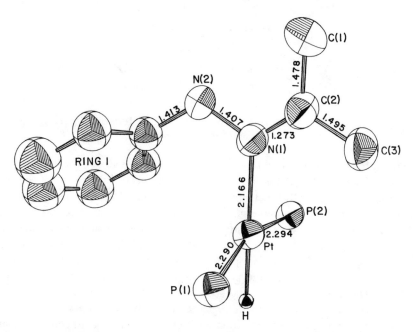

Figure 4.   The inner coordination sphere of $PtH(PhHNNCMe_2)(PPh_3)_2{}^+$. The hydride ligand was not located but is shown at an assumed position.

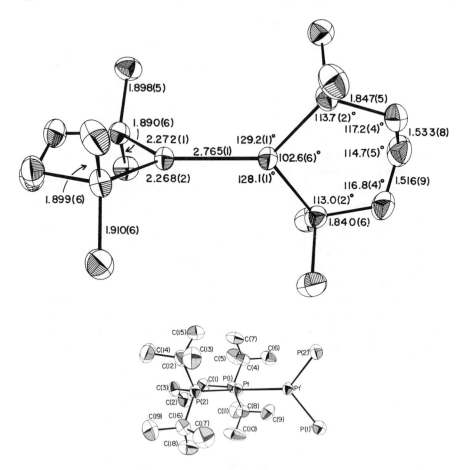

Figure 5.   The inner coordination sphere of [Pt((tert-Bu)₂P(CH₂)₃P(tert Bu)₂)]₂

bridging hydrides was found there.   But there was ir evidence for Pt–H bonding—evidence that was later corroborated by the successful location of the hydride signal in the NMR spectrum.   There is x-ray, NMR, and ir evidence to
support bridging hydrides in the similar compound [Ni(μ-H)(Cy₂P(CH₂)₃PCy₂)]₂
(19).   In both of these verified bridging hydride structures, the LML/LML interplanar angles are near 23°.   In our dimer, this angle is 82°.   In essence then,
there is negative diffraction and spectroscopic evidence for the presence of hydride bridges in the present complex, but negative evidence is never fully satisfactory.   Here we must appeal to stereochemical arguments that if the compound
were a hydride, the PPtP/PPtP interplanar angle would be close to 0°.   This
example illustrates, particularly for transition metal complexes containing or
possibly containing bridging hydride ligands, that x-ray diffraction and spectroscopic results may not always lead to a totally satisfying conclusion.

Conclusions

In this short chapter, I have attempted to provide an overview of the types of results that one can expect to obtain from x-ray diffraction studies of transition metal hydride complexes. The salient points seem to me to be: (1) one cannot predict at the outset whether one will be successful in locating the hydride position or, in fact, a given feature of a structure owing to the possibility of disorder and, at times, to unknown causes; (2) in favorable cases, the hydride position can be located to an accuracy commensurate with or exceeding our ability to use such information in theoretical models; (3) our expectations concerning the stereochemistry of transition metal hydrides that contain terminal hydride ligands have changed little in the 12 years since the determination of the structure of $RhH(CO)(PPh_3)_3$; hence it is unlikely that the study of such structures in themselves is a fertile field of structural chemistry.

Acknowledgment

Most of the work discussed here was kindly supported by the National Science Foundation.

Literature Cited

1. Ibers, J. A., *Ann. Rev. Phys. Chem.* (1965) **16**, 375.
2. La Placa, S. J., Ibers, J. A., *J. Am. Chem. Soc.* (1963) **85**, 3501.
3. La Placa, S. J., Ibers, J. A., *Acta Crystallogr.* (1965) **18**, 511.
4. La Placa, S. J., Hamilton, W. C., Ibers, J. A., Davison, A., *Inorg. Chem.* (1969) **8**, 1928.
5. Brown, L. D., Ibers, J. A., *Inorg. Chem.* (1976) **15**, 2794.
6. Brown, L. D., Robinson, S. D., Sahajpal, A., Ibers, J. A., *Inorg. Chem.* (1977) **16**, 2728.
7. Schramm, K. D., Ibers, J. A., *Inorg. Chem.* (1977) **16**, 3287.
8. Waters, J. M., Ibers, J. A., *Inorg. Chem.*, (1977) **16**, 3273.
9. Krogsrud, S., Toniolo, L., Croatto, U., Ibers, J. A., *J. Am. Chem. Soc.* (1977) **99**, 5277.
10. Yoshida, T., Yamagata, T., Tulip, T. H., Ibers, J. A., Otsuka, S., *J. Am. Chem. Soc.*, in press.
11. Nakamura, A., Yoshida, T., Cowie, M., Otsuka, S., Ibers, J. A., *J. Am. Chem. Soc.* (1977) **99**, 2108.
12. Laing, K. R., Robinson, S. D., Uttley, M. F., *J. Chem. Soc., Dalton Trans.* (1974) 1205.
13. Robinson, S. D., Sahajpal, A., *J. Organomet. Chem.* (1976) **117**, C111.
14. Knoth, W. H., Hertler, W. R., Muetterties, E. L., *Inorg. Chem.* (1965) **4**, 280.
15. Knoth, W. H., *J. Am. Chem. Soc.* (1972) **94**, 104.
16. Collins, T. J., Roper, W. R., Town, K. G., *J. Organomet. Chem.* (1976) **121**, C41.
17. Haymore, B. L., "Abstracts of Papers," *Joint Conf. CIC/ACS, 2nd, Montreal, May 29–June 2, 1977,* INDOR 126.
18. Stone, F. G. A., ADV. CHEM. SER. (1978) **167**, 000.
19. Jolly, P. W., Wilke, G., "The Organic Chemistry of Nickel," Vol. 1, p. 145, Academic, London, 1974.

RECEIVED July 19, 1977.

4

Molecules with Bridging Hydride Ligands

Direct Comparisons of M(μ-H)M Bonds with M–M or M(μ-Cl)M Bonds

MELVYN ROWEN CHURCHILL
Department of Chemistry, State University of New York, Buffalo, NY 14214

A series of trinuclear osmium carbonyl derivatives that contain single unsupported equatorial μ_2-hydride ligands, including $(\mu_2\text{-}H)(H)Os_3(CO)_{11}$, $(\mu_2\text{-}H)(H)Os_3(CO)_{10}(PPh_3)$, "$Os_3(CO)_9$-$(EtC\equiv CH)_2$", and $(\mu_2\text{-}H)_2Os_3Re_2(CO)_{20}$, have been investigated. The Os(μ-H)Os distances range from 2.989(1) to 3.083(3) Å, as opposed to a normal Os–Os bond length of 2.877(3) Å in $Os_3(CO)_{12}$. In $(\mu_2\text{-}H)_2Os_3(CO)_{10}$, the dihydrido-bridged Os–Os bond is 2.681(1) Å in length while in $(\mu_2\text{-}H)$-$Os_3(CO)_{10}(\mu_2\text{-}CH\cdot CH_2\cdot PMe_2Ph)$, the dibridged Os–Os bond distance is 2.800(1) Å. Similar effects are noted for ruthenium complexes based upon studies on $Ru_3(CO)_{12}$, $(\mu_2\text{-}H)Ru_3$-$(CO)_{10}(\mu_2\text{-}C=NMe_2)$, and $(\mu_2\text{-}H)_4Ru_4(CO)_{10}(diphos)$. In the dinuclear species $[(\eta^5\text{-}C_5Me_5)MCl]_2(\mu\text{-}X)(\mu\text{-}Cl)$ (M = Rh, Ir; X = Cl, H), the μ_2-hydrido-μ_2-chloro-bridged complexes have M–M separations of 2.906(1) Å (M = Rh) and 2.903(1) Å (M = Ir) whereas the di-μ_2-chloro-bridged species have increased nonbonding M$\cdots$M distances of 3.719(1) Å (M = Rh) and 3.769(1) Å (M = Ir).

Hydride complexes of the transition metals occupy a central role in contemporary chemistry, both because of their importance as catalytic or stoichiometric reagents for fundamental organic transformations (e.g., the catalytic hydrogenation of unsaturated systems) and because of their chemical interest per se.

The structural chemistry of these species has been the subject of a remarkable number of misconceptions and only now is beginning to emerge as a coherent, ordered discipline. Thus, prevailing attitudes on the nature of even the simplest

0-8412-0390-3/78/33-167-036/$06.25/0

system, the terminal hydride ligand, have undergone almost an entire cycle of change, as illustrated by the following approximate chronology.

(1) Prior to 1960. The metal–hydrogen bond was believed to be very short (*1*). As such, the hydrogen atom was viewed as being buried in the metal *d*-orbitals and having, in essence, a negative or close-to-zero covalent radius. As a result of electron diffraction studies on $HCo(CO)_4$ and $H_2Fe(CO)_4$ (*2*), in which the carbonyl ligands were shown to have a tetrahedral disposition around the central metal atom, the hydride ligand also was believed to exert no stereochemical influence (i.e., it was thought not to occupy a regular coordination site on the metal atom).

(2) ca. 1960–1970. Because of a series of x-ray diffraction studies beginning with $HPtBr(PEt_3)_2$ (in which the hydrogen atom was not located directly) (*3*) and $HRh(CO)(PPh_3)_3$ (in which the hydrogen atom was located directly) (*4, 5*) and a unique early neutron diffraction study of $K_2[ReH_9]$ (*6*), attitudes changed drastically. The revised credo was that the metal–hydrogen bond was entirely normal (i.e., could be predicted safely to be close to $(r(M) + 0.3)$ Å in length, where $r(M)$ is the covalent radius appropriate to the metal under consideration, and 0.3 Å is the approximate covalent radius for hydrogen) and that the hydride ligand occupied a regular stereochemical site in the coordination sphere of the transition metal.

(3) ca. 1970–present. The view currently accepted is that the metal–hydrogen distance will be normal unless otherwise affected by external factors (e.g., a ligand of extremely high trans effect). The categorical statement as to the hydride ligand occupying a regular stereochemical site should be viewed circumspectly. This is usually the case, but a caveat must be issued that bulky ligands may encroach upon the cone-angle of space formally allotted to the coordination site of the hydride ligand. Perhaps the most flagrant examples of this violation occur in the $HRh(PPh_3)_4$ (*7*) and $HRh(PPh_3)_3(AsPh_3)$ (*8*) molecules; here, the triphenylpnicogen ligands occupy essentially regular tetrahedral sites around the central Rh(I) atom, but the hydride ligands can be detected spectroscopically and are believed to occupy sites directly trans to PPh_3 or $AsPh_3$ ligands (two rhodium–hydrogen stretches are observed for the mixed ligand complex). A similar, but far less severe, case occurs in the $HMn(CO)_5$ molecule; here the equatorial ligands are displaced by only 6–8° toward the terminal hydride ligand (*9*).

The structural chemistry of species containing bridging hydride ligands has lagged behind that of the terminal hydride complexes, partly because of the later realization of their synthesis and partly because of the greater complexity of the molecules involved. [A consistent problem for x-ray studies is that the electron density for a hydrogen atom ($Z = 1$) is far smaller than that for a transition metal ($Z = 21–29$ for a first-row transition metal, 39–47 for a second-row metal, and 71–79 for a third-row metal). The hydride ligand thus may be obscured by background noise on an electron density map.] The first x-ray structural analysis on one of these species was that by Doedens and Dahl (*10*) on the molecule $[(\eta^5\text{-}C_5H_5)Mo(CO)_2]_2(\mu_2\text{-}H)(\mu_2\text{-}PMe_2)$. Although the bridging hydride ligand was not located in this study, its location was deduced by comparison with related structures. The structure was described at the time as containing a localized, bent, three-center metal–hydrogen–metal bond without a metal–metal bond. Thus, despite a molybdenum–molybdenum distance of only 3.262(7) Å [cf. the molybdenum–molybdenum single bond distance of 3.235(1) Å in [(η^5-C_5H_5)-$Mo(CO)_3]_2$ (*12*)], the core of this μ_2-hydrido species was drawn originally as that in Structure **1**; i.e., without any direct metal–metal interaction.   A similar situation arose in the duo of complexes $Fe_3(CO)_{12}$ (*13, 14*) and $[HFe_3(CO)_{11}{}^-]$ (*15*). While these were formally illustrated as Structures **2** and **3**, the iron–iron distances of note are 2.558(1) Å for the $Fe(\mu_2$-$CO)_2Fe$ system in $Fe_3(CO)_{12}$ and 2.577(3) Å for the $Fe(\mu_2$-$H)(\mu_2$-$CO)Fe$ system in the $[HFe_3(CO)_{11}{}^-]$ ion.   Thus, within this formalism, a change from a formal metal–metal single bond to a formal no-bond situation is accompanied by a change in metal–metal distance of only 0.019 Å.   Clearly, the original description of these bent metal–hydrogen–metal bonds needed some modification.

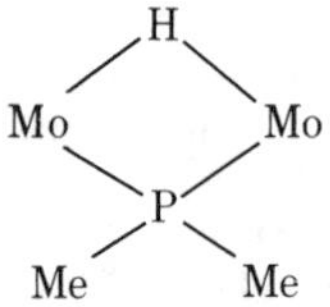

A final important structure in the early days was $[HCr_2(CO)_{10}{}^-]$.   The Cr···Cr distance here was 3.406(9) Å, i.e., some 0.44 Å longer than in the conjugate base, $[Cr_2(CO)_{10}{}^{2-}]$, where the chromium–chromium bond length is 2.97(1) Å (*16, 17*).   A unique feature of the $[HCr_2(CO)_{10}{}^-]$ ion was the eclipsed arrangement of equatorial carbonyl ligands on the two chromium atoms.   It is interesting to note that the $[HCr_2(CO)_{10}{}^-]$ ion lies on a crystallographic center of symmetry. For some ten years this molecule was believed to be the archetypal example of a molecule with a truly linear metal–hydrogen–metal system.   This now has been shown not to be the case.   A neutron diffraction study shows the bridging hydride ligand to be disordered (*18*) and displaced from co-linearity with the Cr···Cr vector.

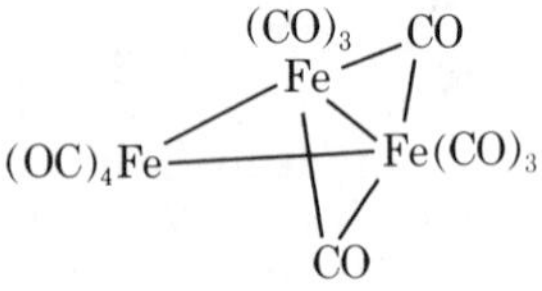

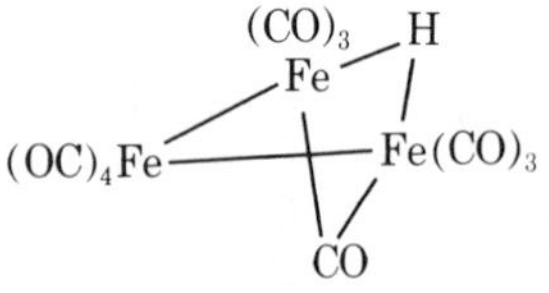

Early Studies on Rhenium–Carbonyl–Hydride Clusters and Related Species

Our initiation into the field of transition metal hydride chemistry resulted from collaborative research with H. D. Kaesz on the structures of a series of rhenium–carbonyl–hydride complexes.   Single-crystal x-ray diffraction studies of $HRe_2Mn(CO)_{14}$ (*19, 20*) and derivatives of the $[H_2Re_3(CO)_{12}{}^-]$ (*21*), $[HRe_3(CO)_{12}{}^{2-}]$ (*22, 23*), $[H_6Re_4(CO)_{12}{}^{2-}]$ (*24*), and $[Re_4(CO)_{16}{}^{2-}]$ (*25, 26*) anions led to our proposing some rules whereby the positions of bridging hydride ligands could be determined indirectly.   (We note parenthetically here that our equipment and data were not, at this time, good enough to enable us to locate hydride ligands directly.)   These simple, empirical rules were as follows:

(1)   A normal nonbridged rhenium–rhenium linkage was about 3.00 Å in length in a rhenium carbonyl cluster.   This distance expanded by about 0.15 Å

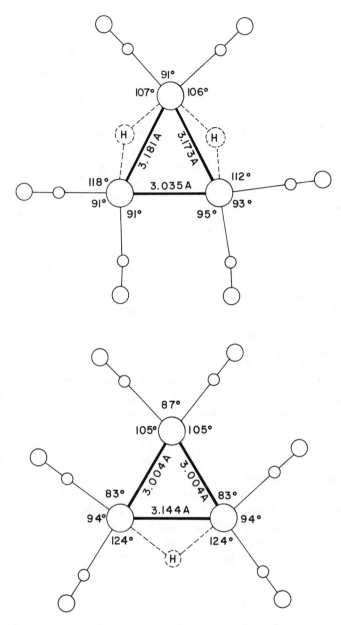

Figure 1.   Interatomic distances and angles within the equatorial planes of (top) $[H_2Re_3(CO)_{12}^-]$ (see Ref. 21) and (bottom) $[HRe_3(CO)_{12}^{2-}]$ (see Ref. 22, 23).   Approximate locations of the hydride ligands are shown.   Note the increased rhenium–rhenium distances and the displacement of equatorial carbonyl ligands associated with the presence of the μ_2-hydride ligands.

Figure 2.   The $H_2Ru_6(CO)_{18}$ mol-ecule, projected onto one of its large open faces (cf. Ref. 27,28).

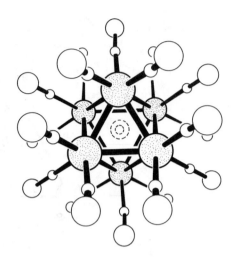

Six of the eight faces of the Ru_6 octahe-dron each have three axial carbonyl li-gands associated with them.   The other two faces (those lying in the plane of the figure) have no truly axial carbonyl li-gands and are defined by ruthenium–ruthenium distances of 2.952(3)–2.959(3) Å as opposed to ruthenium–ruthenium distances of 2.858(3)–2.874(3) Å for the remaining metal–metal bond lengths. The hydride ligands are believed to oc-cupy μ_3-bridging positions over the large open faces—they are shown as dashed circles.

to ca. 3.15 Å for a mono-μ_2-hydrido-bridged rhenium–rhenium vector in a tri-angulated cluster complex and to about 3.40 Å for a close-to-linear rhenium–hydrogen–rhenium system.

(2)   In addition to (1), it was found that ligands adjacent to bridging hydride ligands suffered repulsive displacements such that the locations of the bridging hydride ligands were perceivable as holes in the coordination surface of the cluster.   The species $[H_2Re_3(CO)_{12}^-]$ and $[HRe_3(CO)_{12}^{2-}]$ effectively demon-strate this and are illustrated in Figure 1.

These studies also were extended successfully to the cases of $H_2Ru_6(CO)_{18}$ (believed to contain two μ_3-bridging hydride ligands—*see* Figure 2) (27, 28) and $H_6Cu_6(PPh_3)_6$ (believed to contain six μ_2-bridging hydride ligands—*see* Figure 3) (29, 30).

It should be noted here that our work to this point led only to the identifi-cation of the metal–metal vector (or triangular cluster face) with which the hy-dride ligand was associated.   Dahl and co-workers extended this idea to estimate the actual position of the bridging hydride ligand (31).   Their work was based on a consideration of metal cluster complexes in which each metal atom had octahedral coordination geometry.   The missing bridging hydride ligand then could be assigned to the point of intersection of the OC → M vectors from car-bonyl ligands trans to the hydride location.   This, while valuable, has been shown

to be an over simplification of the true state of affairs.   Thus, the estimated tungsten–hydrogen–tungsten angle for $(\mu_2\text{-H})W_2(CO)_9(NO)$ is 159°, whereas the angle determined by neutron diffraction is only 125.5° (32).   Thus, the hy-dride ligand apparently is displaced outward from the position pointed to by the axial ligands; i.e., it is displaced further away from the metal–metal vector.   This strongly suggests the the metal–hydrogen–metal system is a closed rather than an open, two-electron, three-center bond (i.e., Structure **4** rather than Structure **5**).

Current Studies on Species with Bridging Hydride Ligands

Our current research work is proceeding along two principal routes.

(1) Characterization of trinuclear and larger polynuclear osmium (and to a lesser extent, ruthenium) carbonyl derivatives containing bridging hydride ligands; this work is a result of an on-going collaborative effort with J. R. Shapley. Our principal aims here are: (a) to identify the sites of the bridging hydride ligands directly (by difference-Fourier maps and least-squares refinement of hydrogen atom positions) if possible and indirectly (as outlined above for rhenium clusters) when the direct approach fails; (b) to determine the effect of bridging hydride ligands and other bridging ligands on the length of an osmium–osmium linkage.

(2) Characterization of binuclear species containing bridging hydride ligands and comparison of their geometry with that of related species in which the hydride ligand is replaced with an electron-precise ligand such as a bridging chloride ligand; our efforts have concentrated on rhodium and iridium complexes, with a brief excursion into tungsten chemistry. These studies are detailed below.

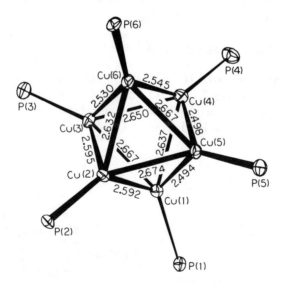

Inorganic Chemistry

Figure 3.   Bond lengths within the P_6Cu_6 core of $H_6Cu_6(PPh_3)_6$ (Ref. 29, 30).   The six hydride ligands are believed to occupy μ_2-bridging positions over the six long copper–copper vectors.

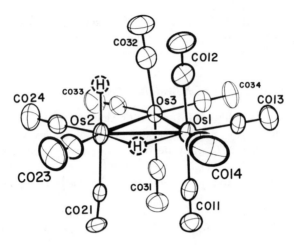

Inorganic Chemistry

Figure 4. Molecular geometry of $(\mu_2\text{-}H)(H)$-$Os_3(CO)_{11}$ (see Refs. 33 and 35) with hydride ligands in their deduced positions

Osmium–Carbonyl–Hydride Clusters and Related Ruthenium Complexes. Our investigation of these species began with a study of the species $(\mu_2\text{-}H)(H)Os_3(CO)_{11}$, prepared from $Os_3(CO)_{12}$ via the unsaturated species $(\mu_2\text{-}H)_2Os_3(CO)_{10}$ (33) (*see* Reaction 1).

$$Os_3(CO)_{12} \xrightarrow[H_2]{\Delta} (\mu_2\text{-}H)_2Os_3(CO)_{10} \xrightarrow{+\,CO} (\mu_2\text{-}H)(H)Os_3(CO)_{11} \qquad (1)$$

The structure of $(\mu_2\text{-}H)(H)Os_3(CO)_{11}$ is, unfortunately, subject to some disorder and it was not possible to locate the hydride ligands by direct methods. However, their locations could be determined unambiguously from the geometry of the cluster; the derived ordered structure is illustrated in Figure 4. The position

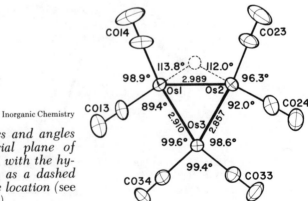

Inorganic Chemistry

Figure 5. Distances and angles within the equatorial plane of $(\mu_2\text{-}H)(H)Os_3(CO)_{11}$, with the hydride ligand shown as a dashed circle in its probable location (see Ref. 35)

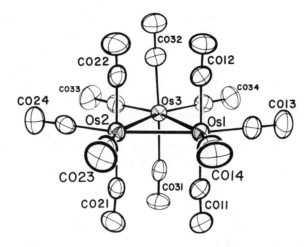

Inorganic Chemistry

Figure 6.   Molecular geometry of $Os_3(CO)_{12}$ (see Ref. 35)

of the bridging hydride ligand was determined indirectly from the longer Os(1)–Os(2) distance and from the abnormally large Os(2)–Os(1)–CO(14) and Os(1)–Os(2)–CO(23) angles adjacent to this vector (*see* Figure 5). A pleasant bonus was in store for us, since we found the $(\mu_2\text{-H})(\text{H})Os_3(CO)_{11}$ molecule to be isomorphous with $Os_3(CO)_{12}$ (previously studied by Corey and Dahl (34)). We now accurately redetermined the structure of $Os_3(CO)_{12}$ (35) (*see* Figures 6 and 7). The isomorphous nature of $(\mu_2\text{-H})(\text{H})Os_3(CO)_{11}$ and $Os_3(CO)_{12}$ means

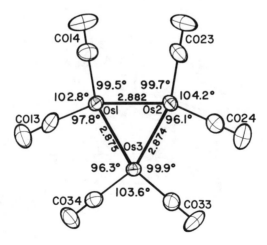

Inorganic Chemistry

Figure 7.   Distances and angles within the equatorial plane of $Os_3(CO)_{12}$ (see Ref. 35)

that external effects on geometry are cancelled. Any differences in the observed intramolecular geometry therefore are attributable to the substitution of one axial carbonyl ligand in $Os_3(CO)_{12}$ by one terminal and one bridging hydride ligand to form $(\mu_2\text{-}H)(H)Os_3(CO)_{11}$. Furthermore, since the terminal hydride ligand occupies the same stereochemical site as the replaced axial carbonyl ligand, any changes in osmium–osmium bond length or Os–Os–CO angles between the two complexes may be attributed primarily to the presence or absence of a bridging hydride ligand!

A detailed comparison of Figures 5 and 7 thus provides perhaps the most thorough assessment available for the stereochemical influence of a single equatorial μ_2-bridging hydride ligand.

We now turned our attention to the complex $(\mu_2\text{-}H)(H)Os_3(CO)_{10}(PPh_3)$, prepared as indicated in Reaction 2. This structure was found to be ordered, with the PPh_3 ligand occupying an equatorial site (36). The overall geometry is shown in Figure 8. Both hydride ligands were located directly by difference-Fourier methods, and their positions were refined by the method of least squares. The Os–H(T) distance is 1.52(7) Å, while the bridging hydride ligand is characterized by bond lengths of Os(1)–H(B) = 1.74(6) Å and Os(2)–H(B) = 2.00(6) Å and an angle of Os(1)–H(B)–Os(2) = 107.5(28)°. Distances and angles in the equatorial plane are illustrated in Figure 9.

$$(\mu_2\text{-}H)_2Os_3(CO)_{10} + PPh_3 \rightarrow (\mu_2\text{-}H)(H)Os_3(CO)_{10}(PPh_3) \tag{2}$$

We note here that the normal osmium–osmium distance is taken as 2.877(3) Å (the average of the three osmium–osmium bond lengths in $Os_3(CO)_{12}$ (35)),

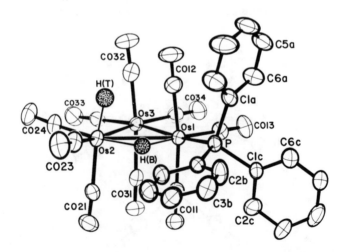

Inorganic Chemistry

Figure 8.   The overall molecular geometry of $(\mu_2\text{-}H)$-$(H)Os_3(CO)_{10}(PPh_3)$ (see Ref. 36). Both hydride ligands were located directly.

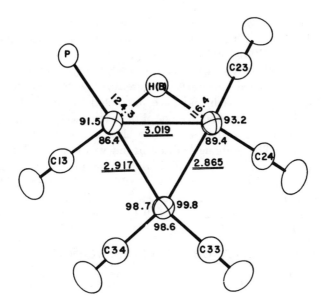

Inorganic Chemistry

Figure 9.   Distances and angles within the equatorial plane of $(\mu_2\text{-}H)(H)Os_3(CO)_{10}(PPh_3)$ (see Ref. 36)

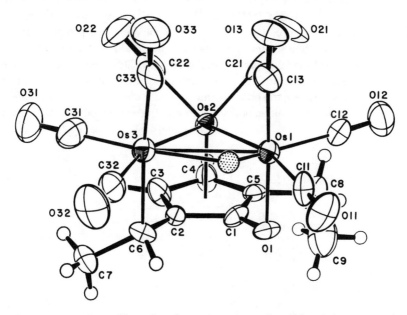

Figure 10.   Overall molecular geometry of $Os_3(CO)_9(EtC{\equiv}CH)_2$ showing the equatorial location of the μ_2-bridging hydride ligand (see Ref. 37)

and that a singly hydrido-bridged Os⋯Os distance (with the hydride ligand in the equatorial plane) is increased to 2.9886(9) Å in $(\mu_2\text{-H})(\text{H})\text{Os}_3(\text{CO})_{11}$ and 3.0185(6) Å in $(\mu_2\text{-H})(\text{H})\text{Os}_3(\text{CO})_{10}(\text{PPh}_3)$.

We next investigated the species $\text{Os}_3(\text{CO})_9(\text{EtC}{\equiv}\text{CH})_2$ prepared as in Reaction 3. The determined structure (*see* Figure 10) shows that two ethylacetylene moieties and one carbonyl ligand have interacted to form a trisubstituted η^5-cyclopentadienyl system, and that one of these ethylacetylene fragments also has become involved in oxidative addition of a carbon–hydrogen bond to a transition metal. The bridging hydride ligand was located (albeit not very precisely); it lies essentially in the plane of the triosmium cluster. Despite being bridged also by the four-membered $\text{C}(6)\text{–C}(2)\text{–C}(1)\text{–O}(1)$ system, the μ_2-hydrido-bridged $\text{Os}(1)\text{–Os}(3)$ bond length here is 3.007(1) Å as opposed to $\text{Os}(1)\text{–Os}(2)$ and $\text{Os}(2)\text{–Os}(3)$ distances of 2.819(1) Å and 2.889(1) Å, respectively (37).

$$(\mu_2\text{-H})_2\text{Os}_3(\text{CO})_{10} + 2\text{EtC}{\equiv}\text{CH} \rightarrow \text{Os}_3(\text{CO})_{10}(\text{EtC}{\equiv}\text{CH})_2$$

$$\xrightarrow{\Delta} \text{Os}_3(\text{CO})_9(\text{EtC}{\equiv}\text{CH})_2 \quad (3)$$

Very recently we have become interested in the structure, stereochemistry, and connectivity of a series of mixed osmium–rhenium carbonyl hydride cluster complexes. These were first reported by Knight and Mays (38) as part of a study on the reaction of $\text{M}_3(\text{CO})_{12}(\text{M} = \text{Fe, Ru, Os})$ clusters with the pentacarbonylmetallate anions, $[\text{M}'(\text{CO})_5{}^-]$ ($\text{M}' = \text{Mn, Re}$). While very few complexes were obtained analytically pure, the stoichiometry of the identified products suggests that the overall reaction sequence is something like that represented by Scheme 1. The structure of the 62-electron complex $\text{HOs}_3\text{Re}(\text{CO})_{15}$ has been determined (39) and is shown in Figure 11. The structure suffers from a fourfold pattern of disorder; nevertheless, the observed metal–metal distances (2.944(1)–2.957(1) Å) and the angular distribution of ligands around the metal atoms militate against

Scheme I

$$\text{Os}_3(\text{CO})_{12} + [\text{Re(CO)}_5{}^-]$$
$$\downarrow -\text{CO}$$
$$[\text{Os}_3\text{Re(CO)}_{16}{}^-] \xrightarrow{\text{H}^+} \text{HOs}_3\text{Re(CO)}_{16}$$
$$\downarrow -\text{CO}$$
$$[\text{Os}_3\text{Re(CO)}_{15}{}^-] \xrightarrow{\text{H}^+} \text{HOs}_3\text{Re(CO)}_{15}$$
$$\downarrow -2\text{CO}, +2\text{e}^-$$
$$[\text{Os}_3\text{Re(CO)}_{13}{}^{3-}] \xrightarrow{3\text{H}^+} \text{H}_3\text{Os}_3\text{Re(CO)}_{13} \; (+ \text{intermediate anionic hydrides?})$$

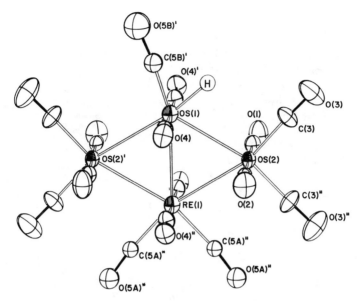

Inorganic Chemistry

Figure 11.   The molecular geometry of $HOs_3Re(CO)_{15}$ (39) as deduced from deconvolution of a fourfold disordered structure. (The hydride ligand was not located.)

the presence of a bridging hydride ligand and in favor of the presence of a terminal hydride ligand on a bridgehead osmium atom.

A second route to this class of complexes recently has been discovered by Shapley and co-workers (*40*). In particular, the species $H_2Os_3Re_2(CO)_{20}$ can be synthesized as shown in Reaction 4. A structural analysis of this complex has been undertaken (*41*). There is a strange crystallographic complication in that the complex crystallizes in the noncentrosymmetric space group Cc with two molecules in the asymmetric unit; furthermore, these molecules are interrelated by a local center of symmetry, but there is no general crystallographic center of symmetry. The overall molecular geometry of this species is indicated in Figure 12. The hydride ligands were not located directly, but their positions are defined unambiguously both by the metal–metal distances [Os(1)–Os(2) and Os(2)–Os(3) distances in the two independent molecules range from 3.058(3) to 3.083(3) Å while the nonhydrido-bridged Os(1)–Os(3) distances are each 2.876(3) Å, and osmium–rhenium bondlengths range from 2.946(4) Å to 2.982(3) Å] and by the angular arrangement of ligands in the equatorial belt around the central triosmium triangle.

$$Os_3(CO)_{10}(C_8H_{14})_2 + 2HRe(CO)_5 \xrightarrow{\text{benzene}} H_2Os_3Re_2(CO)_{20} + 2C_8H_{14} \quad (4)$$

The geometry of triosmium derivatives containing a single unsupported (equatorial) bridging hydride ligand thus appears to be self consistent and pre-

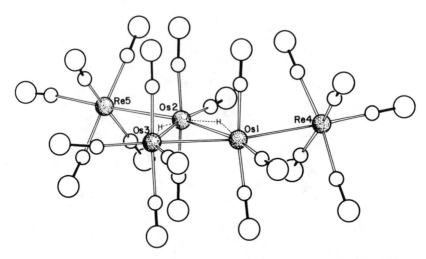

Journal of the American Chemical Society

Figure 12.   Molecular geometry of $H_2Os_3Re_2(CO)_{20}$ (40) with hydride ligands shown in their deduced positions

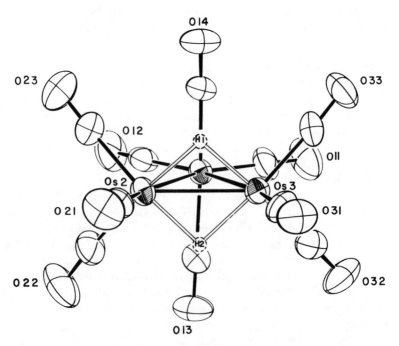

Inorganic Chemistry

Figure 13.   The overall molecular geometry of $(\mu_2\text{-}H)_2Os_3(CO)_{10}$ (45). Hydride ligands were not located reliably and are shown in their probable positions.

$$\begin{array}{ccc}
\text{(CO)}_4 & \text{(CO)}_4 & \text{(CO)}_4 \\
\text{Os} & \text{Os} & \text{Os}
\end{array}$$

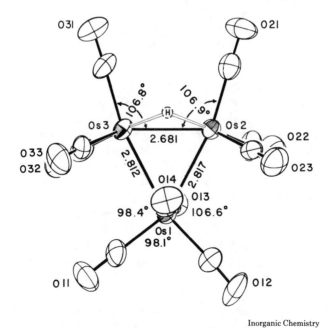

$$\begin{array}{ccc}
\text{(OC)}_3\text{Os} \text{====} \text{Os(CO)}_3 & \text{(OC)}_3\text{Os} \text{----} \text{Os(CO)}_3 & \text{(OC)}_3\text{Os} \quad \text{Os(CO)}_3
\end{array}$$

(Os===Os = 2.670 Å)   (Os---Os = 2.863 Å)   (Os   Os = 3.078 Å)

6   **7**   **8**

dictable. We now turn our attention to related species in which a given osmium–osmium vector of a triangular array of osmium atoms is associated with two simple bridging ligands.   The trio of complexes $(\mu_2\text{-H})_2\text{Os}_3(\text{CO})_{10}$ (Structure 6), $(\mu_2\text{-H})(\mu_2\text{-SEt})\text{Os}_3(\text{CO})_{10}$ (Structure 7), and $(\mu_2\text{-OMe})_2\text{Os}_3(\text{CO})_{10}$ (Structure 8) were initially studied by Mason and co-workers (*42, 43*).   The pattern of distances for the dibridged osmium–osmium vectors was of very great interest, for it was the first clear demonstration that a bridging hydride ligand had a net bonding effect on the appropriate metal–metal vector.   While there has been considerable disagreement as to the precise nature of this bonding interaction

Inorganic Chemistry

Figure 14.   Dimensions within the equatorial plane of
$(\mu_2\text{-H})_2\text{Os}_3(\text{CO})_{10}$ *(45)*

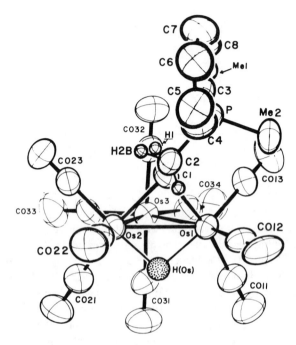

Inorganic Chemistry

Figure 15. An overall view of $(\mu_2\text{-}H)Os_3\text{-}$ $(CO)_{10}(\mu_2\text{-}CH\cdot CH_2\cdot PMe_2Ph)$ (46) showing the location of the bridging hydride ligand

(43, 44), one may, to a first approximation, view a $M_A(\mu_2\text{-}H)M_B$ system as containing a protonated metal–metal bond. Since the hydrogen atom uses only its spherically symmetric $1s$ orbital in bonding, the orbital on M_A that interacts with the hydrogen atom also must be of appropriate symmetry to interact with the orbital on M_B that also is interacting with the hydrogen atom. (The only unknown is the $M_A - M_B$ overlap integral!) Such a system thus contains a closed two-electron, three-center bond and as such will be associated with a metal–metal distance that must by definition be longer than a nonprotonated, two-electron (two-center) metal–metal σ-bond in an otherwise identical chemical environment. A $M(\mu_2\text{-}H)_2M$ system may well have a metal–metal distance that is shorter than a normal metal–metal single bond. However, such a system is viewed formally as a doubly protonated double bond and as such is correctly compared only with a metal–metal double bond. Similar arguments apply, mutatis mutandis, to $M(\mu_2\text{-}H)_3M$, $M(\mu_2\text{-}H)_4M$, and $(\mu_3\text{-}H)M_3$ systems.

While admitting that a full molecular orbital treatment may be necessary to describe accurately the bonding in these species, our simple arguments above provide (at least) an accurate prediction of trends in hydrido-bridged metal–metal distances.

We very recently have reexamined the structure of $(\mu_2\text{-H})_2\text{Os}_3(\text{CO})_{10}$, and the detailed molecular geometry is illustrated in Figure 13.   Dimensions of interest within the equatorial plane are shown in Figure 14 (*45*).   One point worthy of note is that there is no observable repulsion of ligands adjacent to the di-μ_2-hydrido-bridged osmium–osmium vector.   This is presumably because neither the μ_2-hydride ligands nor the adjacent carbonyl ligands lie directly in the equatorial plane of the molecule.   Interligand repulsions are then relatively unimportant.

The final osmium hydride complex that we have examined is the species $(\mu_2\text{-H})\text{Os}_3(\text{CO})_{10}(\mu_2\text{-C}^-\text{H·CH}_2\text{·P}^+\text{Me}_2\text{Ph})$ (*46*) that is prepared as shown in Reaction 5 (*47*).   The molecular stereochemistry of this interesting 1,3-dipolar species is shown in Figures 15 and 16.   A particular point of interest is that the dibridged osmium–osmium vector, Os(1)–Os(2), is only 2.8002(6) Å in length whereas the nonbridged osmium–osmium bond lengths are Os(1)–Os(3) = 2.8688(6) Å and Os(2)–Os(3) = 2.8729(6) Å.   At first sight, this would appear to contradict our arguments above.   However, this apparent discrepancy is interpreted easily.   The second bridging ligand, the $\mu_2\text{-C}^-\text{H·CH}_2\text{·P}^+\text{Me}_2\text{Ph}$ ligand, also clearly exerts an influence on the osmium–osmium distance.   In fact, in common with all bridging ligands with carbon as the single bridge atom, the $\mu_2\text{-C}^-\text{H·CH}_2\text{·P}^+\text{Me}_2\text{Ph}$ ligand exerts a pronounced bond-shortening influence.

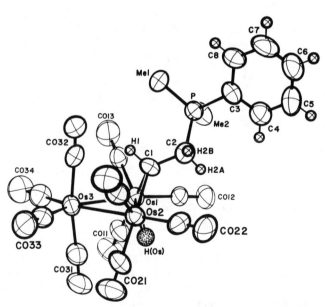

Inorganic Chemistry

Figure 16.   A further view of $(\mu_2\text{-H})\text{Os}_3(\text{CO})_{10}(\mu_2\text{-CH·}$ $CH_2\text{·}PMe_2Ph)$ (46) showing the relative juxtaposition of the μ_2-bridging ligands

$$(\mu_2\text{-H})_2\text{Os}_3(\text{CO})_{10} \xrightarrow{\text{C}_2\text{H}_2} (\mu_2\text{-H})\text{Os}_3(\text{CO})_{10}(\sigma,\pi\text{-CH}\!=\!\text{CH}_2)$$

$$\xrightarrow{\text{PMe}_2\text{Ph}} (\mu_2\text{-H})\text{Os}_3(\text{CO})_{10}(\mu_2\text{-C}^-\text{H}\cdot\text{CH}_2\cdot\text{P}^+\text{Me}_2\text{Ph}) \quad (5)$$

Our investigations of trinuclear and tetranuclear ruthenium complexes began with a study of $(\mu_2\text{-H})\text{Ru}_3(\text{CO})_{10}(\mu_2\text{-C}\!=\!\text{NMe}_2)$, prepared by Abel and co-workers (48) as indicated in Reaction 6. This molecule nominally contains a 1,2 dipolar (or ylid) $>\text{C}^-\!=\!\text{N}^+\text{Me}_2$ ligand. There are two crystallographically independent (but chemically equivalent) molecules in the asymmetric unit. The geometry of one such molecule is illustrated in Figure 17. As with the $(\mu_2\text{-H})\text{-}$ $\text{Os}_3(\text{CO})_{10}(\mu_2\text{-C}^-\text{H}\cdot\text{CH}_2\cdot\text{P}^+\text{Me}_2\text{Ph})$ molecule above, the dibridged ruthenium–ruthenium linkages are slightly, but significantly, shorter than the nonbridged metal–metal bonds (49). Precise distances are 2.7997(5) and 2.8016(6)

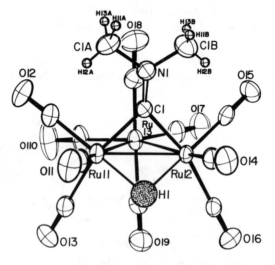

Inorganic Chemistry

Figure 17.   A view of a $(\mu_2\text{-H})\text{Ru}_3(\text{CO})_{10}(\mu_2\text{-}$ $\text{C}\!=\!\text{NMe}_2)$ molecule (49), showing the location of the μ_2-hydride ligand

Å for the dibridged vectors and from 2.8316(6) to 2.8336(6) Å for the normal nonbridged ruthenium–ruthenium bond lengths. Once again, the explanation is that the bond-lengthening influence of the μ_2-bridging hydride ligand is more than counterbalanced by the bond-shortening effect of the bridging $>\text{C}\!=\!\text{NMe}_2$ ligand. It should be noted that the bridging hydride ligands were located, with individual ruthenium–hydrogen distances ranging from 1.80(3) to 1.93(5) Å and averaging 1.85 Å. The ruthenium–hydrogen–ruthenium bridge angles in the two independent molecules were 95(2)° and 101(2)°.

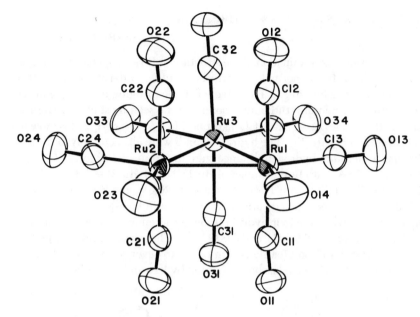

Figure 18. *The overall molecular structure of* $Ru_3(CO)_{12}$ *(50)*

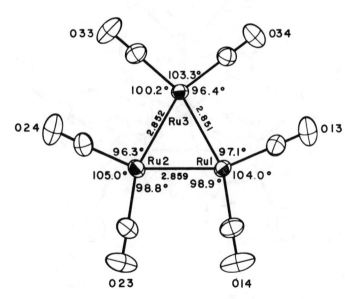

Figure 19. *Dimensions within the equatorial plane of* $Ru_3(CO)_{12}$ *(50)*

$$Ru_3(CO)_{12} + Me_3Sn-CH_2NMe_2 \rightarrow HRu_3(CO)_{10}(C{=}NMe_2)$$
$$+ (Me_3Sn)_2Ru(CO)_4 + \dots \quad (6)$$

Our recent redetermination of the structure of $Ru_3(CO)_{12}$ (50) (see Figures 18 and 19) leads to the normal (unperturbed) ruthenium–ruthenium distance being defined as 2.854(5) Å. The effect of single bridging hydride ligands on the length of ruthenium–ruthenium bonds can be seen most clearly from our study of the species $(\mu_2\text{-}H)_4Ru_4(CO)_{10}(diphos)$ (51). The core of this molecule is illustrated in Figure 20. The hydride ligands were all accurately located, individual ruthenium–hydrogen distances ranging from 1.64(6) to 1.81(4) Å and averaging 1.76 Å, with bridge angles of Ru(1)–H(12)–Ru(2) = 116(3)°, Ru(1)-–H(13)–Ru(3) = 115(2)°, Ru(1)–H(14)–Ru(4) = 120(3)°, and Ru(2)–H(23)–Ru(3) = 110(2)°. The nonbridged ruthenium–ruthenium bondlengths are 2.785(1) and 2.796(1) Å while the four hydrido-bridged ruthenium–ruthenium distances are 2.931(1), 2.946(1), 2.998(1), and 3.006(1) Å. These last data also suggest that the M–H–M system is soft (easily deformed). However, a far more striking example of this phenomenon has been reported by Bau and co-workers (52) from

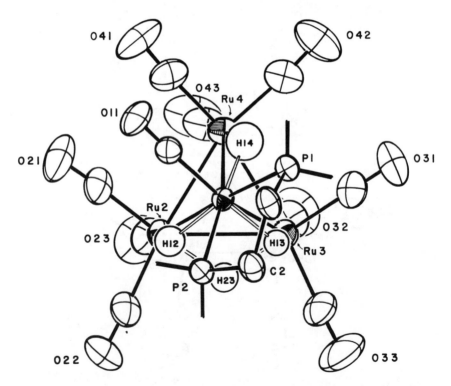

Journal of the American Chemical Society

Figure 20. A portion of the structure of $(\mu_2\text{-}H)_4Ru_4(CO)_{10}(diphos)$ (51) showing the positions of the hydride ligands

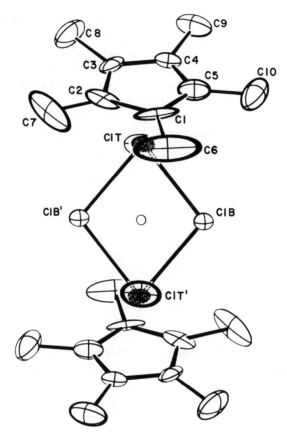

*Figure 21.   The [(η⁵-C₅Me₅)IrCl]₂(μ-Cl)₂ mole-
cule, projected onto its Ir(μ-Cl)₂Ir plane (see Ref.
58).   Note the obtuse Ir–Cl–Ir angles.*

studies of [(PPh₃)₂N⁺] and [Et₄N⁺] salts of the anion [(μ₂-H)W₂(CO)₁₀⁻]. [The
all-trans OC–W⋯W–CO system is linear with d(W⋯W) = 3.504(1) Å in the
[Et₄N⁺] salt and is bent by 15° with d(W⋯W) = 3.391(1) Å in the [(PPh₃)₂N⁺]
salt.]

Binuclear Complexes.   While the work on (principally) trinuclear complexes described above has, indeed, enabled us to locate and to characterize
structurally the hydride ligand in a variety of chemical environments, it is apparent that the M(μ₂-H)ₓM system under investigation is also subject to an additional restraint that essentially has been ignored.   Thus, for example, the
Os(μ₂-H)₂Os bridge in (μ₂-H)₂Os₃(CO)₁₀ suffers from the restriction that it is
bridged by a further group, namely the -Os(CO)₄- entity.   Clearly then, if one
is to see the full geometric effect of bridging hydride ligands, attention must be
restricted to binuclear species.

We have begun recently a series of studies on the species $[(\eta^5\text{-}C_5Me_5)\text{-}MX]_2(\mu_2\text{-}X)_2$ and $[(\eta^5\text{-}C_5Me_5)MX]_2(\mu_2\text{-}H)(\mu_2\text{-}X)$ (M = Rh, Ir; X = halogen). Some of these species originally were synthesized by Maitlis and co-workers (53), as shown in Reactions 7 and 8, and are of additional interest because of their efficacy as catalysts for the homogeneous hydrogenation of olefins etc. (54, 55). The active catalyst is, in fact, the μ-hydrido–μ-chloro species. However, this can be generated in situ from the di-μ-chloro species by the heterolytic cleavage of dihydrogen under the experimental conditions for hydrogenation.

$$2(\text{Dewar-}C_6Me_6) + 2MCl_3 \cdot 3H_2O + 4MeOH \rightarrow [(\eta^5\text{-}C_5Me_5)MCl]_2(\mu_2\text{-}Cl)_2$$
$$+ 2MeCH(OMe)_2 + 2HCl + 6H_2O \quad (7)$$

$$[(\eta^5\text{-}C_5Me_5)MCl]_2(\mu_2\text{-}Cl)_2 + H_2 \rightarrow [(\eta^5\text{-}C_5Me_5)MCl]_2(\mu_2\text{-}Cl)(\mu_2\text{-}H) + HCl$$
$$(8)$$

To date, we have completed single-crystal x-ray diffraction studies on four of these species, $[(\eta^5\text{-}C_5Me_5)RhCl]_2(\mu_2\text{-}Cl)_2$ (56), $[(\eta^5\text{-}C_5Me_5)RhCl]_2(\mu_2\text{-}H)(\mu_2\text{-}Cl)$ (57), $[(\eta^5\text{-}C_5Me_5)IrCl]_2(\mu_2\text{-}Cl)_2$ (58), and $[(\eta^5\text{-}C_5Me_5)IrCl]_2(\mu_2\text{-}H)(\mu_2\text{-}Cl)$ (58). The two iridium species are pictured in Figures 21 and 22. (The rhodium complexes are isomorphous with their appropriate iridium complex.) Geometric details of these four structures are compiled in Table I. There are some remarkable changes as one replaces a bridging chloride ligand with a bridging hydride ligand. Thus (using the parlance of Mason (43)), the nonbonding Rh···Rh and Ir···Ir distances in the electron-precise di-μ-chloro complexes are 3.719(1) and 3.769(1) Å, respectively. The corresponding rhodium–rhodium and iridium–iridium distances in the electron-deficient, μ-hydrido–μ-chloro species are contracted from these previous values by close to one Ångström unit, the resulting bonding distances being 2.903(1) and 2.906(1) Å, respectively. This contraction

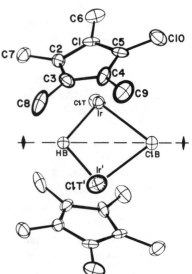

Inorganic Chemistry

Figure 22.   The $[(\eta^5\text{-}C_5Me_5)IrCl]_2\text{-}$ ($\mu\text{-}H$)($\mu\text{-}Cl$) molecule, projected onto its Ir($\mu\text{-}H$)($\mu\text{-}Cl$)Ir plane (see Ref. 58). Note the acute Ir–Cl–Ir angle.   The hydride ligand was located.

Table I.   Comparative Geometry of M(μ-X)(μ-Cl)M Bridges in [(η^5-C$_5$Me$_5$)MCl]$_2$(μ-X)(μ-Cl) Molecules (M = Ir, Rh; X = H, Cl)

	Ir(μ-H)(μ-Cl)Ir Complex	Rh(μ-H)(μ-Cl)Rh Complex	Ir(μ-Cl)$_2$Ir Complex	Rh(μ-Cl$_2$)Rh Complex
(A) Distances, Å				
M$\cdots$M	2.903(1)	2.906(1)	3.769(1)	3.719(1)
M-Cl(B)	2.451(4)	2.437(2)	2.453(5)	2.459(9)
M-Cl(T)	2.397(4)	2.393(2)	2.387(4)	2.397(1)
M-H(B)	1.939(65)	1.849(47)	—	—
M-C, av	2.155	2.151	2.132	2.128
(B) Angles, deg				
M-Cl(B)-M'	72.65(8)	73.20(6)	100.45(12)	98.29(3)
M-H(B)-M'	96.9(25)	103.6(37)	—	—
Cl(B)-M-H(B)	95.2(17)	91.6(10)	—	—
Cl(B)-M-Cl(B)'	—	—	79.55(12)	81.71(3)

in metal–metal distance is associated with numerous other geometric changes; in particular, there is a significant decrease (i.e., more than 25°) in the rhodium–chlorine–rhodium and iridium–chlorine–iridium bond angles. This can be seen clearly in Figures 21 and 22 and is indicated quantitatively by the data in Table I. It is interesting to note that a structural study on the related species [(μ_2-H)$_3$Ir$_2$(H)$_2$(PPh$_3$)$_4^+$] has been reported (59); here the tri-μ-hydrido-bridged iridium–iridium distance is reduced to 2.518 Å. Such a linkage can (in keeping

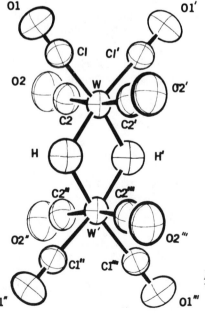

Inorganic Chemistry

Figure 23.   Geometry of the [(μ_2-H)$_2$-W$_2$(CO)$_8^{2-}$] ion (see Ref. 61). Hydride ligands were located.

with our discussion above) be viewed in a formal sense as a triply protonated triple bond.

The final structure with which we have been concerned is the anionic species $[(\mu_2\text{-H})_2W_2(CO)_8^{2-}]$. This was isolated as the $[Et_4N^+]$ salt by Davison and co-workers *(60)*, and the crystal structure was determined by our research group *(61)*. The hydride ligands were located directly (W–H = 1.86(6) Å and the W–H–W angles are 108.6(52)°). The molecule is shown in Figure 23. The W–W bond length is 3.0162(11) Å and is some 0.2 Å shorter than the W–W single-bond length of 3.222(1) Å in $[(\eta^5\text{-}C_5H_5)W(CO)_3]_2$ *(12)*. The $[(\mu_2\text{-H})_2\text{-}W_2(CO)_8^{2-}]$ anion is isoelectronic with the neutral $(\mu_2\text{-H})_2Re_2(CO)_8$ molecule that Bennett and co-workers found *(62)* to have a Re–Re bond length of 2.896(3) Å (i.e., some 0.1 Å shorter than a Re–Re single bond). Of course, each of these dihydrido-bridged systems can be regarded formally as containing a diprotonated double bond.

Finally, we note here that Bau and co-workers have discovered a $Re(\mu_2\text{-}H)_4Re$ system in the $(\mu_2\text{-H})_4Re_2(H)_4(PEt_2Ph)_4$ molecule *(63)*. Such a system can be viewed formally as a quadruply protonated quadruple bond and is characterized by a very short Re–Re separation of 2.538(4) Å.

Acknowledgment

None of this work would have been completed if it were not for collaborative efforts with other research groups. I list the directors of these in the approximate historical order in which our joint efforts took place: H. D. Kaesz, M. J. Mays, J. A. Osborn, P. M. Maitlis, A. Davison, E. W. Abel, and J. R. Shapley.

The crystallographic work was carried out by graduate students and postdoctoral fellows. These include (again, in historical order) R. Bau, P. H. Bird, S. W. Kirtley, J. Wormald, S. A. Bezman, S. W. Y. (Ni) Chang, B. G. DeBoer, F. J. Rotella, R. A. Lashewycz, S. A. Julis, F. J. Hollander, and J. P. Hutchinson. This work has been generously funded by the National Science Foundation, through grants GP-42724-X, CHE76-05564, and CHE77-04981.

Literature Cited

1. Hieber, W., *Die Chemie* (1942) **55**, 24.
2. Ewens, R. V. G., Lister, M. W., *Trans. Faraday Soc.* (1939) **35**, 681.
3. Owston, P. G., Partridge, J. M., Rowe, J. M., *Acta Crystallogr.* (1960) **13**, 246.
4. La Placa, S. J., Ibers, J. A., *J. Am. Chem. Soc.* (1963) **85**, 3501.
5. La Placa, S. J., Ibers, J. A., *Acta Crystallogr.* (1965) **18**, 511.
6. Abrahams, S. C., Ginsberg, A. P., Knox, K., *Inorg. Chem.* (1964) **3**, 558.
7. Baker, R. W., Pauling, P., *J. Chem. Soc., Chem. Commun.* (1969) 1495.
8. Baker, R. W., Ilmaier, B., Pauling, P. J., Nyholm, R. S., *J. Chem. Soc., Chem. Commun.* (1970) 1077.
9. La Placa, S. J., Hamilton, W. C., Ibers, J. A., Davison, A., *Inorg. Chem.* (1969) **8**, 1928.
10. Doedens, R. J., Dahl, L. F., *J. Am. Chem. Soc.* (1965) **87**, 2576.

11. Petersen, J. L., Dahl, L. F., Williams, J. M., *J. Am. Chem. Soc.* (1974) **96**, 6610.
12. Adams, R. D., Collins, D. M., Cotton, F. A., *Inorg. Chem.* (1974) **13**, 1086.
13. Cotton, F. A., Troup, J. M., *J. Am. Chem. Soc.* (1974) **96**, 4155.
14. Wei, C. H., Dahl, L. F., *J. Am. Chem. Soc.* (1969) **91**, 1351.
15. Dahl, L. F., Blount, J. F., *Inorg. Chem.* (1965) **4**, 1373.
16. Handy, L. B., Treichel, P. M., Dahl, L. F., Hayter, R. G., *J. Am. Chem. Soc.* (1966) **88**, 366.
17. Handy, L. B., Ruff, J. K., Dahl, L. F., *J. Am. Chem. Soc.* (1970) **92**, 7312.
18. Roziere, J., Williams, J. M., Stewart, R. P., Petersen, J. L., Dahl, L. F., *J. Am. Chem. Soc.* (1977) **99**, 4497.
19. Kaesz, H. D., Bau, R., Churchill, M. R., *J. Am. Chem. Soc.* (1967) **89**, 2775.
20. Churchill, M. R., Bau, R., *Inorg. Chem.* (1967) **6**, 2086.
21. Churchill, M. R., Bird, P. H., Kaesz, H. D., Bau, R., Fontal, B., *J. Am. Chem. Soc.* (1968) **90**, 7135.
22. Kirtley, S. W., Kaesz, H. D., Churchill, M. R., Knobler, C., in preparation.
23. Kirtley, S. W., Ph.D. dissertation, UCLA, 1972.
24. Kaesz, H. D., Fontal, B., Bau, R., Kirtley, S. W., Churchill, M. R., *J. Am. Chem. Soc.* (1969) **91**, 1021.
25. Bau, R., Fontal, B., Kaesz, H. D., Churchill, M. R., *J. Am. Chem. Soc.* (1967) **89**, 6374.
26. Churchill, M. R., Bau, R., *Inorg. Chem.* (1968) **7**, 2606.
27. Churchill, M. R., Wormald, J., Knight, J., Mays, M. J., *J. Chem. Soc., Chem. Commun.* (1970) 458.
28. Churchill, M. R., Wormald, J., *J. Am. Chem. Soc.* (1971) **93**, 5670.
29. Churchill, M. R., Bezman, S. A., Osborn, J. A., Wormald, J., *Inorg. Chem.* (1972) **11**, 1818.
30. Bezman, S. A., Churchill, M. R., Osborn, J. A., Wormald, J., *J. Am. Chem. Soc.* (1971) **93**, 2063.
31. White, R. P., Block, T. E., Dahl, L. F., reported by B. A. Frenz and J. A. Ibers, "Transition Metal Hydrides," E. L. Muetterties, Ed., p 61, Marcel Dekkar, Inc., 1971.
32. Olsen, J. P., Koetzle, T. F., Kirtley, S. W., Andrews, M., Tipton, D. L., Bau, R., *J. Am. Chem. Soc.* (1974) **96**, 6621.
33. Shapley, J. R., Keister, J. B., Churchill, M. R., DeBoer, B. G., *J. Am. Chem. Soc.* (1975) **97**, 4145.
34. Corey, E. R., Dahl, L. F., *Inorg. Chem.* (1962) **1**, 521.
35. Churchill, M. R., DeBoer, B. G., *Inorg. Chem.* (1977) **16**, 878.
36. Ibid. (1977) **16**, 2397.
37. Churchill, M. R., Lashewycz, R. A., Tachikawa, M., Shapley, J. R., *J. Chem. Soc., Chem. Commun.* (1977) 699.
38. Knight, J., Mays, M. J., *J. Chem. Soc., Dalton Trans.* (1972) 1022.
39. Churchill, M. R., Hollander, F. J., *Inorg. Chem.* (1977) **16**, 2493.
40. Shapley, J. R., Pearson, G. A., Tachikawa, M., Schmidt, G. D., Churchill, M. R., Hollander, F. J., *J. Am. Chem. Soc.* (1977) **99**, 8064.
41. Churchill, M. R., Hollander, F. J., unpublished data.
42. Mason, R., *IUPAC Congress, 23rd, Massachusetts, 1971*, **6**, 31.
43. Mason, R., Mingos, D. M. P., *J. Organomet. Chem.* (1973) **50**, 53.
44. Teo, B. K., Hall, M. B., Fenske, R. F., Dahl, L. F., *J. Organomet. Chem.* (1974) **70**, 413.
45. Churchill, M. R., Hollander, F. J., Hutchinson, J. P., *Inorg. Chem.* (1977) **16**, 2697.
46. Churchill, M. R., DeBoer, B. G., *Inorg. Chem.* (1977) **16**, 1141.
47. Churchill, M. R., DeBoer, B. G., Shapley, J. R., Keister, J. B., *J. Am. Chem. Soc.* (1976) **98**, 2357.
48. Churchill, M. R., DeBoer, B. G., Rotella, F. J., Abel, E. W., Rowley, R. J., *J. Am. Chem. Soc.* (1975) **97**, 7158.
49. Churchill, M. R., DeBoer, B. G., Rotella, F. J., *Inorg. Chem.* (1976) **15**, 1843.
50. Churchill, M. R., Hollander, F. J., Hutchinson, J. P., *Inorg. Chem.* (1977) **16**, 2655.

51. Shapley, J. R., Richter, S. I., Churchill, M. R., Lashewycz, R. A., *J. Am. Chem. Soc.* (1977) **99**, 7384.
52. Wilson, R. D., Graham, S. A., Bau, R., *J. Organomet. Chem.* (1975) **91**, C49.
53. Kang, J. W., Moseley, K., Maitlis, P. M., *J. Am. Chem. Soc.* (1969) **91**, 5970.
54. White, C., Gill, D. S., Kang, J. W., Lee, H. B., Maitlis, P. M., *J. Chem. Soc., Chem. Commun.* (1971) 734.
55. White, C., Oliver, A. J., Maitlis, P. M., *J. Chem. Soc., Dalton Trans.* (1973) 1901.
56. Churchill, M. R., Julis, S. A., Rotella, F. J., *Inorg. Chem.* (1977) **16**, 1137.
57. Churchill, M. R., Ni, S. W.-Y., *J. Am. Chem. Soc.* (1973) **95**, 2150.
58. Churchill, M. R., Julis, S. A., *Inorg. Chem.* (1977) **16**, 1488.
59. Crabbee, R. H., Felkine, H., Morris, G. E., King, T. J., Richards, J. A., *J. Organomet. Chem.* (1976) **113**, C7.
60. Churchill, M. R., Chang, S. W.-Y., Berch, M. L., Davison, A., *J. Chem. Soc., Chem. Commun.* (1973) 691.
61. Churchill, M. R., Chang, S. W.-Y., *Inorg. Chem.* (1974) **13**, 2413.
62. Bennett, M. J., Graham, W. A. G., Hoyano, J. K., Hutcheon, W. L., *J. Am. Chem. Soc.* (1972) **94**, 6232.
63. Bau, R., Carroll, W. E., Teller, R. G., Koetzle, T. F., *J. Am. Chem. Soc.* (1977) **99**, 3872.

RECEIVED July 19, 1977.

Neutron Diffraction Studies of Tetrahedral Cluster Transition Metal Hydride Complexes: $HFeCo_3(CO)_9$-$(P(OCH_3)_3)_3$ and $H_3Ni_4(C_5H_5)_4$

THOMAS F. KOETZLE and RICHARD K. McMULLAN–Department of Chemistry, Brookhaven National Laboratory, Upton, NY 11973

ROBERT BAU, DONALD W. HART, RAYMOND G. TELLER, DONALD L. TIPTON and ROBERT D. WILSON[1]–Department of Chemistry, University of Southern California, Los Angeles, CA 90007

Structures of the tetrahedral cluster transition metal hydride complexes $HFeCo_3(CO)_9(P(OCH_3)_3)_3$ and $H_3Ni_4(C_5H_5)_4$ have been investigated by low-temperature neutron diffraction techniques. Both complexes have approximate C_{3v} symmetry. In $HFeCo_3(CO)_9(P(OMe)_3)_3$, the hydride ligand is found outside the $FeCo_3$ cluster, 0.978(3) Å from the Co_3 face and essentially on the molecular threefold axis, triply bridging the cobalt atoms. In $H_3Ni_4(Cp)_4$, the three hydride ligands are face bridging, and their mean displacement from the faces of the cluster is 0.90(3) Å. The H_3Ni_4 core may be envisaged as a distorted cube with one vertex unoccupied. The observed geometries of the two clusters considered here suggest a plausible model for chemisorption of hydrogen on {111} or {001} surfaces of ccp or hcp metals, respectively, in which hydrogen atoms are located approximately 1 Å above the centers of triangles of metal atoms.

A variety of factors contribute to the great current interest in polynuclear metal hydride complexes. These include the novel geometries found in these systems and their usefulness as models for the bonding of hydrogen to metals, such as may occur in catalysis (1) or hydrogen-storage applications (2). A comprehensive review of metal hydride complexes, in which polynuclear species are included, has been published by Kaesz and Saillant (3).

[1] Present address: Department of Chemistry, Northwestern University, Evanston, IL 60201.

0-8412-0390-3/78/33-167-061/$05.00/0

Direct location of hydride ligands in metal complexes by x-ray diffraction may be difficult, especially in the case of bridging hydrides commonly occurring in polynuclear systems. Several cases have been reported where the hydride was found successfully despite this difficulty. For example, Churchill and De-Boer have located the bridging hydride in $HOs_3(CO)_{10}(CHCH_2PMe_2Ph)$ (4) while we have used a Fourier-averaging technique to locate the face-bridging hydrogen atoms in $H_4Re_4(CO)_{12}$ (5). However, x-ray diffraction studies cannot, in any event, be expected to provide metal–hydrogen bond lengths of accuracy much better than ±0.1 Å. Thus, precise information on geometries of metal hydride complexes has depended on neutron diffraction. The sensitivity of neutron diffraction to light atoms in general and hydrogen in particular is caused by the large relative cross sections of these atoms compared with those of heavy atoms. For example, the ratio $\sigma(H):\sigma(Os)$, is 0.12 for neutrons and 1.7×10^{-4} for x-rays ($2\theta = 0°$). Thus, the relative contribution of hydrogen in a structure containing osmium will be roughly three orders of magnitude greater in neutron than in x-ray diffraction.

In this chapter we briefly review some results of prior neutron diffraction studies and present new results for two tetrahedral cluster complexes with face-bridging hydride ligands: $HFeCo_3(CO)_9(P(OMe)_3)_3$ and $H_3Ni_4(Cp)_4$. (Abbreviations used in this paper are as follows: Me–methyl, Et–ethyl, Cp–cyclopentadienyl, and Ph–phenyl.) In addition to these tetrahedral complexes, single-crystal, neutron diffraction data currently are available for the triangular ruthenium and osmium species $HRu_3(CO)_9$ (C≡C–C(Me)_3) (6), $HOs_3(CO)_9R$, R = H, vinyl (7), and $HDOs_3(CO)_{10}·(CHD)$ (8), as well as for the dodecanickel cluster anions $[HNi_{12}(CO)_{21}]^{3-}$ and $[H_2Ni_{12}(CO)_{21}]^{2-}$ (9). A neutron powder diffraction study of HNb_6I_{11} also has been reported (10). The hydride ligand was located at the center of the octahedral Nb_6 cluster, similar to the situation in the dodecanickel anions mentioned above where the hydride ligands occur in octahedral sites in the nickel framework.

Prior Neutron Diffraction Work on Transition Metal Hydride Complexes

The first neutron diffraction study of a transition metal hydride complex was that of K_2ReH_9 (11), reported in 1964, that showed that the $[H_9Re]^{2-}$ dianion forms a tricapped trigonal prism with a mean rhenium–hydrogen bond distance of 1.68(1) Å. This investigation, together with subsequent x-ray (12) and neutron (13) diffraction studies of $HMn(CO)_5$, demonstrated unequivocally that hydrogen is a stereochemically active ligand in the latter complex, and that terminal metal–hydrogen distances correspond to those expected for normal covalent bonds. More recently, a substantial body of accurate data on terminal and bridging metal–hydrogen bonds has emerged based on neutron diffraction studies of 23 complexes, listed in Table I. We have published a review covering this work up to 1976 (26), and results for polyhydride complexes are discussed in an accompanying article (27).

Table I.   Transition Metal Hydride Complexes Studied by Neutron Diffraction[a]

A. *Mononuclear*

$HMn(CO)_5$ (*13*)
$HZnN(Me)C_2H_4N(Me)_2$ (*14*)
$D_2Mo(Cp)_2$ (*15*)
$H_2Mo(Cp)_2$ (*16*)
$H_3Ta(Cp)_2$ (*17*)
$K_2^+[H_9Re]^{2-}$ (*11*)
$H_4Os(PMe_2Ph)_3$ (*18*)

B. *Binuclear*

$[Et_4N]^+[HCr_2(CO)_{10}]^-$ (*19*)
$[(Ph_3P)_2N]^+[HCr_2(CO)_{10}]^-$ (*20*)
$HMo_2(Cp)_2(CO)_4(PMe_2)$ (*21*)
$HW_2(CO)_9(NO)$ (*22*)
$HW_2(CO)_8(NO)(P(OMe)_3)$ (*23*)
$H_8Re_2(PEt_2Ph)_4$ (*24*)
$[H_3Ir_2(C_5Me_5)_2]^+BF_4^-$ (*25*)

C. *Polynuclear*

$HFeCo_3(P(OMe)_3)_3$
$H_3Ni_4(Cp)_4$
$[(Ph_3P)_2N]_3^+]HNi_{12}(CO)_{21}]^{3-}$ (*9*)
$[(Ph_3P)_2N]_2^+[H_2Ni_{12}(CO)_{21}]^{2-}$ (*9*)
HNb_6I_{11} (*10*)
$HRu_3(CO)_9(C{\equiv}C-C(Me)_3)$ (*6*)
$H_2Os_3(CO)_{10}$ (*7*)
$HOs_3(CO)_{10}(C_2H_3)$ (*7*)
$HDOs_3(CO)_{10}(CHD)$ (*8*)

[a] Me: methyl; Et: ethyl; Cp: cyclopentadienyl; Ph: phenyl.

The studies of $HMo_2(Cp)_2(CO)_4(PMe_2)$ (*21*), $HW_2(CO)_9(NO)$ (*22*), and $HW_2(CO)_8(NO)(P(OMe)_3)$ (*23*) are of particular significance and provide definitive evidence that metal–hydrogen–metal bridges in these binuclear species are best described as closed, three-center bonds with significant metal–metal interaction (*22*). It is not surprising that this is the case since metal orbitals of proper symmetry that interact with H(1s) also can interact with one another, as has been pointed out by Hoffmann (*28*). In $HW_2(CO)_8(NO)(P(OMe)_3)$ (Figure 1), the tungsten–hydrogen–tungsten bridge was found to be slightly asymmetric with the hydride ligand displaced toward the $W(CO)_5$ group, as could be predicted on electron-counting considerations. It is likely that such asymmetry also exists in $HW_2(CO)_9(NO)$, but the effect could not be measured since both crystalline forms of this latter complex exhibit disorder with rotation of the molecule around the pseudo twofold axis passing through the hydrogen atom.

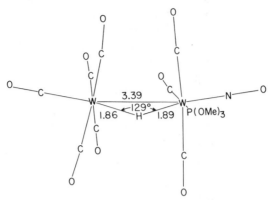

Figure 1.   View of HW$_2$(CO)$_8$(NO)(P(OMe)$_3$)

Experimental

A powdered sample of HFeCo$_3$(CO)$_9$(P(OMe)$_3$)$_3$ was supplied by H. D. Kaesz and B. T. Huie of the University of California, Los Angeles, and recrystallized from a hexane:CH$_2$Cl$_2$ (6:1) mixture.   Single crystals of H$_3$Ni$_4$(Cp)$_4$ were supplied by J. Müller of The Technical University of Berlin.   Large single crystals of both compounds were affixed to aluminum pins and mounted in cryostats on an automated four-circle diffractometer (29, 30) at the Brookhaven High Flux Beam Reactor.   HFeCo$_3$(CO)$_9$(P(OMe)$_3$)$_3$ was studied at 90°K, and H$_3$Ni$_4$(Cp)$_4$ at 81°K.   Crystal data and experimental parameters are summarized in Table II.

Table II.   Crystal Data and Experimental Parameters

	HFeCo$_3$(CO)$_9$-(P(OMe)$_3$)$_3$	H$_3$Ni$_4$(Cp)$_4$
Space group	P2$_1$/c	C2/c
Cell parameters a	15.957(8) Å	28.312(13) Å
b	10.611(5)	9.234(5)
c	18.383(9)	14.783(7)
β	98.70(2)°	103.35(2)°
Cell vol	3077(3) Å^3	3760(3) Å^3
No. of molecules per unit cell (Z)	4	8
Mol wt	858.0	498.2
Calc. density	1.85 g/cm^3	1.76 g/cm^3
Absorption coefficient (μ)[a]	1.54 cm^{-1}	1.94 cm^{-1}
Wavelength	1.1598(1) Å	1.0183(1) Å
Sample vol	31.2 mm^3	12.5 mm^3
Data collection temp	90.0(4)°K	81(1)°K
Data collection limit (sinθ/λ)	0.68 Å^{-1}	0.68 Å^{-1}
No. of reflections used in structure analysis	8229	2656
Final agreement factors[b]	$R_F = 0.070$	$R_F = 0.107$
	$R_{wF} = 0.035$	$R_{wF} = 0.067$

[a] Calculated assuming an incoherent scattering cross section for hydrogen of 40 barn.
[b] $R_F = \Sigma|F_o - |F_c||/\Sigma F_o$ $R_{wF} = \{\Sigma w|F_o - |F_c||^2/\Sigma w F_o{}^2\}^{1/2}$.

For HFeCo$_3$(CO)$_9$(P(OMe)$_3$)$_3$, starting phases were calculated based on the positions of nonhydrogen atoms determined from a prior x-ray analysis (*31*), and all hydrogen atoms were then located in a difference-Fourier synthesis. Initial refinement was carried out with an automated procedure using differential-Fourier syntheses (*32*), followed by full-matrix least-squares based upon F_0^2, including reflections with $F_0^2 < 0$. Parameters were blocked into groups of ca. 250, and anisotropic thermal factors were used for all atoms. Satisfactory convergence was achieved, and all bond distances were determined with precision better than 0.004 Å.

For H$_3$Ni$_4$(Cp)$_4$, the initial phasing model consisted of the nickel and carbon atoms at positions determined in an earlier x-ray study (*33, 34*), with cyclopentadienyl hydrogen atoms in calculated positions. The hydride ligands were located in a difference-Fourier synthesis, and the structure was refined by least-squares procedures, including only reflections with $F_0^2 > 1.5\sigma(F_0^2)$. The relatively high discrepancy between calculated and observed structure factors (*see* Table II) results from the fact that a large fraction of the reflections were measured to have very low intensity, i.e. 3478 of a total of 5633 unique reflections were observed with $F_0^2 < 3\sigma(F_0^2)$. However, chemically equivalent bond lengths in the H$_3$Ni$_4$ core agree to within 0.04 Å. Anisotropic thermal factors refined to quite large values for certain atoms in the Cp rings, as might be expected if the barrier to rotation of the rings in the solid state is assumed to be low.

Results

HFeCo$_3$(CO)$_9$(P(OMe)$_3$)$_3$. This complex is found to possess essentially C_{3v} symmetry, with the geometry shown schematically in Figure 2. Figure 3 illustrates the molecular structure with thermal ellipsoids and gives the atomic numbering scheme. The hydride ligand is located outside the FeCo$_3$ cluster, 0.978(3) Å from the Co$_3$ face, triply bridging the cobalt atoms. Excluding the cobalt–cobalt bonds, the environment of each cobalt atom is approximately octahedral, and the position of the hydride ligand might be inferred by the presence of one vacancy common to all three octahedra, trans to the terminal carbonyl on each cobalt atom. These results confirm the findings of Huie et al. (*31*) based on their x-ray diffraction study at 134°K. The bridging hydride is found es-

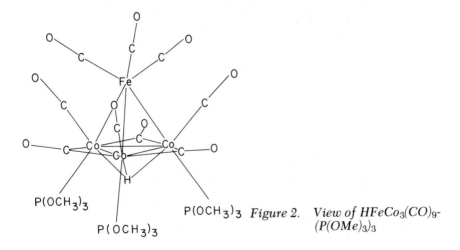

Figure 2. View of HFeCo$_3$(CO)$_9$-(P(OMe)$_3$)$_3$

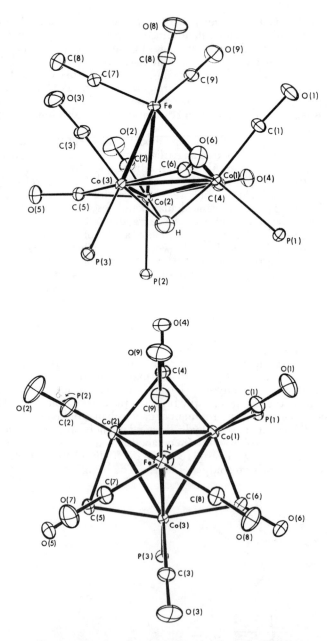

Figure 3. Molecular structure of HFeCo₃(CO)₉-(P(OMe)₃)₃ with thermal ellipsoids drawn to enclose 50% probability (Ref. 35). Methoxy groups have been removed for clarity. (Top) View normal to the three-fold molecular axis; (bottom) view approximately along the three fold axis.

Table III. Selected Bond Distances and Angles in HFeCo₃-(CO)₉(P(OMe)₃)₃ᵃ

Distances (Å)		*Angles (°)*	
Co(1)–H	1.742(3)	Co(1)–H–Co(2)	92.1(1)
Co(2)–H	1.731(3)	Co(1)–H–Co(3)	91.7(1)
Co(3)–H	1.728(3)	Co(2)–H–Co(3)	91.5(1)
Mean	1.734(4)	Mean	91.8(2)
Co(1)–Fe	2.563(2)	Fe–Co(1)–H	89.5(1)
Co(2)–Fe	2.556(2)	Fe–Co(2)–H	90.0(1)
Co(3)–Fe	2.558(2)	Fe–Co(3)–H	90.0(1)
Mean	2.559(2)	Mean	89.8(2)
Co(1)–Co(2)	2.501(2)	Fe–Co(1)–Co(2)	60.6(1)
Co(1)–Co(3)	2.489(3)	Fe–Co(1)–Co(3)	60.8(1)
Co(2)–Co(3)	2.477(3)	Fe–Co(2)–Co(1)	60.9(1)
Mean	2.489(7)	Fe–Co(2)–Co(3)	61.1(1)
		Fe–Co(3)–Co(1)	61.0(1)
Mean values		Fe–Co(3)–Co(2)	61.0(1)
Co–P	2.175(4)	Mean	60.9(1)
Co–C (terminal CO)	1.756(4)		
Co–C (bridging CO)	1.953(6)	Co(1)–Fe–Co(2)	58.5(1)
Fe–C	1.798(2)	Co(1)–Fe–Co(3)	58.2(1)
C–O (terminal)	1.147(1)	Co(2)–Fe–Co(3)	58.0(1)
C–O (bridging)	1.165(1)	Mean	58.2(1)
		Mean values	
		P–Co–H	91.6(39)
		C–Co–H (terminal CO)	170.5(14)
		C–Co–H (bridging CO)	83.8(6)
		Co–C–Co	79.2(1)

ᵃ Standard deviations of mean values are calculated as $(\Sigma(x_i - \bar{x})^2/n(n-1))^{1/2}$, where n is the number of observations. The resulting deviations are to be regarded as rough estimates of uncertainty, in cases where $n = 3$.

sentially on the molecular threefold axis, as illustrated in Figure 3 (bottom). Selected bond distances and angles are presented in Table III. It is interesting to note that the mean carbon–cobalt–hydrogen angle (terminal cobalt) is 170(1)°, so that the hydride ligand lies about 0.3 Å farther from the Co₃ face than would be predicted on the basis of undistorted octahedral coordination around the cobalt atoms. In HRu₃(CO)₉(C≡C–C(Me)₃), the analogous mean carbon–ruthenium–hydrogen angle involving the carbonyls trans to the doubly bridging hydride ligand is 176.4(5)° (6).

H₃Ni₄(Cp)₄. The structure of H₃Ni₄(Cp)₄, shown schematically in Figure 4, consists of a tetrahedral nickel cluster with each nickel atom π-bonded to a Cp ring. The three hydride ligands are face bridging, as deduced on the basis of x-ray data (33, 34). Figure 5 gives a close-up view of the H₃Ni₄ core, which may be envisaged as a tricapped tetrahedron or equivalently as a distorted cube with alternate corners occupied by Ni and H atoms and one corner vacant. Selected bond distances and angles are presented in Table IV. The mean displacement of the hydride ligands from the faces of the Ni₄ cluster is 0.90(3) Å.

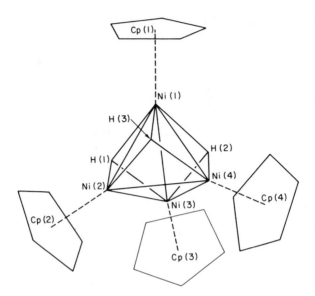

Figure 4.   View of $H_3Ni_4(Cp)_4$

Discussion

In 1968, an unusual structure for $HFeCo_3(CO)_{12}$ with the hydride ligand located inside the cage was proposed by Mays (36) on the basis of mass spectral evidence and electron-counting considerations.   However, this model was disproved by the x-ray work of Huie et al. on the tris(trimethyl phosphite) derivative (31), in which the hydride ligand was located in a difference-Fourier synthesis and shown to bridge the Co_3 face.   The present neutron diffraction study has allowed definitive placement of the hydride ligand and has yielded more accurate bond distances and angles.

One motivation to carry out a neutron diffraction investigation of $H_3Ni_4(Cp)_4$ was to check the possibility of disorder of the hydride ligands over all four faces of the Ni_4 tetrahedron.   The hydrides were not located from the x-ray data (33, 34).   Rather, their positions were inferred from the deviations of the structure from strict tetrahedral symmetry.   The observed $Cp(i)–Cn–Cp(j)$ angles (*see* Table IV) are distorted from the tetrahedral value such that $Cp(2)$, $Cp(3)$, and $Cp(4)$ are bent away from $Cp(1)$.   The face defined by $Ni(2)$, $Ni(3)$, and $Ni(4)$ therefore could be expected to be vacant.   Our neutron results indicate that this is indeed the case, with no evidence for disorder of the hydride ligands on the nuclear density maps.

The metal clusters in $HFeCo_3(CO)_9(P(OMe)_3)_3$ and $H_3Ni_4(Cp)_4$ contain different numbers of electrons.   The former cluster is a closed-shell structure (60 electrons) while the latter contains 63 electrons and is paramagnetic with S

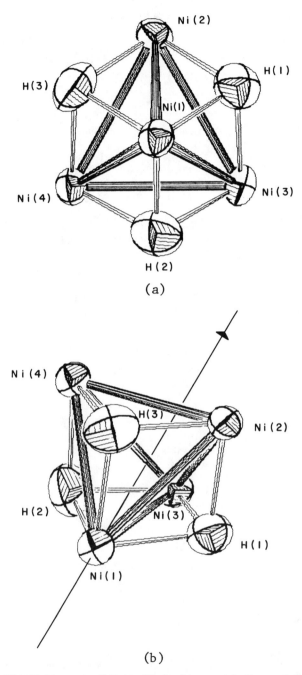

(a)

(b)

Figure 5.   The H_3Ni_4 core of $H_3Ni_4(Cp)_4$, drawn with thermal ellipsoids enclosing 50% probability.   (a) View approximately along the threefold molecular axis; (b) view approximately normal to the Ni(1)–Ni(2) bond.

Table IV.   Selected Distances and Angles in $H_3Ni_4(Cp)_4$[a,b]

Distances (Å)		Angles (°)	
Ni(1)–H(1)	1.720(8)	Ni(1)–H(1)–Ni(2)	94.0(4)
Ni(1)–H(2)	1.718(9)	Ni(1)–H(1)–Ni(3)	93.1(4)
Ni(1)–H(3)	1.711(7)	Ni(1)–H(2)–Ni(3)	93.6(4)
Mean	1.716(3)	Ni(1)–H(2)–Ni(4)	93.9(4)
		Ni(1)–H(3)–Ni(2)	94.7(4)
Ni(2)–H(1)	1.684(7)	Ni(1)–H(3)–Ni(4)	93.0(3)
Ni(2)–H(3)	1.674(8)	Mean	93.7(3)
Ni(3)–H(1)	1.674(8)		
Ni(3)–H(2)	1.661(9)	Ni(2)–H(1)–Ni(3)	94.1(4)
Ni(4)–H(2)	1.672(8)	Ni(3)–H(2)–Ni(4)	95.7(4)
Ni(4)–H(3)	1.704(8)	Ni(2)–H(3)–Ni(4)	93.1(4)
Mean	1.678(6)	Mean	94.3(8)
Ni(1)–Ni(2)	2.490(3)	Cp(1)[c]–Cn[d]–Cp(2)	117.5(2)
Ni(1)–Ni(3)	2.464(3)	Cp(1)–Cn–Cp(3)	112.3(2)
Ni(1)–Ni(4)	2.478(3)	Cp(1)–Cn–Cp(4)	112.3(2)
Mean	2.477(8)	Mean	114.0(17)
Ni(2)–Ni(3)	2.458(3)	Cp(2)–Cn–Cp(3)	105.0(2)
Ni(2)–Ni(4)	2.454(3)	Cp(2)–Cn–Cp(4)	103.6(2)
Ni(3)–Ni(4)	2.471(3)	Cp(3)–Cn–Cp(4)	105.1(2)
Mean	2.461(5)	Mean	104.6(5)
H(1)···H(2)	2.317(11)		
H(1)···H(3)	2.305(10)		
H(2)···H(3)	2.326(9)		
Mean	2.316(6)		
Ni(1)–Cp[c](1)	1.758(2)		
Ni(2)–Cp(2)	1.761(2)		
Ni(3)–Cp(3)	1.763(2)		
Ni(4)–Cp(4)	1.764(2)		
Mean	1.763(1)		
Cn[d]–Cp(1)	3.279(2)		
Cn–Cp(2)	3.258(2)		
Cn–Cp(3)	3.262(2)		
Cn–Cp(4)	3.268(2)		
Mean	3.263(3)		

Overall Mean Values

Ni–Ni	2.469(6)	Ni–H–Ni	93.9(3)
Ni–H	1.691(8)	H–Ni–H	86.1(6)
Ni–Cp	1.762(1)	Ni–Ni–Ni	60.0(2)
Cn–Cp	3.267(5)		
C–C (Cp rings)	1.408(5)		
C–H	1.076(8)		

[a] Standard deviations of mean values calculated as in Table III.
[b] The numbering of Ni(2) and Ni(4), as well as their attached Cp rings, has been interchanged compared with that given in Ref. 30.   This has been done to ensure that atom numbers increase upon clockwise rotation when viewed along the threefold axis, with Ni(1) pointing up.
[c] Cp = ring centroid.
[d] Cn = Ni_4 centroid.

= 3/2 (33). (For a description of the electron-counting procedure as applied to metal clusters, *see* Ref 37.) The paramagnetism of the nickel cluster, in principle, could be detected directly by neutron diffraction with a polarized beam and an external magnetic field. However, such measurements were not undertaken, and the effects of paramagnetism on the observed diffraction intensities, that are small in the present experiment, were ignored.

Mean cobalt–cobalt and nickel–nickel distances observed in these complexes are very close to interatomic distances determined at ambient temperatures in cobalt and nickel metals (Co–Co: 2.489(7) Å vs. 2.507 Å in α-cobalt (38); Ni–Ni: 2.469(6) Å vs. 2.492 Å in the metal (39)). The mean M–H bond lengths, as well as hydride displacements from M_3 faces, are less for nickel in $H_3Ni_4(Cp)_4$ than for cobalt in $HFeCo_3(CO)_9(P(OMe)_3)_3$. Although the differences are marginally significant within error limits (Ni–H 1.691(8) Å vs. Co–H 1.734(4) Å; displacements from plane: Ni_3 0.90(3) Å vs. Co_3 0.978(3) Å), they are in the expected direction since the covalent radius should vary inversely with atomic number within a transition series. However, other effects such as the number of electrons in the cluster also can influence these dimensions.

The geometries of the two clusters considered here suggest a plausible model for chemisorption of hydrogen on close-packed metals. Thus hydrogen atoms might be placed roughly 1 Å above the centers of triangles of metal atoms, such as occur on {111} or {001} surfaces of ccp or hcp metals, respectively. In this model, adjacent hydrogen atoms are roughly 1.4 Å apart and therefore separated well beyond bonding distance. It is widely believed that hydrogen chemisorbed on metals is bound in the monatomic form.

Acknowledgment

We wish to thank Herbert D. Kaesz and Jörn Müller for generously providing the chemical samples used in this work, and Joseph Henriques for technical assistance. We are grateful to Sax A. Mason, Arthur J. Schultz, John R. Shapley, and Jack M. Williams for communicating results to us prior to publication. We acknowledge the National Science Foundation and the Petroleum Research Fund (administered by the American Chemical Society) for financial support through grants CHE-77-00360 and 7800-AC3,6 respectively, and the W. C. Hamilton Memorial Fund for a scholarship awarded to R.G.T. Research at Brookhaven National Laboratory was performed under contract with the U.S. Department of Energy and supported by its Division of Basic Energy Sciences.

Literature Cited

1. Muetterties, E. L., *Science* (1977) **196**, 839.
2. Winsche, W. E., Hoffman, K. C., Salzano, F. J., *Science* (1973) **180**, 1325.
3. Kaesz, H. D., Saillant, R. B., *Chem. Revs.* (1972) **72**, 231.
4. Churchill, M. R., DeBoer, B. G., *Inorg. Chem.* (1977) **16**, 1141.
5. Wilson, R. D., Bau, R., *J. Am. Chem. Soc.* (1976) **98**, 4687.

6. Catti, M., Gervasio, G., Mason, S. A., *J. Chem. Soc., Dalton Trans.* (1977) 2260.
7. Rivera, V., Orpen, G., Bryan, E. G., Pippard, J., Sheldrick, G., *European Crystallogr. Meeting, 4th, Oxford, Collected Abstracts, p. 232, 1977.*
8. Shapley, J. R., private communication of work with Schultz, A. J., Williams, J. M., Suib, S., Stucky, G. D., Calvert, R. B., 1977.
9. Dahl, L. F., Broach, R. W., Longoni, G., Chini, P., Schultz, A. J., Williams, J. M., ADV. CHEM. SER. (1978) **167**, 93.
10. Simon, A., Z. *Anorg. Allg. Chem.* (1967) **355**, 311.
11. Abrahams, S. C., Ginsberg, A. P., Knox, K., *Inorg. Chem.* (1964) **3**, 558.
12. La Placa, S. J., Hamilton, W. C., Ibers, J. A., *Inorg. Chem.* (1964) **3**, 1491.
13. La Placa, S. J., Hamilton, W. C., Ibers, J. A., Davison, A., *Inorg. Chem.* (1969) **8**, 1928.
14. Moseley, P. T., Shearer, H. M. M., Spencer, C. B., *Acta Crystallogr.* (1969) **A25**, S169.
15. Cheetham, A. K., private communication, 1976. Referenced in Prout, K., Cameron, T. S., Forder, R. A., Critchley, S. R., Denton, B., Rees, G. V., *Acta Crystallogr.* (1974) **B30**, 2290.
16. Schultz, A. J., Stearley, K. L., Williams, J. M., Mink, R., Stucky, G. D., *Inorg. Chem.* (1977) **16**, 3303.
17. Wilson, R. D., Koetzle, T. F., Hart, D. W., Kvick, Å., Tipton, D. L., Bau, R., *J. Am. Chem. Soc.* (1977) **99**, 1775.
18. Hart, D. W., Bau, R., Koetzle, T. F., *J. Am. Chem. Soc.* (1977) **99**, 7557.
19. Roziere, J., Williams, J. M., Stewart, R. P., Petersen, J. L., Dahl, L. F., *J. Am. Chem. Soc.* (1977) **99**, 4497.
20. Petersen, J. L., Dahl, L. F., Williams, J. M., ADV. CHEM. SER. (1978) **167**, 11.
21. Petersen, J. L., Dahl, L. F., Williams, J. M., *J. Am. Chem. Soc.* (1974) **96**, 6610.
22. Olsen, J. P., Koetzle, T. F., Kirtley, S. W., Andrews, M., Tipton, D. L., Bau, R., *J. Am. Chem. Soc.* (1974) **96**, 6621.
23. Love, R. A., Chin, H. B., Koetzle, T. F., Kirtley, S. W., Whittlesey, B. R., Bau, R., *J. Am. Chem. Soc.* (1976) **98**, 4491.
24. Bau, R., Carroll, W. E., Teller, R. G., Koetzle, T. F., *J. Am. Chem. Soc.* (1977) **99**, 3872.
25. Bau, R., Carroll, W. E., Teller, R. G., Koetzle, T. F., unpublished data. Referenced in *24*.
26. Koetzle, T. F., Bau, R., *Proc. Conf. Neutron Scattering, Tennessee, p. 507, 1976.*
27. Bau, R., Carroll, W. E., Hart, D. W., Teller, R. G., Koetzle, T. F., ADV. CHEM. SER., (1978) **167**, 73.
28. Hoffmann, R., private communication, 1977.
29. Dimmler, D. G., Greenlaw, N., Kelley, M. A., Potter, D. W., Rankowitz, S., Stubblefield, F. W., *IEEE Trans. Nucl. Sci.* (1976) **NS-23**, 398.
30. McMullan, R. K., and in part, Andrews, L. C., Koetzle, T. F., Reidinger, F., Thomas, R., Williams, G. J. B., *NEXDAS.* Neutron and X-ray Data Acquisition System, unpublished work (1976).
31. Huie, B. T., Knobler, C. B., Kaesz, H. D., *J. Chem. Soc., Chem. Commun.* (1975) 684.
32. McMullan, R. K., unpublished work (1977).
33. Müller, J., Dorner, H., Huttner, G., Lorenz, H., *Angew. Chem., Int. Ed. Engl.* (1973) **12**, 1005.
34. Huttner, G., Lorenz, H., *Chem. Ber.* (1974) **107**, 996.
35. Johnson, C. K., ORTEP-II, Report ORNL-5138, Oak Ridge National Laboratory, Tennessee, 1976.
36. Mays, M. J., Simpson, R. N. F., *J. Chem. Soc. A* (1968) 1444.
37. Kaesz, H. D., *Chem. Br.* (1973) **9**, 344.
38. Jette, E. R., Foote, F., *J. Chem. Phys.* (1935) **3**, 605.
39. Anantharaman, T. R., *Curr. Sci.* (1958) **27**, 51.

RECEIVED July 19, 1977.

Crystallographic Investigations on Polyhydride Metal Complexes

ROBERT BAU, W. EAMON CARROLL, DONALD W. HART[1], and RAYMOND G. TELLER—Department of Chemistry, University of Southern California, Los Angeles, CA 90007

THOMAS F. KOETZLE—Department of Chemistry, Brookhaven National Laboratory, Upton, NY 11973

The molecular structures of four transition metal polyhydride complexes are reported: $H_3Ir(PMe_2Ph)_3$ and $H_7Re(PMe_2Ph)_2$ by x-ray diffraction analysis and $H_4Os(PMe_2Ph)_3$ and H_8Re_2-$(PEt_2Ph)_4$ by neutron diffraction analysis. Although the hydride ligands were not located in $H_3Ir(PMe_2Ph)_3$, the arrangement of phosphorus atoms about the iridium atom suggests a distorted octahedral geometry for this molecule. X-ray analysis of $H_7Re(PMe_2Ph)_2$ reveals a bent phosphorus–rhenium–phosphorus backbone for the molecule, that is consistent with a tricapped trigonal prism with phosphorus atoms in equatorial positions. Neutron diffraction analysis of $H_4Os(PMe_2Ph)_3$ shows that the molecule is a distorted pentagonal bipyramid, with one equatorial and two axial phosphine ligands. The neutron diffraction analysis of the $H_8Re_2(PEt_2Ph)_4$ dimer reveals a molecule with four terminal and four bridging hydride ligands. This molecule provides the first example of a metal–metal bond bridged by four hydrogen atoms.

During the 1960s, J. Chatt, B. L. Shaw, and co-workers synthesized numerous mixed hydride–phosphine complexes of the third-row transition metals tungsten (*1, 2, 3*), rhenium (*4, 5, 6, 7*), osmium (*8, 9, 10, 11*), and iridium (*12, 13, 14, 15*) (*see* Table I). These unusual covalent compounds, called polyhydride complexes, were found to be remarkably stable and may contain up to seven hydride ligands per metal atom. The general method of preparation involves reaction of the metal chlorides with tertiary phosphines under refluxing conditions (in alcohol) to yield complexes of the type $MCl_xL_{6-x}(L = PR_3)$, followed by reduction to the corresponding polyhydride with $LiAlH_4$ (in THF) or

[1]Present address: Department of Chemistry, University of Arkansas, Fayetteville, AR 72701.

Table I.   Schematic of Polyhydride Complexes

	H_7ReL_2	H_6OsL_2	H_5IrL_2
H_6WL_3	H_5ReL_3	H_4OsL_3	H_3IrL_3
H_4WL_4	H_3ReL_4	H_2OsL_4	$HIrL_4$

(L = tertiary phosphine)

$NaBH_4$ (in ethanol). The original number of phosphines usually remains constant during the hydrogenation step. The synthetic approach to the rhenium series of polyhydride complexes (Scheme 1) (7) is a representative example. The resulting polyhydride has all chloride ligands replaced by a sufficient number of hydrogen atoms to achieve a closed-shell (18 electron) configuration. The compounds are generally colorless (or light yellow), diamagnetic, and soluble in many organic solvents. Most are air stable in the solid state. The fact that phosphine ligands stabilize these molecules can be attributed to the softness (i.e., π-accepting ability) and steric bulk of these ligands.

The polyhydride series (Table I) later was extended by Tebbe to include $H_5Ta(Me_2PCH_2CH_2PMe_2)_2$ (16) and by Ginsberg to include the anionic species $[H_8Re(PR_3)]^-$ (17). (Abbreviations used in this paper are as follows: Me–methyl, Et–ethyl, Cp–cyclopentadienyl, and Ph–phenyl.) Technically, the classic $[ReH_9]^=$ anion (18) also can be considered a member of this series. Many analogous compounds involving second-row and first-row transition metals are known (19); however, these are generally less stable than those involving third-row transition metals. Hydride–phosphine complexes of first-row transition elements are limited mainly to compounds having few hydrogen atoms, such as $HCoL_4$, H_2FeL_4, and H_3CoL_3 (19).

The subject of polyhydride complexes was included as part of a larger, comprehensive review of metal hydride complexes by Kaesz and Saillant (19), and their interesting NMR behavior has been summarized by Jesson (20). The

Scheme 1.   Synthesis of rhenium polyhydride complexes (4, 5, 6, 7, 17)

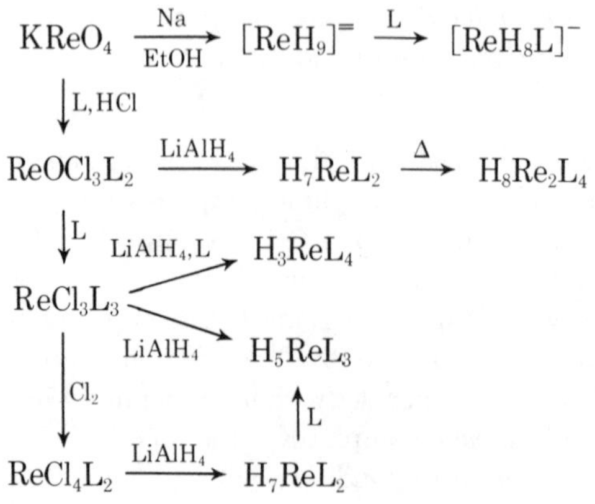

$$KReO_4 \xrightarrow[\text{EtOH}]{\text{Na}} [ReH_9]^= \xrightarrow{L} [ReH_8L]^-$$

$$\downarrow L, HCl$$

$$ReOCl_3L_2 \xrightarrow{LiAlH_4} H_7ReL_2 \xrightarrow{\Delta} H_8Re_2L_4$$

$$\downarrow L$$

$$ReCl_3L_3 \underset{\overbrace{}}{\overset{LiAlH_4, L}{\nearrow}} H_3ReL_4$$

$$\xrightarrow{LiAlH_4} H_5ReL_3$$

$$\Big\downarrow Cl_2 \qquad \qquad \uparrow L$$

$$ReCl_4L_2 \xrightarrow{LiAlH_4} H_7ReL_2$$

(L = tertiary phosphine)

rhenium hydride complexes, in particular, have been reviewed by Giusto (*21*). In this chapter, we will focus on structural aspects; background material will be reviewed briefly, and some recent results will be presented.

Prior Crystallographic Work on Polyhydride Complexes

The general historical background of structural work on hydride complexes has been covered very well in review articles by Ginsberg (*22*) and by Frenz and Ibers (*23*). The latter article catalogs all crystallographic investigations of metal hydrides up to 1970. Since many of the problems involved in structurally characterizing the hydride ligand have been discussed in these reviews, they will not be repeated here. Suffice it to say that from x-ray data often it is not possible to locate hydrogen atoms attached to third-row transition metals, and that neutron

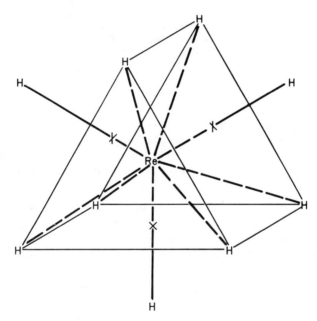

Figure 1. The tricapped trigonal prismatic geometry of the [ReH₉]⁼ anion (18)

diffraction methods are preferable for the accurate determination of hydrogen positions in such compounds. (There are, however, examples of third-row complexes in which hydride ligands have been located and refined from x-ray data, as have been reported by Churchill and Chang (*24, 25*) and by Ibers (*26*).)

One of the first metal polyhydride complexes to be thoroughly studied was K_2ReH_9, the nature of which was established firmly only after a neutron diffraction study reported in 1964 by Abrahams, Ginsberg, and Knox (*18*). At that

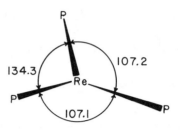

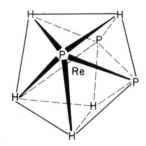

Figure 2. (Top) A sketch of the nonplanar ReP_3 skeleton in $H_5Re(PPh_3)_3$ (28). (Bottom) A possible structure of $H_5Re(PPh_3)_3$ based on the dodecahedron.

time, this compound was the only known example of a binary transition metal polyhydride anion (i.e., with no phosphine ligands); somewhat later, the technetium analog was prepared (27). The geometry of the $[ReH_9]^=$ dianion (see Figure 1) is that of a tricapped trigonal prism with an average Re–H distance of 1.68(1) Å.

Ginsberg and co-workers also investigated the eight-coordinate complex $H_5Re(PPh_3)_3$ (28). Decomposition of the crystal during x-ray data collection prevented a precise structure determination, but the rhenium and three phosphorus atoms were unambiguously located. The ReP_3 skeleton, which is non-

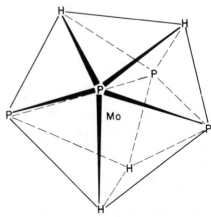

Figure 3. A representation of the dodecahedral structure of $H_4Mo(P-Me_2Ph)_4$, as derived from x-ray diffraction (29)

planar, is illustrated in Figure 2 (top).   The phosphorus–rhenium–phosphorus
bond angles (134.3°, 107.2°, and 107.1°) are consistent with coordination based
on either the dodecahedron (Figure 2 (bottom)), truncated octahedron, or bicapped octahedron.   Since the hydride ligands could not be located, the exact
geometry of this complex remains in doubt.

Another eight-coordinate polyhydride structure that was studied is that of
$H_4Mo(PPhMe_2)_4$ (*29*).   The x-ray data collected for this complex were of sufficient quality to permit location of the four hydride ligands.   The geometry
(Figure 3) is that of a dodecahedron comprising two distorted tetrahedra: one
composed of four hydrogen atoms and the other of four phosphorus atoms.   The
average Mo–H distance is 1.70(3) Å, which is in agreement with other reported
M–H (terminal) distances.

X-ray structural data also exist for two seven-coordinate polyhydrides,
$H_3Re(DPPE)(PPh_3)_2$ and $H_3Re(DPPE)_2$ (*30*) (DPPE is bisdiphenylphosphi-
noethane, also known as diphos).   Analysis of the data for the latter compound

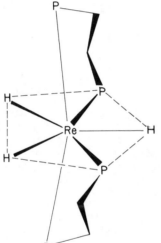

*Figure 4.   A sketch of the
postulated   structure   of
$H_3Re(Ph_2PCH_2CH_2PPh_2)_2$
(21, 30)*

revealed the following arrangement of phosphorus atoms about the metal: two
axial phosphines [P–Re–P angle = 167.4(5)°] and two equatorial phosphines with
a P–Re–P angle of 151.5(5)°.   A nonbonding interaction energy map calculated
about the rhenium atom, assuming a Re–H bond distance of 1.68 Å, showed three
energy minima that were assigned to the three hydrogen ligand positions (*30*,
31).   The geometry of the resultant molecule is that of a distorted pentagonal
bipyramid (as illustrated in Figure 4) in which the axial phosphine ligands are
bent away from the two equatorial phosphorus atoms and toward the two cis
hydrogen ligands.

The structure determination of *mer*-$H_3Ir(PPh_3)_3$ is an example of a situation
in which hydrogen atoms bonded to third-row transition elements could be lo-

cated but not refined (32). A series of difference electron density syntheses revealed three peaks that were assigned to the hydrogen positions. The geometry of this molecule is approximately octahedral with phosphorus–iridium–phosphorus angles of 153° (trans) and 103° (cis) (see Figure 5). The average iridium–hydrogen bond distance is 1.60 Å. Again note the deviation from perfect octahedral geometry that is caused by the displacement of the phosphine ligands away from each other and toward the hydrogen atoms. Because of the small size of hydrogen atoms, it seems reasonable that steric interactions would cause larger ligands to distort toward them. However, Elian and Hoffmann have presented theoretical evidence that this type of distortion also can be rationalized in terms of electronic effects (33).

About two years ago we began a systematic investigation of the structures of polyhydride complexes. In this article we present some x-ray diffraction results on $fac\text{-}H_3Ir(PMe_2Ph)_3$ and $H_7Re(PMe_2Ph)_2$ and neutron diffraction results on $H_4Os(PMe_2Ph)_3$ and $H_8Re_2(PEt_2Ph)_4$. The latter compound, which is a pyrolysis product of $H_7Re(PEt_2Ph)_2$, is the only dimeric member of the polyhydride series

Table II.

	$H_3Ir(PMe_2Ph)_3$ (X-ray)	$H_7Re(PMe_2Ph)_2$ (X-ray)
Crystal type	Triclinic	Monoclinic
Space group	$P\bar{1}$	$P2_1/n$
Cell constants a	6.454(2) Å	19.038(17) Å
b	16.061(7) Å	6.337(4) Å
c	13.303(6) Å	15.234(13) Å
α	106.38(2)°	90.0
β	103.48(2)°	93.72(4)°
γ	78.13(2)°	90.0
Cell vol	1272.0 Å^3	1834.0 Å^3
No. of molecules in the unit cell	2	4
Calc. density	1.59 g cm^{-3}	1.70 g cm^{-3}
Obs. density	—	1.70 g cm^{-3}
Absorption coefficient	36.0 cm^{-1}	71.6 cm^{-1}
Wavelength used in data collection	Mo Kα x-rays $\lambda = 0.71069$ Å	Mo Kα x-rays $\lambda = 0.71069$ Å
Data collection T	Room Temp	Room Temp
Data collection limit (sin θ/λ)	0.54 Å^{-1}	0.54 Å^{-1}
No. of reflections used in structure analysis	3417 (I > 3σ)	1672 (I > 3σ)
Final agreement factors[a]	$R_F = 0.045$ $R_{wF} = 0.064$	$R_F = 0.050$ $R_{wF} = 0.055$

[a] $R_F = \Sigma|F_0 - |F_c||/\Sigma F_0$; $R_{wF} = \{\Sigma w|F_0 - |F_c||^2/\Sigma w F_0^2\}^{1/2}$

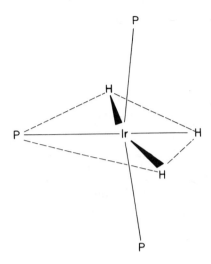

Figure 5. The distorted octahedral structure of mer-$H_3Ir(PPh_3)_3$ *(32)*

Crystal Data

$H_8Re_2(PEt_2Ph)_4$		$H_4Os(PMe_2Ph)_3$	
(*X-ray*)	(*Neutron*)	(*X-ray*)	(*Neutron*)
Monoclinic		Triclinic	
$C2/c$		$P\bar{1}$	
23.309(7) Å	23.137(7) Å	11.489(4) Å	11.409(2) Å
12.353(4) Å	12.276(4) Å	12.441(4) Å	12.388(2) Å
19.634(6) Å	19.438(5) Å	11.103(4) Å	11.098(2) Å
90.0	90.0	90.54(2)°	90.36(5)°
129.38(1)°	129.51(2)°	124.63(2)°	125.07(1)°
90.0	90.0	89.93(2)°	90.06(4)°
4369.8 Å³	4259.5 Å³	1306.8 Å³	1283.8 Å³
4	4	2	2
1.59 g cm⁻³	1.64 g cm⁻³	1.55 g cm⁻³	1.57 g cm⁻³
1.60 g cm⁻³	—	1.51 g cm⁻³	—
60.4 cm⁻¹	2.01 cm⁻¹	53.6 cm⁻¹	2.23 cm⁻¹
{Mo Kα x-rays λ = 0.71069 Å}	{thermal neutrons λ = 1.1598 Å}	{Mo Kα x-rays λ = 0.71069 Å}	{thermal neutrons λ = 1.01939 Å}
Room Temp	80° K	Room Temp	90° K
0.54 Å⁻¹	0.62 Å⁻¹	0.59 Å⁻¹	0.63 Å⁻¹
2367	2729	3486	3381
(I > 2σ)	(I > 2σ)	(I > 3σ)	(I > 3σ)
$R_F = 0.053$	$R_F = 0.086$	$R_F = 0.055$	$R_F = 0.044$
$R_{wF} = 0.057$	$R_{wF} = 0.049$	$R_{wF} = 0.060$	$R_{wF} = 0.042$

known to exist. This compound originally was referred to as an agnohydride complex by Chatt and Coffey (7), who were at the time unaware of the exact number of hydrogen atoms in the molecule.

Experimental

The compounds fac-$H_3Ir(PMe_2Ph)_3$ (13), $H_7Re(PMe_2Ph)_2$ (7), H_4Os-$(PMe_2Ph)_3$ (10), and $H_8Re_2(PEt_2Ph)_4$ (7) all were prepared using standard literature methods. The rhenium complexes were recrystallized from n-hexane and the others from absolute ethanol.

Unit cell parameters, originally obtained photographically and later redetermined accurately from diffractometer settings, are listed with other relevant crystal data in Table II. X-ray diffraction data were collected on a Nonius CAD-3 diffractometer in the manner described in an earlier publication (34). Reflec-

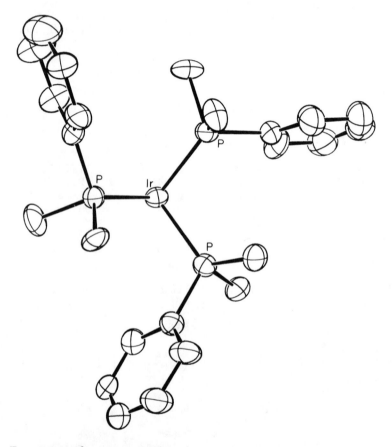

Figure 6.   The structure of fac-$H_3Ir(PMe_2Ph)_3$, *viewed approximately along the noncrystallographic threefold axis of the molecule. This structure is also based on the distorted octahedron.   The hydride ligands were not located.*

Table III. Selected Distances and Angles in H₄Os(PMe₂Ph)₃ [a]

Distances (in Å)

Os–H(1)	1.663(3)	Os–P(1)	2.317(2)
Os–H(2)	1.648(3)	Os–P(2)	2.307(2)
Os–H(3)	1.644(3)	Os–P(3)	2.347(2)
Os–H(4)	1.681(3)		

Angles (in degrees)

H(1)–Os–H(2)	69.4(2)	P(1)–Os–H(1)	92.9(1)
H(2)–Os–H(3)	67.9(2)	P(1)–Os–H(2)	83.7(1)
H(3)–Os–H(4)	70.0(2)	P(1)–Os–H(3)	83.9(1)
H(1)–Os–H(3)	137.3(2)	P(1)–Os–H(4)	91.3(1)
H(2)–Os–H(4)	137.9(2)	P(2)–Os–H(1)	89.6(1)
H(1)–Os–H(4)	152.7(2)	P(2)–Os–H(2)	84.4(1)
P(3)–Os–H(1)	79.7(1)	P(2)–Os–H(3)	84.9(1)
P(3)–Os–H(2)	149.1(2)	P(2)–Os–H(4)	92.7(1)
P(3)–Os–H(3)	143.0(2)		
P(3)–Os–H(4)	73.0(1)		

[a] From neutron diffraction analysis.

tions were measured at room temperature with Mo Kα radiation to a $(\sin\theta/\lambda)$ limit of 0.54 Å⁻¹. Analysis of the data yielded the positions of all the nonhydrogen atoms in the molecules. Full-matrix, least-squares refinement resulted in the final agreement factors given in Table II.

Neutron diffraction data for H₄Os(PMe₂Ph)₃ and H₈Re₂(PEt₂Ph)₄ were collected at the Brookhaven High Flux Beam Reactor under operating conditions (*35, 36*) specified in Table II. Nonhydrogen atom positions obtained from an x-ray analysis were used to phase the neutron data, and subsequent difference syntheses revealed the positions of all hydrogen atoms in both molecules. Least-squares refinements were carried out with anisotropic temperature factors included for all atoms to yield the agreement factors listed in Table II.

Least-squares computations were performed with a local version of ORFLS (*37*), with the X-RAY system (*38*), or with the CRYM system (*39*). Molecular plots were produced with ORTEP (*40*). Calculations for the neutron structures were carried out on CDC 7600 and CDC 6600 computers at Brookhaven National Laboratory, making use of programs described by Berman et al. (*41*).

Results

fac-H₃Ir(PMe₂Ph)₃. This six-coordinate molecule, which is the facial analog of the known *mer*-H₃Ir(PPh₃)₃ (*32*), exhibits noncrystallographic, threefold symmetry (Figure 6). Although the hydride ligands have not been located in this study, the arrangement of phosphorus atoms [iridium–phosphorus distances are 2.296(3), 2.296(3), and 2.291(3) Å; phosphorus–iridium–phosphorus angles are 101.4(1)°, 102.1(1)°, and 99.5(1)°] leaves little doubt that the geometry of the molecule is that of a trigonally distorted octahedron with a facial (cis) arrangement of ligands. As in other transition metal hydride complexes, this distortion is related to the modest steric requirements of the hydrogen atoms.

H₄Os(PMe₂Ph)₃. A few years ago, Mason studied the PEt₂Ph analog of this compound with x-ray techniques and found its OsP₃ skeleton to be a planar, distorted T-shaped unit (*42*). These unpublished results have been mentioned briefly by Aslanov et al. (*43*). In our neutron diffraction analysis of H₄Os-(PMe₂Ph)₃, (*44*), we confirmed this basic geometry and determined the positions of the four hydride ligands. Distances and angles about the osmium atom in this compound are given in Table III. The geometry that is illustrated in Figure 7 is reminiscent of that of $H_3Re(DPPE)_2$ discussed earlier (*see* Figure 4). It is a distorted pentagonal bipyramid with two phosphine ligands in axial positions. The equatorial H_4OsP fragment is planar within ± 0.01 Å. Osmium–hydrogen distances are 1.663(3), 1.648(3), 1.644(3), and 1.681(3) Å, and nonbonding H···H contact distances are 1.883(5), 1.840(6), and 1.909(5) Å. The hydrogen–os-

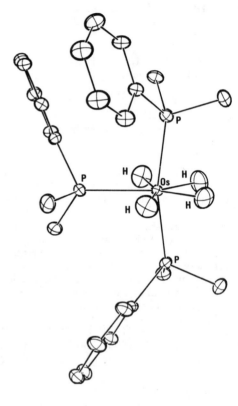

Figure 7. (Top) *The pentagonal bipyramidal geometry of $H_4Os(PMe_2Ph)_3$ as determined by neutron diffraction. Hydrogen atoms on the methyl and phenyl groups have been omitted for clarity.* (Bottom) *The central core of the molecule, viewed normal to the equatorial H_4OsP plane. Note the slight bending of the phosphorus–osmium–phosphorus axis* (69)

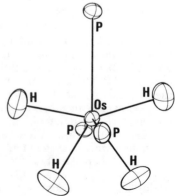

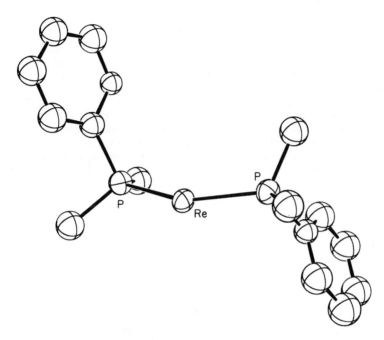

Figure 8. The structure of $H_7Re(PMe_2Ph)_2$ as derived from x-ray diffraction, showing the nonlinearity of the main phosphorus–rhenium–phosphorus axis of the molecule

mium–hydrogen angles [67.9(2)°, 69.4(2)°, and 70.0(2)°] are somewhat compressed from the normal pentagonal value of 72° because of the steric influence of the equatorial phosphine. In H_3TaCp_2 (45), an even more compact arrangement of hydrides around the central metal was found [H–Ta = 1.769(8), 1.775(9), 1.777(9) Å; H–Ta–H = 62.8(5)°, 63.0(4)°; H⋯H = 1.846(9), 1.856(9) Å]. At room temperature, the NMR spectrum of $H_4Os(PMe_2Ph)_3$ in the hydridic region consists of a symmetrical quartet.

$H_7Re(PMe_2Ph)_2$. This is a particularly interesting compound because it has the largest number of hydride ligands per metal in the known series of neutral polyhydride complexes (*see* Table I). The phosphorus–rhenium–phosphorus axis in this molecule is nonlinear (Figure 8). Although it was not possible to locate the hydride ligands in our x-ray study, we suggest a geometry similar to that of the $[ReH_9]^=$ anion with phosphine ligands capping two of the square faces of the trigonal prism (Figure 9). The large phosphorus–rhenium–phosphorus angle [146.8(1)°, as compared with 120° for an ideal tricapped trigonal prism] could be attributed to steric crowding of the two phosphine ligands. Alternatively, one could assign one phosphine ligand on a corner of the prism and the other ligand on a remote equatorial position. The rhenium–phosphorus bond lengths in this molecule are 2.396(4) and 2.395(4) Å. At room temperature, this compound is also fluxional on the NMR time scale; its spectrum consists of a simple triplet.

$H_8Re_2(PEt_2Ph)_4$. As mentioned earlier in this article, pyrolysis of H_7ReL_2 leads to the formation of a dimeric molecule which is undoubtedly the most unusual compound in the series discussed here. The exact nature of this dimer previously had not been established fully because of the uncertainty over the number of hydride ligands (7). The results of our neutron diffraction analysis (46) establish this number as eight (four bridging and four terminal hydrogen atoms). The molecular configuration is shown in Figure 10, and selected distances and angles are given in Table IV. A crystallographic center of inversion exists at the midpoint of the rhenium–rhenium bond.

The molecule contains the first example of a metal–metal bond bridged by four hydrogen atoms. The coordination around each rhenium atom can be envisaged as a distorted trigonal prism (six hydrogen ligands) capped on the three square faces by two phosphine ligands and the rhenium–rhenium bond (Figure 10). An alternative view of the molecule is presented in Figure 11, which shows the terminal H_2P_2 units and the bridging H_4 group in a mutually staggered ar-

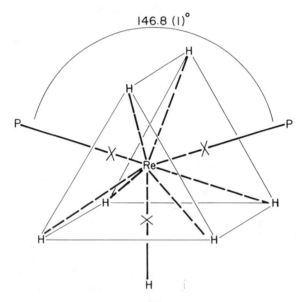

Figure 9. A postulated structure of $H_7Re(P\text{-}Me_2Ph)_2$, based on the tricapped trigonal prism, with the phosphine ligands assigned to equatorial positions

rangement. This unusual molecule formally possesses a rhenium–rhenium triple bond (via standard electron-counting procedures) and hence, the bonding can be described loosely as a quadruply protonated, metal–metal triple bond. The rhenium–rhenium distance of 2.538(4) Å is intermediate between that of the quadruply bonded anion $[Re_2Cl_8]^=$ (47, 48, 49), 2.241(7) Å, and the species $H_2Re_2(CO)_8$ (50) that has a formal double bond of length 2.896(3) Å.

Terminal rhenium–hydrogen bond lengths in $H_8Re_2(PEt_2Ph)_4$ [average 1.669(7) Å] agree well with those found in $[ReH_9]^=$ [1.68(1) Å] and are 0.21 Å shorter than the bridging rhenium–hydrogen distances [average 1.878(7) Å]. This lengthening of metal–hydrogen bonds from the terminal to the bridging mode has been noted previously (with increases ranging from 0.1 to 0.2 Å), but this is the first time that an accurate comparison could be made within the same mol-

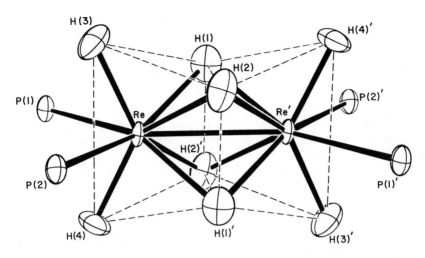

Journal of the American Chemical Society

Figure 10.   The central core of the $H_8Re_2(PEt_2Ph)_4$ molecule, as determined by neutron diffraction (70)

ecule.   The central H_4 unit is planar, as required by crystallographic symmetry, and defines a distorted square [H···H = 1.870(8), 2.042(8) Å] normal to the rhenium–rhenium bond.   An interesting question associated with the structure of $H_8Re_2(PEt_2Ph)_4$ concerns the eclipsed conformation of the phosphine ligands. Intuitively, one might expect that for steric reasons the phosphines would be staggered (i.e., with one terminal H_2P_2 unit turned 90° relative to the other).

$[(C_5Me_5)Ir(H)_3Ir(C_5Me_5)]^+BF_4^-$.   A related compound that we recently have analyzed by neutron diffraction is $[(C_5Me_5)Ir(H)_3Ir(C_5Me_5)]^+BF_4^-$ (Figure 12) (51), that contains a metal–metal triple bond surrounded by three hydrogen atoms rather than four.   The bridging hydrogen atoms essentially form an equilateral triangle normal to the iridium–iridium bond, with the following mean molecular parameters: Ir–Ir = 2.458(6) Å; Ir–H = 1.745(12) Å; H···H = 2.142(24)

Table IV.   Selected Distances and Angles in $H_8Re_2(PEt_2Ph)_4{}^a$

Distances (in Å)

Re–Re'	2.538(4)		
Re–P(1)	2.336(4)	Re–P(2)	2.333(5)
Re–H(1)	1.882(7)	Re–H(2)	1.885(7)
Re–H(1)'	1.862(7)	Re–H(2)'	1.883(8)
Re–H(3)	1.682(7)	Re–H(4)	1.656(6)

Angles (in degrees)

Re'–Re–P(1)	129.2(1)	Re'–Re–H(3)	115.4(3)
Re'–Re–P(2)	128.0(1)	Re'–Re–H(4)	116.3(3)
P(1)–Re–P(2)	102.7(2)	H(3)–Re–H(4)	128.3(4)
H(1)–Re–H(1)'	94.6(3)	H(2)–Re–H(2)'	95.3(3)
H(1)–Re–H(2)	65.6(3)	H(1)–Re–H(2)'	59.6(3)
Re–H(1)–Re'	85.4(3)	Re–H(2)–Re'	84.7(3)

^{a} From neutron diffraction analysis.

Å; Ir–H–Ir = 89.5(8)°; and H–Ir–H = 75.8(9)°. As expected, the bridging hydrogen atoms in this compound are not as closely packed as those in H_8Re_2-$(PEt_2Ph)_4$ since $M(\mu\text{-}H)_3M$ represents a less crowded arrangement than $M(\mu\text{-}H)_4M$.

Discussion

From the evidence presented in this chapter, there appear to be enough data to make some tentative statements about the geometries of mononuclear polyhydride complexes.

Six-coordinate complexes are octahedral, as expected. X-ray results on *mer*-$H_3Ir(PPh_3)_3$ (*see* Figure 5) (*32*) and *fac*-$H_3Ir(PMe_2Ph)_3$ (*see* Figure 6) leave little doubt of this. What also is evident in these compounds is a pronounced bending of the phosphine ligands away from each other and toward the hydrogen atoms; this recurring feature is found in all polyhydride complexes. Unlike the other polyhydrides, six-coordinate complexes are stereochemically rigid molecules (*15*).

Seven-coordinate complexes that have been analyzed so far show structures consistent with the pentagonal bipyramid, with hydrogen atoms in equatorial positions. The structures of $H_3Re(DPPE)(PPh_3)_2$ (*31*), $H_3Re(DPPE)_2$ (*see* Figure

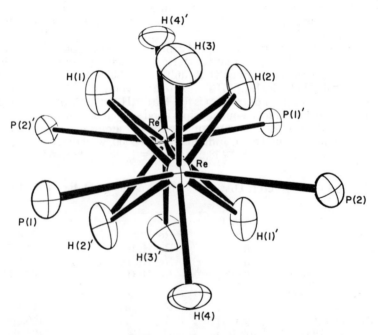

Journal of the American Chemical Society

Figure 11.   Another view of the skeleton of $H_8Re_2(PEt_2Ph)_4$ showing the mutually staggered arrangement of the terminal H_2P_2 units and the bridging H_4 square (70)

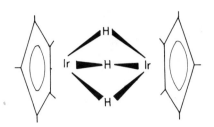

Figure 12. A schematic sketch of the cation in [$H_3Ir_2(C_5Me_5)_2$]- $^+BF_4^-$

4) (*30*), $H_4Os(PEt_2Ph)_3$ (*42*), and $H_4Os(PMe_2Ph)_3$ (*see* Figure 7) (*44*) all seem to fit this pattern. Very preliminary x-ray results on $H_5Ir(PEt_2Ph)_2$ (*52*) indicate the presence of a linear phosphorus–iridium–phosphorus backbone, that again is consistent with the pentagonal bipyramidal geometry. It may be that the small steric bulk of the hydride ligand is particularly suited for the crowded pentagonal arrangement associated with this geometry.

There is less information for eight-coordinate complexes than for seven-coordinate complexes. Both $H_4Mo(PMe_2Ph)_4$ (*29*) (*see* Figure 3) and $H_5Re(PPh_3)_3$ (*28*) (*see* Figure 2) are compatible with the dodecahedral arrangement, but it has been pointed out that the geometry of the latter compound also can be fitted into a bicapped octahedral or truncated octahedral framework (*28*). Clearly, a few definitive neutron-diffraction structure determinations are needed here. Eight-coordinate complexes exhibit an interesting NMR behavior; some are partially fluxional (*29*) and seem to bridge the gap between the rigid six-coordinate and the totally fluxional seven- and nine-coordinate complexes. NMR studies on complexes of the type H_4ML_4 indicate that these are rigid on the NMR time scale at or just below room temperature but become fluxional when the temperature is raised (*29*).

The overall geometry of nine-coordinate complexes is, like that of six-coordinate molecules, readily predictable. Nine-coordination corresponds to sp^3d^5 hybridization which, in turn, indicates the tricapped trigonal prism (*see* Figure 1). X-ray results on $H_7Re(PMe_2Ph)_2$ are consistent with this geometry (*see* Figure 9), and it is likely that other nine-coordinate compounds such as $[H_8ReL]^-$ and H_6WL_3 also are based on the tricapped trigonal prism with phosphine groups in equatorial positions.

The only binuclear compound discussed here is the complex $H_8Re_2(PEt_2Ph)_4$ (*see* Figure 10) (*46*). The four bridging hydrogen atoms found in this compound may represent an upper limit; that is, it is possible that a metal–metal bond can accommodate no more than four hydride bridges. The $M(\mu\text{-}H)_4M$ arrangement formally corresponds to a situation in which hydrogen $1s$ orbitals are associated with each of the four lobes that comprise the π bonds of a metal–metal triple bond. The shorter of the two intramolecular H⋯H nonbonding distances in $H_8Re_2(PEt_2Ph)_4$ [H(1)⋯H(2)′ = 1.870(8) Å] is close to minimum H⋯H contacts found in other molecules [such as 1.872(15) Å in K_2ReH_9 (*18*), 1.840(6) Å in $H_4Os(PMe_2Ph)_3$ (*44*), and 1.846(9) Å in H_3TaCp_2 (*45*)] and probably represents the minimum distance beyond which nonbonding hydrogen atoms cannot be

Table V.   Compilation of M–H Bond Lengths Determined by
Single-Crystal Neutron Diffraction[a]

(A) Terminal (M–H) Bonds

Compound	M	M–H (Å)	Ref.
HMn(CO)$_5$	Mn	1.601(16)	53
[HZnN(Me)C$_2$H$_4$NMe$_2$]$_2$	Zn	1.60	54
H$_2$MoCp$_2$	Mo	1.685(3)	55
H$_3$TaCp$_2$	Ta[c]	1.769(8)	45
		1.775(9)	
		1.777(9)	
K$_2$[ReH$_9$]	Re	1.68(1)[b]	18
H$_8$Re$_2$(PEt$_2$Ph)$_4$	Re	1.682(7)	46
		1.656(7)	
H$_4$Os(PMe$_2$Ph)$_3$	Os	1.663(3)	44
		1.648(3)	
		1.644(3)	
		1.681(3)	

(B) Bridging (M–H–M) Bonds

Compound	M	M–H (Å)	M–H–M (deg)	M–M (Å)	Ref.
[HCr$_2$(CO)$_{10}$]$^-$[Et$_4$N]$^+$	Cr[c]	1.737(19)	158.9(6)	3.386(6)	56
		1.707(21)			
HMo$_2$(C$_5$H$_5$)$_2$(CO)$_4$(P-Me$_2$)	Mo	1.851(4)	122.9(2)	3.267(2)	57
		1.869(4)			
HRu$_3$(CO)$_9$(C≡C-CMe$_3$)	Ru	1.789(5)	102.3(2)	2.792(3)	58
		1.796(5)			
α-HW$_2$(CO)$_9$(NO)	W	1.875(4)	125.0(2)	3.328(3)	59
		1.876(4)			
β-HW$_2$(CO)$_9$(NO)	W	1.870(4)	125.9(4)	3.330(3)	59
HW$_2$(CO)$_8$(NO)(P(O-Me)$_3$)	W	1.859(6)	129.4(3)	3.393(4)	34
		1.894(6)			
H$_2$Os$_3$(CO)$_{10}$	Os	1.86	92.6	2.69	60
H$_8$Re$_2$(PEt$_2$Ph)$_4$	Re	1.862(7)	85.4(3)	2.538(4)	46
		1.883(8)	84.7(3)		
		1.882(7)			
		1.885(7)			
[H$_3$Ir$_2$(C$_5$Me$_5$)$_2$]$^+$BF$_4$$^-$	Ir	1.745(12)[b]	89.5(8)[b]	2.458(6)	51

(C) Triply bridging (M$_3$H) Bonds

Compound	M	M–H (Å)	M–H–M (deg)	M–M (Å)	Ref.
HFeCo$_3$(CO)$_9$(P(O-Me)$_3$)$_3$	Co	1.742(3)	92.1(1)	2.501(2)	61
		1.731(3)	91.7(1)	2.489(3)	
		1.728(3)	91.5(1)	2.477(3)	
H$_3$Ni$_4$Cp$_4$	Ni	1.691(8)[b]	93.9(3)[b]	2.469(6)[b]	62

Table V.   (*Continued*)

(D) M–H–B Bonds

Compound	M	M–H (Å)	M–H–B (deg)	M–B (Å)	Ref.
U(BH$_4$)$_4$	U^c	$\begin{Bmatrix} 2.36(2) \\ 2.34(2) \\ 2.33(2) \end{Bmatrix}^d$	$\begin{Bmatrix} 82(1) \\ 86(1) \\ 82(1) \end{Bmatrix}^d$	2.52(1)^d	63
		$\begin{Bmatrix} 2.44(3) \\ 2.46(2) \\ 2.36(2) \\ 2.36(2) \end{Bmatrix}^e$	$\begin{Bmatrix} 97(2) \\ 99(1) \\ 99(1) \\ 96(2) \end{Bmatrix}^e$	$\begin{Bmatrix} 2.90(1) \\ 2.82(2) \end{Bmatrix}^e$	63
Hf(BH$_4$)$_4$	Hf^c	2.06(2)	83(1)	2.25(3)	64
Hf(C$_5$H$_4$Me)$_2$(BH$_4$)$_2$	Hf^c	2.069(7) 2.120(8)	96.3(5) 97.3(5)	2.553(6)	65

^{a} Not included in this table are the results from [(Ph$_3$P)$_2$N]$^+$[HCr$_2$(CO)$_{10}$]$^-$, [(Ph$_3$P)$_2$N]$_2$$^+$ [H$_2$Ni$_{12}$(CO)$_{21}$]$^{2-}$, and [(Ph$_3$P)$_2$N]$_3$$^+$[HNi$_{12}(CO)_{21}$]$^{3-}$, which will be reported elsewhere in this volume *(66, 67)*.
^{b} Average values.
^{c} For comparison, the M–H distances found in binary metal hydrides are as follows: Cr–H = 1.67 Å in CrH, Ta–H = 1.74 Å in TaH, Hf–H = 2.05 Å in HfH$_2$, U–H = 2.32 Å in UH$_3$. Values taken from Ref. *68*.
^{d} Distances and angles associated with terminal BH$_4$ groups [i.e., M(μ-H)$_3$BH].
^{e} Distances and angles associated with bridging BH$_4$ groups [i.e., M(μ-H)$_2$B(μ-H)$_2$M'].

compressed further.   Interestingly, the hydridic region of the NMR spectrum of H$_8$Re$_2$(PEt$_2$Ph)$_4$ at room temperature is that of a simple quintet, indicating that the hydride ligands are equivalent on the NMR time scale and are being influenced equally by the phosphorus atoms in the molecule.   At $-120°$C, the quintet begins to broaden appreciably.   Perhaps one might envisage simple rearrangement processes {e.g., [H$_2$P$_2$Re(μ-H)$_4$ReH$_2$P$_2$ $\leftrightarrow$ H$_3$P$_2$Re(μ-H)$_2$ReH$_3$P$_2$] or [H$_2$P$_2$Re(μ-H)$_4$ReH$_2$P$_2$ $\leftrightarrow$ H$_4$P$_2$Re$\equiv$ReH$_4$P$_2$]} to account for this fluxional behavior.

One rationale for studying metal hydride complexes with neutron diffraction is that the metal–hydrogen bond, which is free from π-bonding effects and other complicating features, should serve as a reliable source of information from which accurate values of the covalent radii of the heavy elements can be obtained.   In Table V we summarize terminal and bridging metal–hydrogen bond lengths derived from single-crystal neutron diffraction data.   Most of the results were obtained from complexes of third-row transition elements.

Based on the data of Table V, we can make two general statements:

(a) bridging metal–hydrogen bonds are longer than terminal bonds by about 0.10–0.20 Å.   As mentioned earlier, the best example of this effect comes from H$_8$Re$_2$(PEt$_2$Ph)$_4$, in which a meaningful internal comparison can be made.   One also can compare distances involving different elements [e.g., the terminal manganese–hydrogen bond in HMn(CO)$_5$ vs. the triply bridging cobalt–hydrogen bond in HFeCo$_3$(CO)$_9$(P(OMe)$_3$)$_3$] after making suitable adjustments to account for the different sizes of the metal atoms.

(b) the expected decrease in covalent radii as we move from left to right across a row of the periodic table is reflected in the metal–hydrogen distances. For example, one can see this trend by comparing the following mean terminal metal–hydrogen values: Ta–H (1.774 Å), Re–H (1.675 Å, averaged between two compounds), and Os–H (1.659 Å). It would be desirable to complete this series by making measurements on terminal hafnium–hydrogen, tungsten–hydrogen, iridium–hydrogen, and platinum–hydrogen bond lengths.

Acknowledgment

We are grateful to the National Science Foundation and the Petroleum Research Fund (administered by the American Chemical Society) for financial support through grants CHE-77-00360 and 7800-AC3,6 respectively. We also thank Joseph Henriques for technical assistance in operating the neutron diffractometers, and the W. C. Hamilton Memorial Fund for providing a scholarship to R.G.T. Research at Brookhaven National Laboratory was performed under contract with the Department of Energy and supported by its Division of Basic Energy Sciences.

Literature Cited

1. Moss, J. R., Shaw, B. L., *J. Chem. Soc., Chem. Commun.* (1968) 632.
2. Bell, B., Chatt, J., Leigh, G. J., Ito, T., *J. Chem. Soc., Chem. Commun.* (1972) 34.
3. Moss, J. R., Shaw, B. L., *J. Chem. Soc., Dalton Trans.* (1972) 1910.
4. Chatt, J., Rowe, G. A., *J. Chem. Soc.* (1962) 4019.
5. Chatt, J., Garforth, J. D., Johnson, N. P., Rowe, G. A., *J. Chem. Soc.* (1964) 601.
6. Chatt, J., Leigh, G. J., Mingos, D. M. P., Paske, R. J., *J. Chem. Soc. A* (1968) 2636.
7. Chatt, J., Coffey, R. S., *J. Chem. Soc. A* (1969) 1963.
8. Douglas, P. G., Shaw, B. L., *J. Chem. Soc., Chem. Commun.* (1969) 624.
9. Leigh, G. J., Levinson, J. J., Robinson, S. D., *J. Chem. Soc., Chem. Commun.* (1969) 705.
10. Douglas, P. G., Shaw, B. L., *J. Chem. Soc. A* (1970) 334.
11. Bell, B., Chatt, J., Leigh, G. J., *J. Chem. Soc., Dalton Trans.* (1973) 997.
12. Chatt, J., Shaw, B. L., *J. Chem. Soc.* (1963) 3371.
13. Chatt, J., Coffey, R. S., Shaw, B. L., *J. Chem. Soc.* (1965) 7391.
14. Mann, B. E., Masters, C., Shaw, B. L., *J. Chem. Soc., Chem. Commun.* (1970) 703.
15. Ibid. (1970) 846.
16. Tebbe, F. N., *J. Am. Chem. Soc.* (1973) **95**, 5823.
17. Ginsberg, A. P., *J. Chem. Soc., Chem. Commun.* (1968) 857.
18. Abrahams, S. C., Ginsberg, A. P., Knox, K., *Inorg. Chem.* (1964) **3**, 558.
19. Kaesz, H. D., Saillant, R. B., *Chem. Revs.* (1972) **72**, 231.
20. Jesson, J. P., "Transition Metal Hydrides," E. L. Muetterties, Ed., p. 75, Marcel Dekker, New York, 1971.
21. Giusto, D., *Inorg. Chem. Acta Rev.* (1972) **6**, 91.
22. Ginsberg, A. P., *Transition Metal Chem.* (1965) **1**, 112.
23. Frenz, B. A., Ibers, J. A., "Transition Metal Hydrides," E. L. Muetterties, Ed., p. 33, Marcel Dekker, New York, 1971.
24. Churchill, M. R., Chang, S. W. Y., *Inorg. Chem.* (1974) **13**, 2413.
25. Churchill, M. R., ADV. CHEM. SER. (1978) **167**, 36.
26. Ibers, J. A., ADV. CHEM. SER. (1978) **167**, 26.

27. Ginsberg, A. P., *Inorg. Chem.* (1964) **3**, 567.
28. Ginsberg, A. P., Abrahams, S. C., Jamieson, P. B., *J. Am. Chem. Soc.* (1973) **95**, 4731.
29. Meakin, P., Guggenberger, L. J., Peet, W. G., Muetterties, E. L., Jesson, J. P., *J. Am. Chem. Soc.* (1973) **95**, 1467.
30. Albano, V. G., Bellon, P., *J. Organomet. Chem.* (1972) **37**, 151.
31. Albano, V., Bellon, P., Scatturin, V., *Rend. Ist. Lomb. Accad. Sci. Lett. A.* (1966) **100**, 986; quoted in Ref. *21.*
32. Clark, G. R., Skelton, B. W., Waters, T. N., *Inorg. Chim. Acta.* (1975) **12**, 235.
33. Elian, M., Hoffmann, R., *Inorg. Chem.* (1975) **14**, 1058.
34. Love, R. A., Chin, H. B., Koetzle, T. F., Kirtley, S. W., Whittlesey, B. R., Bau, R., *J. Am. Chem. Soc.* (1976) **98**, 4491.
35. Dimmler, B. G., Greenlaw, N., Kelley, M. A., Potter, D. W., Rankowitz, S., Stubblefield, F. W., *IEEE Trans. Nucl. Sci.* (1976) **NS-23**, 398.
36. McMullan, R. K., and, in part, Andrews, L. C., Koetzle, T. F., Reidinger, F., Thomas, R., Williams, G. J. B., NEXDAS. Neutron and X-ray Data Acquisition System, unpublished work, 1976.
37. Busing, W. R., Martin, K. O., Levy, H. A., Report ORNL-TM-305, Oak Ridge National Laboratory, Tennessee, 1962.
38. J. M. Stewart, Ed., Technical Report TR-446, Computer Science Center, University of Maryland, College Park, 1976.
39. Duchamp, D. J., Abstracts, Am. Crystallogr. Assoc., Montana, paper no. 14, p. 29, 1964.
40. Johnson, C. K., ORTEP-II, Report ORNL-5138, Oak Ridge National Laboratory, Tennessee, 1976.
41. H. M. Berman, F. C. Bernstein, H. J. Bernstein, T. F. Koetzle, G. J. B. Williams, Eds., *Brookhaven National Laboratory, Chemistry Department, CRYSNET Manual,* Informal Report BNL 21714, Brookhaven National Laboratory, New York, 1976.
42. Mason, R., private communication, 1974.
43. Aslanov, L., Mason, R., Wheeler, A. G., Whimp, P. O., *J. Chem. Soc., Chem. Commun.* (1970) 30.
44. Hart, D. W., Bau, R., Koetzle, T. F., *J. Am. Chem. Soc.* (1977) **99**, 7557.
45. Wilson, R. D., Koetzle, T. F., Hart, D. W., Kvick, Å., Tipton, D. L., Bau, R., *J. Am. Chem. Soc.* (1977) **99**, 1775.
46. Bau, R., Carroll, W. E., Teller, R. G., Koetzle, T. F., *J. Am. Chem. Soc.* (1977) **99**, 3872.
47. Cotton, F. A., *Chem. Soc. Rev.* (1975) **4**, 27.
48. Cotton, F. A., Harris, C. B., *Inorg. Chem.* (1965) **4**, 330.
49. Kusnetzov, B. G., Koz'min, P. A., *Zhur. Strukt. Khim.* (1963) **4**, 55.
50. Bennett, M. J., Graham, W. A. G., Hoyano, J. K., Hutcheon, W. L., *J. Am. Chem. Soc.* (1972) **94**, 6232.
51. Bau, R., Carroll, W. E., Teller, R. G., Koetzle, T. F., unpublished data, included in the Ph.D. dissertation of R. G. Teller, University of Southern California (1978).
52. Hart, D. W., Bau, R., unpublished data.
53. La Placa, S. J., Hamilton, W. C., Ibers, J. A., Davison, A., *Inorg. Chem.* (1969) **8**, 1928.
54. Moseley, P. T., Shearer, H. M. M., Spencer, C. B., *Acta Crystallogr.* (1969) **A25**, S169.
55. Schultz, A. J., Stearley, K. L., Williams, J. M., Mink, R., Stucky, G. D., *Inorg. Chem.* (1977) **16**, 3303.
56. Roziere, J., Williams, J. M., Stewart, R. P., Jr., Petersen, J. L., Dahl, L. F., *J. Am. Chem. Soc.* (1977) **99**, 4497.
57. Petersen, J. L., Dahl, L. F., Williams, J. M., *J. Am. Chem. Soc.* (1974) **96**, 6610.
58. Catti, M., Gevasio, G., Mason, S. A., *J. Chem. Soc., Dalton Trans.* (1978).
59. Olsen, J. P., Koetzle, T. F., Kirtley, S. W., Andrews, M., Tipton, D. L., Bau, R., *J. Am. Chem. Soc.* (1974) **96**, 6621.
60. Sheldrick, G. M., private communication, 1977.

61. Teller, R. G., Wilson, R. D., McMullan, R. K., Koetzle, T. F., Bau, R., *J. Am. Chem. Soc.* (1978) **100**, 0000.
62. Koetzle, T. F., McMullan, R. K., Bau, R., Hart, D. W., Teller, R. G., Tipton, D. L., Wilson, R. D., ADV. CHEM. SER. (1978) **167**, 61.
63. Bernstein, E. R., Hamilton, W. C., Keiderling, T. A., La Placa, S. J., Lippard, S. J., Mayerle, J. J., *Inorg. Chem.* (1972) **11**, 3009.
64. Bernstein, E. R., Hamilton, W. C., Keiderling, T. A., Kennelly, W. J., La Placa, S. J., Marks, T. J., Mayerle, J. J., unpublished data, included in the Ph.D. dissertation of Keiderling, T. A., p. 174–188, Princeton University, 1974.
65. Johnson, P. L., Cohen, S. A., Marks, T. J., Williams, J. M., submitted for publication.
66. Dahl, L. F., Broach, R. W., Longoni, G., Chini, P., Schultz, A. J., Williams, J. M., ADV. CHEM. SER. (1978) **167**, 93.
67. Petersen, J. L., Dahl, L. F., Williams, J. M., ADV. CHEM. SER. (1978) **167**, 11.
68. Gibb, T. R. P., Jr., Schumacher, D. P., *J. Phys. Chem.* (1960) **64**, 1407.
69. Hart, D. W., Bau, R., Koetzle, T. F., *J. Am. Chem. Soc.* (1977) **99**, 7557.
70. Bau, R., Carroll, W. E., Teller, R. G., Koetzle, T. F., *J. Am. Chem. Soc.* (1977) **99**, 3872.

RECEIVED July 19, 1977.

Stereochemistry of Monohydrido- and Dihydrido-Dodecanickel Carbonyl Clusters Containing a Hexagonal Close Packed Nickel Fragment

A Model System for Hydrogen Interaction with a Close Packed Metal Lattice

ROBERT W. BROACH and LAWRENCE F. DAHL—Department of Chemistry, University of Wisconsin–Madison, Madison, WI   53706

GIULIANO LONGONI and PAOLO CHINI—Instituto di Chimica Generale dell' Universita di Milano, 20133 Milano, Italy

ARTHUR J. SCHULTZ and JACK M. WILLIAMS—Chemistry Division, Argonne National Laboratory, Argonne, IL   60439

Single-crystal x-ray diffraction studies of the $[Ph_4As]^+$, $[PPN]^+$, and $[Ph_4P]^+$ salts (Compounds 1, 2, and 3) of the $[Ni_{12}(CO)_{21}H_2]^{2-}$ dianion and of $[Ph_4As]_3^+[Ni_{12}(CO)_{21}H]^{3-}\cdot$ (acetone) (Compound 4) and subsequent neutron diffraction investigations of Compounds 3 and 4 provided the first detailed crystallographic description of the interaction of interstitial hydrogens with a close-packed fragment of metal atoms. X-ray diffraction results revealed the common geometry of the anions as a 12-atom nickel fragment of a hcp metal lattice (containing two octahedral and six tetrahedral holes) surrounded by nine terminal and 12 doubly bridging carbonyl ligands. Neutron diffraction results showed hydrogen occupation of one octahedral hole in the trianion and both octahedral holes in each dianion. Stereochemical implications are discussed with respect to possible proton migration within these nickel clusters and to their observed reversible protonation-deprotonation reactions.

X-ray diffraction studies on salts of the nickel and platinum carbonyl dianions $[M_3(CO)_3(\mu_2\text{-}CO)_3]_n^{2-}$ (M = Ni, Pt) showed that these

0-8412-0390-3/78/33-167-093/$05.00/0
© American Chemical Society

complexes are based on the stacking of a triangular $M_3(CO)_3(\mu_2\text{-}CO)_3$ building block (1–7). The platinum carbonyl dianions, where the stacking of the triangular trimetal units is in a nearly eclipsed fashion, have been isolated (8) for n = 1 to 6 and $n \sim$10 (and characterized ($4, 5$) structurally for n = 2, 3, 4, 5); the nickel analogues, where the triangular trimetal units are staggered, have been obtained ($9, 10$) (and structurally analyzed ($3, 7$)) only for n = 2 and n = 3. These latter dianions were found by Chini and Longoni ($9, 10$) to give rise by acidic hydrolysis to a new series of nickel carbonyl anions, for which analytical data indicated 12 nickel atoms in each anion. A trianion, which was prepared by hydrolysis of the $[Ni_3(CO)_3(\mu_2\text{-}CO)_3]_2{}^{2-}$ dianion at a buffered pH of $\sim$5 to 6 in THF, gave an ir spectrum containing carbonyl bands at 1990(s) and 1830(s) cm^{-1} while its ^{1}H NMR spectrum in acetone-d^6 exhibited a high-field resonance for hydrogen at 34τ. A dianion, which was prepared from either the $[Ni_3(CO)_3$ $(\mu_2\text{-}CO)_3]_2{}^{2-}$ or $[Ni_3(CO)_3(\mu_2\text{-}CO)_3]_3{}^{2-}$ dianions by hydrolysis at a lower buffered pH of $\sim$3 to 4, showed ir carbonyl bands in THF at 2020(s) and 1860(s) cm^{-1} and a NMR proton resonance in acetone-d^6 at 28τ. Relative integration of the NMR peaks of the cation's phenyl hydrogens to the anion's hydrogens indicated that the trianion contained one hydrogen atom while the dianion contained two. Furthermore, it was found that the monohydrido trianion could be protonated by acetic acid in THF to give the dihydrido dianion, and conversely the dianion could be deprotonated by potassium *tert*-butoxide in acetone or THF to give the trianion.

The presence of hydrogen atoms in this new series of anions indicated that the anions' stereochemistry was markedly different from those of the previously characterized $[M_3(CO)_3(\mu_2\text{-}CO)_3]_n{}^{2-}$ dianions, and, moreover, the highly shielded nature of the hydrogens indicated from their NMR resonances suggested (11) the distinct possibility that they were interstitial. This indication of interstitial metal–hydrogen clusters intensified our interest in their stereochemistry since previously there have been only two substantiated examples of transition metal polyhedral cluster compounds containing interstitial hydrogen atoms. Based upon prior independent x-ray diffraction analyses ($12, 13$) which showed that Nb_6I_{11} contains $[Mo_6Cl_8]^{4+}$ type clusters (14) (with each of the other iodine atoms bridging two such $[Nb_6I_8]$ groups) in accord with the compound's formulation as $[Nb_6I_8]^{3+}(I^-)_{6/2}$, powder neutron diffraction measurements (15) of HNb_6I_{11} and DNb_6I_{11} indicated that the hydrogen atoms occupy the octahedral sites of the $[Nb_6I_{11}]$ groups. The interstitial nature of the hydrogen atoms in the $[Rh_{13}(CO)_{24}H_{5-n}]^{n-}$ (n = 2,3,4) anions (16) was clearly disclosed from an analysis (17) of their ^{1}H NMR spectra.

To determine both the stoichiometry and atomic arrangement of these new nickel carbonyl cluster anions (including an assessment of the geometrical effects of the hydrogen atoms), x-ray diffraction studies of $[Ph_4As]_2{}^+[Ni_{12}(CO)_{21}H_2]^{2-}$ (**1**), $[PPN]_2{}^+[Ni_{12}(CO)_{21}H_2]^{2-}$ (**2**) (where PPN denotes the bis(triphenylphosphine)iminium cation), $[Ph_4P]_2{}^+[Ni_{12}(CO)_{21}H_2]^{2-}$ (**3**), and $[Ph_4As]_3{}^+\text{-}$ $[Ni_{12}(CO)_{21}H]^{3-}\cdot Me_2CO$ (**4**), were carried out, followed by neutron diffraction analyses of Compounds 3 and 4. These investigations have ascertained the correct formulation to be $[Ni_{12}(CO)_{21}H_{4-n}]^{n-}$ (n = 2,3) and have revealed a highly unusual geometry for the dodecanickel framework based on a hcp array of metal atoms. The subsequent neutron diffraction studies not only have substantiated the proposed presence of the interstitial hydrogen atoms but also have allowed the first detailed crystallographic examination of the interaction of interstitial hydrogen atoms with a close-packed fragment of metal atoms. This work, which is of interest with respect to the behavior of hydrogen in metals (*18, 19*), was found to be of particular pertinence in connection with the observed structural differences between the interstitial face-centered nickel–hydrogen compound, $NiH_{0.6}$ (*20*), and ccp nickel (*21*).

Experimental

Since the crystallographic analyses of these four compounds were not at all straightforward, an outline of the approaches used to obtain their crystal structures is given below. Details of the x-ray and neutron diffraction work (*22*), including the extensive refinements, will be published elsewhere.

X-ray Diffraction Studies. Initial attempts to solve the triclinic crystal structures of the PPN^+ and the Ph_4As^+ salts of the $[Ni_{12}(CO)_{21}H_2]^{2-}$ dianions by trial-and-error models founded on a considerable number of "heavy atom" interpretations of the calculated Patterson maps and by direct methods, via the use of MULTAN (*23*), were unsuccessful. Several models based on 13-atom nickel fragments geometrically similar to the 13-atom rhodium fragment previously determined (*16, 17*) in the $[Rh_{13}(CO)_{24}H_3]^{2-}$ dianion gave limited agreement, but no such model under least-squares refinements converged with an unweighted $R_1(F)$ discrepancy index lower than 30%. The correct crystal structure containing the dodecanickel dianion was ultimately unraveled for the Ph_4As^+ salt **1** by "brute force," involving an initial phasing under $P\overline{1}$ symmetry of a Fourier map with two tentative arsenic positions, followed by a recognition on this map of low-density peaks as a possible dodecanickel framework together with a disregarding of much higher density peaks produced by pseudo-nickel imagery. This knowledge of the dianion's architecture then enabled one to obtain the correct solution of the crystal structure of the PPN^+ salt **2** from a wrong one generated on an E-map by MULTAN (*23*).

The crystallographic determination of the tetraphenylphosphonium salt **3**, crystals of which were subsequently furnished by Longoni and Chini, was necessitated from its crystals being more suitable for a neutron diffraction investigation (vide infra) of the $[Ni_{12}(CO)_{21}H_2]^{2-}$ dianion. The solution of this structural problem was based on the initial assumption that the space group of the compound conforms to triclinic $P1$ symmetry rather than to the true monoclinic $P2_1/c$ symmetry. Since the choice of origin is arbitrary under noncentrosymmetric $P1$ symmetry, a nickel triangle of side 2.5 Å was placed in the unit cell in the orientation determined from the computed Patterson map with the origin situated at one of the nickel atoms. A Fourier map phased on these three nickel atoms gave, under $P1$ symmetry, four separate assemblages of peaks. Sets of 12 peaks (corresponding to the known geometry of the dodecanickel fragment determined from Compound 1) were chosen from the distributions of peaks in each assemblage in accord with monoclinic $P2_1/c$ symmetry, and the coordinates for 12 independent nickel atoms were shifted to what was assumed to be the true

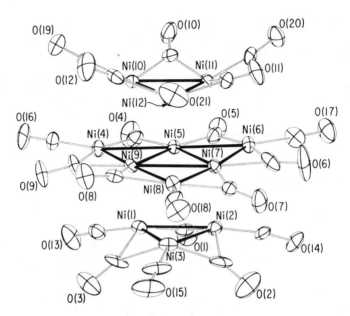

Figure 1. Architecture of the $[Ni_{12}(CO)_{21}H]^{3-}$ trianion in Compound 4, as determined from the x-ray diffraction refinement, showing the anisotropic thermal ellipsoids of 20% probability and the atom labeling scheme common to the trianion and three dianions of Compounds 1, 2, 3, and 4. All four anions are similarly constructed from a planar Ni_6-$(CO)_3(\mu_2\text{-}CO)_6$ fragment of $D_{3h}\text{-}\overline{6}2m$ symmetry capped by two $Ni_3(CO)_3(\mu_2\text{-}CO)_3$ fragments through Ni–Ni interactions.

center of symmetry. Further Fourier syntheses on atoms from one independent anion under $P2_1/c$ symmetry revealed positions for 38 atoms of the independent anion. At this point it was discovered that the initial choice for the center of symmetry was incorrect. One of the anions was related to the first by the c_b glide plane, but the other two were not related to the first by either the center of symmetry or 2_1 screw axis. Phased on the 38 atoms previously found, a Fourier synthesis calculated under Pc symmetry revealed two anions related by a center of symmetry at approximately 0, 0, 1/4. After the coordinates for the independent anion were shifted to correspond to the true center of symmetry at the origin, further Fourier and difference Fourier syntheses under $P2_1/c$ symmetry exhibited positions for all independent nonhydrogen atoms.

The crystal structure of Compound 4, which was found to possess a centrosymmetric triclinic unit cell containing six tetraphenylarsonium cations, two $[Ni_{12}(CO)_{21}H]^{3-}$ trianions, and two solvent acetone molecules, also was solved by the combined Patterson–Fourier method, based on the assumption of the dodecanickel core of the trianion being analogous to that of the dianion. Peaks corresponding to all of the nickel and arsenic atoms were first located from successive Fourier syntheses in the unit cell under noncentrosymmetric $P1$ symmetry, after which initial atomic coordinates for the 12 independent nickel and three independent arsenic atoms were obtained by an origin shift to an approximate center of symmetry relating pairs of these peaks to one another.

Neutron Diffraction Studies.   The two interstitial hydrogen atoms in the dianion of **3** and the one interstitial hydrogen atom in the trianion of **4** were located from single-crystal neutron diffraction data collected at the CP-5 reactor at Argonne National Laboratory.   In each case, the initial positions of the interstitial hydrogen nuclei were obtained from a difference Fourier synthesis of nuclear scattering density based on the coordinates for all nonhydrogen atoms determined from the x-ray refinements and for the phenyl hydrogen atoms fixed at idealized positions.   In Compound **4**, the six methyl hydrogen atoms of the solvent acetone molecule were unresolved in the difference map, presumably because of large librations of the methyl groups and because of errors introduced by uncertainties in the thermal parameters.   As a consequence of the unusually large number of independent atoms in both Compound **3** and Compound **4**, only the positional and isotropic nuclear temperature factors for all atoms in the anion were refined (with the coordinates for all atoms in the cations fixed at their x-ray determined values and with their isotropic nuclear temperature factors fixed at reasonable values).

Results and Discussion

General Description of the Structures.   The x-ray structural determinations of the four compounds revealed discrete anions and cations with normal interionic separations and with stoichiometries consistent with the analytical and spectral data.   Shown in Figures 1 and 2 are ORTEP drawings of the

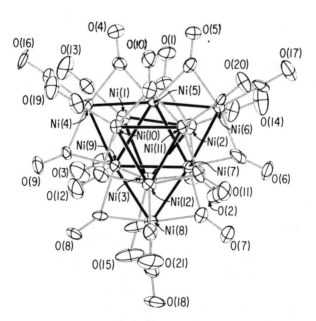

*Figure 2.   A view of the $[Ni_{12}(CO)_{21}H]^{3-}$ trianion of Compound **4** normal to the $Ni_6(CO)_3(\mu_2\text{-}CO)_6$ plane, emphasizing the approximate threefold axis.   The entire trianion ideally possesses C_{3v}-3m symmetry.*

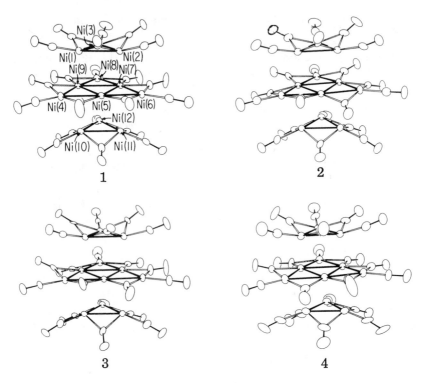

Figure 3. Analogous views with identical atom labelings and with 30 %
anisotropic thermal ellipsoids obtained from the x-ray diffraction refinements of the $[Ni_{12}(CO)_{21}H_2]^{2-}$ dianion for Compounds **1, 2,** and **3,** and
$[Ni_{12}(CO)_{21}H]^{3-}$ trianion for Compound **4.** The dianions ideally conform
to D_{3h}-$\overline{6}2m$ symmetry, while the trianion ideally possess C_{3v}-3m symmetry.

$[Ni_{12}(CO)_{21}H]^{3-}$ trianion (in Compound **4**) which consists of a planar $Ni_6(CO)_3$
$(\mu_2$-CO$)_6$ fragment symmetrically attached to two equivalent $Ni_3(CO)_3(\mu_2$-CO$)_3$
moieties by Ni–Ni interactions. This 12-atom nickel fragment of a hcp metal
lattice surrounded by nine terminal and 12 doubly bridging carbonyl ligands is
common to the anions of all four compounds, as shown in Figure 3. Although
none of the anions has any crystallographic site symmetry constraints, the dihydrido dianions of Compounds **1, 2,** and **3** ideally conform to D_{3h}-$\overline{6}2m$ symmetry
(where the horizontal mirror plane contains the hexanickel fragment and the
threefold axis passes through the centroids of the hexanickel and two trinickel
fragments) while the monohydrido trianion of Compound **4** ideally conforms
to C_{3v}-3m symmetry. The view in Figure 2 of the trianion clearly reveals the
approximate threefold axis common to all of the anions. Geometrical deviations
of the dodecanickel framework of the trianion from C_{3v} symmetry and of the
dianion from D_{3h} symmetry are not appreciable in the four compounds. The
observed angular distortions of the outer two nickel triangles with respect to the

central three nickel atoms differ by only 1° to 7° from the regular staggered value of 60° while the calculated tipping of each outer trinickel plane relative to the central hexanickel plane is slight, with the values ranging from 0.4° to 2.2°. These geometrical distortions can be attributed primarily to intra-anion steric effects, upon which are superimposed asymmetric packing forces. The strong steric interactions between the carbonyl ligands of the central hexanickel fragment and those of each of the trinickel fragments produce a considerable outward bending of the carbonyl groups in the two outer trinickel fragments. The relative constancy of this carbonyl bending in the anions of the four compounds apparently reflects the equilibration achieved between nonbonding repulsive effects of the carbonyl ligands and interplanar bonding interactions of the nickel atoms.

Stereochemistry of the Dianion and Trianion. (a) THE DODECANICKEL CORE. A drawing of the dodecanickel framework for the dianion of Compound 1, showing the three kinds of equivalent nickels under assumed D_{3h} symmetry, is given in Figure 4. The mean Ni–Ni bonding distances (Table I) obtained from the refinements based on the x-ray diffraction data are the averages for each of the five different types of distances—viz., those for the Ni_1–Ni_1 bonds within the two $Ni_3(CO)_3(\mu_2$-CO$)_3$ fragments, for the carbonyl-bridged Ni_2–Ni_3 and noncarbonyl-bridged Ni_2–Ni_2 bonds within the central $Ni_6(CO)_3(\mu_2$-CO$)_6$ fragment, and for the Ni_1–Ni_2 and Ni_1–Ni_3 bonds between the central and outer fragments. A striking feature revealed in Table I is the close similarity of the corresponding Ni–Ni distances in the four anions together with the small, but nevertheless highly significant, differences in the Ni_1–Ni_3 and Ni_1–Ni_2 inter-ring distances between the two halves of the $[Ni_{12}(CO)_{21}H]^{3-}$ trianion. The similarities of the larger values in one of the two halves of the trianion with those in both equivalent halves of the dianion in Compounds 1, 2, and 3 provided the first evidence that the hydrogen atom in the trianion was localized in only one of the two octahedral interstices (rather than disordered in both octahedral interstices) and that the addition of the second proton to the trianion causes a similar inter-

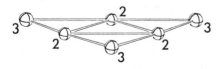

Figure 4. The dodecanickel framework showing the three kinds of nickel atoms under assumed D_{3h}-$\overline{6}2m$ symmetry. This system can be viewed as a hcp array of nickel atoms (with an a:b:a stacking pattern) containing two octahedral and six tetrahedral interstices.

Table I.   Mean Ni–Ni Distances (Å) in the $[Ni_{12}(CO)_{21}H_{4-n}]^{n-}$ Anions (n = 2, 3) Determined from X-ray Diffraction Studies

Anion	$[Ni_{12}(CO)_{21}H_2]^{2-}$			$[Ni_{12}(CO)_{21}H]^{3-}$	
Idealized Geometry	D_{3h}			C_{3v}	
Compound	1	2	3	4	
					Differ-
	w/Hᵃ	w/H	w/H	w/H; wo/Hᵃ	enceᵈ
A. Two trinickel fragments					
Ni_1-Ni_1	[6]ᵇ	2.438	2.442	2.438 [3]	2.435; 2.415	+0.020
B. Central hexanickel fragment					
Ni_2-Ni_3	[6]	2.429	2.429	2.431 [3]	2.416
Ni_2-Ni_2	[3]	2.665	2.654	2.658 [3]	2.682
C. Inter-ring (between fragments)					
Ni_1-Ni_3	[6]	2.807	2.814	2.808 [3]	2.822; 2.750	+0.072
Ni_1-Ni_2	[12]	2.847	2.850	2.847 [6]	2.847; 2.811	+0.036
TriNi(c)–HexaNi(c)ᶜ	[2]	2.43	2.44	2.44 [1]	2.44; 2.38

ᵃw/H and wo/H, which denote with hydrogen and without hydrogen, respectively, are used to distinguish between the two nonequivalent trinickel fragments in the monohydrido trianion caused by hydrogen occupation of only one of the two octahedral interstices. In the dihydrido dianion, both of the two equivalent trinickel fragments are associated with hydrogen because of occupation of both octahedral interstices.

ᵇThe square brackets enclose the number of equivalent Ni–Ni distances (averaged under the idealized geometry) for each value listed in the right columns.

ᶜDesignates the distance between the centroids of the trinickel and hexanickel fragments.

ᵈThis bond-length expansion attributed to the interstitial hydrogen is between the half of the monohydrido trianion containing the interstitial hydrogen (w/H) and the other unoccupied half (wo/H).

ring expansion by 1–2% of the second octahedral hole, in accord with both octahedral holes being occupied in the dianion.

The carbonyl-bridged Ni_1-Ni_1 bonds in the three dihydrido dianions and in the half of the monohydrido trianion containing the interstitial hydrogen possess identical mean lengths of 2.44 Å.   The corresponding mean value of 2.42 Å for the carbonyl-bridged Ni_1-Ni_1 bonds in the half of the trianion not containing the interstitial hydrogen is only slightly smaller.   It is noteworthy that these mean values are significantly longer than the mean values for the corresponding carbonyl-bridged Ni–Ni bonds found in the analogous $Ni_3(CO)_3(\mu_2\text{-}CO)_3$ fragments of the $[Ni_5(CO)_9(\mu_2\text{-}CO)_3]^{2-}$ dianion (2.36 Å) (24), the $[Ni_3(CO)_3(\mu_2\text{-}CO)_3]_2^{2-}$ dianion (2.38 Å) (3), the $[Ni_3(CO)_3(\mu_2\text{-}CO)_3]_3^{2-}$ dianion (2.38 Å) (7), and the $[M_2Ni_3(CO)_{13}(\mu_2\text{-}CO)_3]^{2-}$ dianions (2.34 Å for both M = Mo and M = W) (25).

Table I also shows that the mean lengths of the carbonyl-bridged Ni_2–Ni_3 bonds in the central $Ni_6(CO)_3(\mu_2\text{-}CO)_6$ fragment of the dianions (2.43 Å for each of the Compounds, 1, 2, and 3) and the trianion (2.42 Å) are virtually identical with those of the carbonyl-bridged Ni_1–Ni_1 bonds. Each of these carbonyl-bridged Ni–Ni bonds is considered to correspond to a normal electron-pair bond.

The noncarbonyl-bridged Ni_2–Ni_2 bonds in the central $Ni_6(CO)_3(\mu_2\text{-}CO)_6$ fragment expectedly are considerably longer than the carbonyl-bridged bonds, with the mean intra-triangular distance having a value of 2.68 Å for the trianion and falling within a 2.654–2.665 Å range for the dianions. All of these values are close to the mean of 2.65 Å found for the noncarbonyl-bridged Ni–Ni single-bond lengths in the completely bonding nickel cube of $Ni_8(CO)_8(\mu_4\text{-}PC_6H_5)_6$ (26).

The longer inter-ring Ni_1–Ni_2 and Ni_1–Ni_3 distances between the hexanickel and trinickel fragments are influenced significantly by the intersititial hydrogens. For the three hydrido dianions, the Ni_1–Ni_2 bonds have identical mean values of 2.85 Å and likewise the Ni_1–Ni_3 bonds have identical mean values of 2.81 Å. In the monohydrido trianion, the mean Ni_1–Ni_2 and mean Ni_1–Ni_3 distances about one octahedral site of 2.85 and 2.82 Å, respectively, are similar to those in the dianions, but the corresponding mean Ni_1–Ni_2 and mean Ni_1–Ni_3 distances about the other octahedral hole of 2.81 and 2.75 Å, respectively, are definitely shorter. On the basis of these particular bond-length variations, together with the observation that the above shorter mean Ni_1–Ni_2 distance of 2.81 Å for the presumed unoccupied half of the trianion is nearer to the mean inter-triangular Ni–Ni distance of 2.77 Å found (3) in the $[Ni_3(CO)_3(\mu_2\text{-}CO)_3]_2{}^{2-}$ dianion (which can be considered to possess an unoccupied octahedral hole), it was then concluded and later ascertained from the neutron diffraction results (vide infra) that the dodecanickel dianion and trianion contain interstitial hydrogen atoms that exclusively occupy the octahedral sites. Table I also shows that the separation of the trinickel and hexanickel planes without the presence of the intersitial hydrogen atom is 2.38 Å while the addition of an interstitial hydrogen atom increases this distance to 2.44 Å, in accord with the premise (11) that Ni–H bonding effectively weakens the Ni–Ni bonding interactions. This observed greater expansion of the inter-ring Ni_1–Ni_2 and Ni_1–Ni_3 bonds by 0.04 and 0.07 Å, respectively, relative to the expansion of the intra-ring carbonyl-bridged Ni_1–Ni_1 bonds by 0.02 Å, can be readily attributed to the latter bonds being considerably stronger. In harmony with this rationalization is the qualitative metal-cluster bonding description, first proposed (25) for the $Ni_3(CO)_3(\mu_2\text{-}CO)_3$ units in the $[M_2Ni_3(CO)_{13}(\mu_2\text{-}CO)_3]^{2-}$ dianions (M = Mo, W) and later extended to the $[Ni_5(CO)_9(\mu_2\text{-}CO)_3]^{2-}$ and $[Ni_3(CO)_3(\mu_2\text{-}CO)_3]_2{}^{2-}$ dianions, which suggests normal electron-pair interactions within the intratriangular trinickel fragments of the dodecanickel anions and weaker delocalized, multicentered interactions between the nickel atoms of the hexanickel and trinickel fragments. These bonding conclusions are also consistent with the observed variations in the Ni–Ni distances.

(b) THE CARBONYL LIGANDS.   The Ni–CO(terminal) bond lengths in the dianions and in the trianion appear to be normal ones for the outer triangular fragments, with mean values of range 1.78 ± 0.02 Å for the three dianions and of 1.75 and 1.76 Å for the two nonequivalent halves of the trianion.   These values compare favorably with the mean of 1.75 Å for the Ni–CO(terminal) bond lengths in the $[Ni_3(CO)_3(\mu_2\text{-}CO)_3]_2^{2-}$ dianion.   The corresponding mean values of the Ni–CO(terminal) bond lengths for the hexanickel fragment were found to be slightly shorter in the dianions (1.74–1.76 Å) and in the trianion (1.74 Å).

The bridging carbonyl ligands linking pairs of equivalent Ni_1 atoms in each $Ni_3(CO)_3(\mu_2\text{-}CO)_3$ fragment and the nonequivalent Ni_2 and Ni_3 atoms in the central $Ni_6(CO)_3(\mu_2\text{-}CO)_6$ fragment were determined to be symmetrically coordinated within 0.01 Å in the three dianions and in the trianion, except for a small indicated asymmetry of 0.03 Å between the Ni_2 and Ni_3 atoms in the trianion.   The corresponding mean values of the Ni–CO(bridging) distances, which are expectedly ca. 0.15 Å longer than the Ni–CO(terminal) distances, exhibit a parallel bond-length trend.   The mean values for the $Ni_3(CO)_3(\mu_2\text{-}CO)_3$ fragments in the dianions (1.92–1.93 Å) and the two halves of the trianion (1.89, 1.90 Å) are similar to that of 1.90 Å in the $[Ni_3(CO)_3(\mu_2\text{-}CO)_3]_2^{2-}$ dianion; these values in turn are slightly longer than those for the $Ni_6(CO)_3(\mu_2\text{-}CO)_6$ fragment in the dianions (1.89–1.90 Å) and in the trianion (asymmetrical with 1.86 vs. 1.89 Å).   In the four anions, the individual mean C–O bond lengths of 1.13–1.16 Å (with an overall mean of 1.14 Å) for the terminal carbonyl ligands and of 1.15–1.18 Å (with an overall mean of 1.16 Å) for the bridging carbonyl ligands are normal values.

(c) THE INTERSTITIAL HYDROGEN ATOMS.   The neutron diffraction studies of $[Ph_4P]_2^+[Ni_{12}(CO)_{21}H_2]^{2-}$ (3) and $[Ph_4As]_3^+[Ni_{12}(CO)_{21}H]^{3-}\cdot Me_2CO$ (4) have provided the first detailed stereochemical information on the bonding of an interstitial hydrogen atom.   A summary of the important features for both compounds is given in Table II while ORTEP drawings of the anions of both compounds obtained from the neutron diffraction refinements are shown in Figures 5 and 6.

The dodecanickel core of these anions can be considered as a small fragment of a hcp array of metal atoms stabilized by a surrounding blanket of carbonyl ligands.   The three planar fragments of nickel atoms form an $a:b:a$ stacking pattern, giving rise to two octahedral and six tetrahedral holes.   The results show that the hydrogens in both the dianion and trianion are localized in octahedral sites with no evidence for occupation of tetrahedral sites, which not only is in harmony with the larger size of the octahedral sites compared with the tetrahedral sites but also is consistent with the partial occupation of only the octahedral sites by hydrogens in the ccp $NiH_{0.6}$ (20) (vide infra).

The single hydrogen of the trianion was found to occupy only one of the two octahedral sites rather than being randomly disordered in the crystalline state in both octahedral sites, with an occupancy factor of 0.5 for each position.   This observed crystal ordering of the polar C_{3v} trianions, which are identically oriented

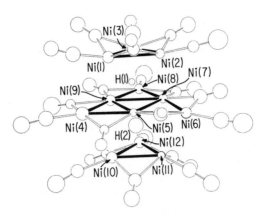

Figure 5. Architecture of the $[Ni_{12}(CO)_{21}$-$H_2]^{2-}$ dianion in Compound 3 with 20% isotropic ellipsoids of nuclear motion as determined from the neutron diffraction refinements. The two hydrogen atoms were found to occupy the two octahedral interstices.

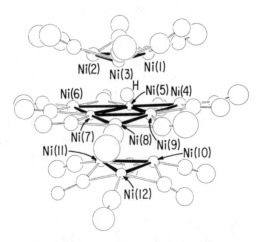

Figure 6. Architecture of the $[Ni_{12}(CO)_{21}$-$H]^{3-}$ trianion in Compound 4 with 20% isotropic ellipsoids of nuclear motion, as determined from the neutron diffraction refinements. The single hydrogen atom was found to be localized in one of the two octahedral interstices rather than crystallographically disordered in both octahedral holes.

Table II. Mean Distances[a] (Å) Involving the Interstitial Hydrogen
Atoms Determined from Neutron Diffraction Studies of
Compounds 3 and 4

	$[Ni_{12}(CO)_{21}H_2]^{2-}$		$[Ni_{12}(CO)_{21}H]^{3-}$	
H–Ni$_1$	[6][b]	2.00[c]	[3]	2.22
H–Ni$_2$	[6]	1.84[d]	[3]	1.72
H$\cdots$H		2.11		—
H$\cdots$(trinickel plane)		1.40		1.69
H$\cdots$(hexanickel plane)		1.04		0.73

[a] Based on assumed D_{3h} and C_{3v} symmetries for the dihydrido dianion and monohydrido trianion, respectively.
[b] The square brackets enclose the number of equivalent distances (averaged under the idealized geometry) for each value listed in the right column.
[c] The average of [3] 2.01 Å in Ni$_6$H fragment (a) and [3] 1.99 Å in Ni$_6$H fragment (b) (See Figure 8).
[d] The average of [3] 1.84 Å in Ni$_6$H fragment (a) and [3] 1.85 Å in Ni$_6$H fragment (b).

by translational symmetry into finite columnar arrays (or strings) along one direction, can be attributed to the resulting asymmetrical crystalline environment about each trianion, as evidenced from the considerably different contact distances of the tetraphenylarsonium cations with the two outer $Ni_3(CO)_3(\mu_2\text{-}CO)_3$ fragments of the trianion.

The local nickel environments of the interstitial hydrogens in the two anions are given in Figure 7. The two hydrogen atoms in the dihydrido dianion are not situated at the centers of the octahedral holes but are distinctly closer to the central hexanickel plane (1.04 Å) than to the two outer trinickel planes (1.40 Å). This shift of the interstitial hydrogens from the centers of the adjacent nickel octahedra toward one another also is reflected in each hydrogen being located at an average Ni–H distance of 1.84 Å from the three Ni$_2$ atoms compared with that of 2.00 Å from the three Ni$_1$ atoms. Figure 7 shows that the geometries of the two crystallographically independent Ni$_6$H fragments (a) and (b), which are equivalent under assumed D_{3h} symmetry, are identical to each other within experimental error with respect to the disposition of the hydrogen nuclei. The indicated difference between the two hydrogen nuclei in their isotropic thermal motion is deemed not to be physically meaningful but instead to be a consequence of the limited refinement model outlined in the experimental section.

The unexpected structural feature elucidated from the neutron diffraction analyses is that the single interstitial hydrogen of the monohydrido trianion in Compound 4 lies much closer to the central hexanickel plane (0.73 Å) than to the outer triangular nickel plane (1.69 Å). In fact, its asymmetrical displacement from the center of the nickel octahedron (Figure 7c) is sufficiently great so the hydrogen could not unreasonably be considered as triply bridging to only the central three Ni$_2$ atoms in the octahedral hole. The three individual Ni–H distances to the Ni$_1$ atoms in the outer nickel triangle are 2.08, 2.27, and 2.32 Å, in contrast to the three Ni–H distances to the Ni$_2$ atoms in the central nickel triangle of 1.54, 1.75, and 1.88 Å. Support for a consideration of this interstitial hydrogen as effectively being coordinated to only three of the six nickel atoms is given by the mean value of 1.72 Å for the three shorter Ni–H distances, being close to the mean value of 1.69 Å for the Ni–H distances determined for the triply bridging hydrogen atoms in $Ni_4(h^5\text{-}C_5H_5)_4(\mu_3\text{-}H)_3$ (27, 28) from a single-crystal neutron diffraction study (29). A rationalization for the preferential interaction of the single hydrogen in the trianion with only three of the six nickel atoms is not apparent.

In the protonation of the trianion to the dianion, the second hydrogen occupies an adjacent face-shared octahedron in the dodecanickel fragment. The

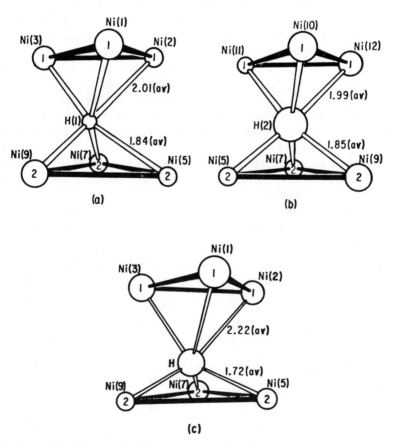

Figure 7. The two analogous Ni_6H fragments (a and b) in $[Ph_4P]_2{}^+$-$[Ni_{12}(CO)_{21}H_2]^{2-}$ (3), and the Ni_6H fragment (c) in $[Ph_4As]_3{}^+$-$[Ni_{12}(CO)_{21}H]^{3-}\cdot Me_2CO$ (4), showing hydrogen atoms in the octahedral interstices. The mean Ni–H distances and 20% isotropic thermal ellipsoids of nuclear motion were obtained from the neutron diffraction experiments.

fact that the two intersitital hydrogens move closer to the centers of the octahedral sites can be ascribed to a composite of two effects. The second hydrogen can be presumed to compete effectively with the first one for electron density on the central nickel triangle such that a decrease in the available electron density at the Ni_2 atoms gives rise to longer Ni–H distances. Alternatively or concurrently, mutual repulsion between the two hydrogen nuclei in neighboring octahedral holes can force them apart to the observed equilibrium distance of 2.11 Å. This distance compares with a mean H–H separation in $Ni_4(h^5-C_5H_5)_4(\mu_3-H)_3$ of 2.325 Å (29).

(d) COMPARISON WITH NICKEL METAL AND $NiH_{0.6}$. Table III provides a comparison of the octahedral and tetrahedral interstices in the monohydrido trianion with those in ccp nickel and in $NiH_{0.6}$. Of prime interest is that this comparison is in accord with the premise that the nature of these interstitial hydrogens in the discrete $[Ni_{12}(CO)_{21}H_{4-n}]^{n-}$ anions ($n = 2, 3$) may be used as a meaningful model with regard to the properties of interstitial metal hydrogen compounds such as $NiH_{0.6}$. In this connection, a powder neutron-diffraction investigation (20) of $NiH_{0.6}$ showed that the partial occupation of only octahedral sites by the intersitial hydrogens resulted in an expansion of the ccp nickel lattice from that in ccp nickel metal (21) such that the average radius of 1.86 Å for the partially filled octahedral holes in $NiH_{0.6}$ compares favorably with the radii of the occupied octahedral holes in the monohydrido and dihydrido anions. Furthermore, the average radius of 1.61 Å for the empty tetrahedral holes in $NiH_{0.6}$ is virtually identical to the average radius of 1.62 Å estimated for the empty tetrahedral holes in the trianion.

(e) CRYSTAL PACKING. Packing diagrams for the four compounds show that the anions are well separated from each other by cations in two unit cell di-

Table III. Comparison of Octahedral and Tetrahedral Interstices in the Monohydrido Trianion of Compound 4 with Those in Nickel Metal and in $NiH_{0.6}$

| | $[Ni_{12}(CO)_{21}H]^{3-}$ | | Ni Metal[a] | $NiH_{0.6}$[b] |
	w/H	wo/H	ccp	ccp
Oct. hole	1.92 (occ.)	1.88 (empty)	1.76 (empty)	1.86 (partially occ.)
Ni–Ni (av)[c]	2.70	2.68	2.49	2.63
Teh. hole	1.63 (empty)	1.62 (empty)	1.52 (empty)	1.61 (empty)
Ni–Ni (av)[d]	2.67	2.64	2.49	2.63

[a] ccp nickel ($a_0 = 3.52$ Å) (21).
[b] $NiH_{0.6}$ ($a_0 = 3.72$ Å). A powder neutron diffraction study (20) of this interstitial compound showed the hydrogen atoms to occupy only octahedral sites of the fcc nickel lattice.
[c] Based on the weighted average of [3] Ni_1-Ni_1, [3] Ni_2-Ni_2, and [6] Ni_1-Ni_2, which comprise the 12 Ni–Ni edges of the octahedral hole either containing the interstitial hydrogen (w/H) or without it (wo/H).
[d] Based on the weighted average of [2] Ni_1-Ni_2, [1] Ni_1-Ni_3, [1] Ni_2-Ni_2, and [2] Ni_2-Ni_3, which comprise six Ni–Ni edges of one of the tetrahedral holes either in the half of the trianion containing the interstitial hydrogen (w/H) or in the other unoccupied half (wo/H).

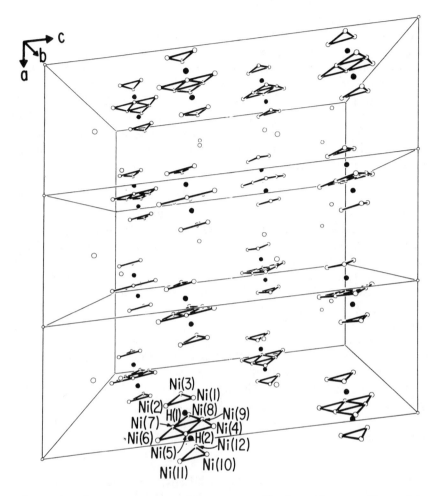

*Figure 8.   Perspective view of three unit cells along the crystallographic
a axis depicting the stacking of the dihydrido dianions in [Ph₄P]₂⁺-
[Ni₁₂(CO)₂₁H₂]²⁻ (3).   For clarity, the positions of the tetraphenylphosphonium cations are indicated by the positions of the phosphorus atoms
which are represented as isolated open circles.*

rections but not in the third direction, such that the anions stack roughly above
one another in a skewed fashion in this third direction.   This stacking, as shown
in Figure 8 for the dianions of $[Ph_4P]_2^+[Ni_{12}(CO)_{21}H_2]^{2-}$ (3), gives rise to a relatively constant closest nonbonding separation of 3.07–3.16 Å between the dianions in Compounds 1, 2, and 3 and a larger separation of 3.45 Å between the
trianions in Compound 4, consistent with a larger electrostatic repulsion for the
more highly charged species.   A different choice of cations or possibly a partial
oxidation of the anions could decrease these anion–anion separations to produce
a one-dimensional system.   Further work is continuing in this direction.

(f) HYDROGEN MIGRATION. It is worthwhile from a structural viewpoint to consider the possible migrational behavior of the interstitial hydrogens in the $[Ni_{12}(CO)_{21}H_{4-n}]^{n-}$ anions ($n = 2, 3$), especially with respect to protonation and deprotonation reactions in solution. A recently reported ^1H, ^1H-$\{^{103}Rh\}$ INDOR, and ^{13}C NMR spectroscopic study (17) of the $[Rh_{13}(CO)_{24}H_{5-n}]^{n-}$ anions ($n = 2, 3$) in solution was consistent with a rapid migration of the interstitial hydrogens as protons inside the hcp rhodium cluster. The stereochemical implications presented below also are completely consistent with proton migration inside as well as outside of this different kind of hcp metal fragment, with concomitant changes in negative charge distribution upon proton migration primarily occurring via the metal-cluster orbitals.

Table IV, which gives the calculated nickel-to-centroid distances for the six kinds of nickel triangles within the anions, provides (under an assumed rigid dodecanickel framework) an indication of the closest Ni–H contacts encountered upon a possible migration of the hydrogen through one of the triangular nickel faces. Based upon these values being used as estimates of the potential barriers to migration in these anions, it appears that migration between the two octahedral sites should be relatively easier than migration out of the octahedral holes.

Migration of a hydrogen as a proton out of the cluster can occur either through a top (1,1,1) nickel triangle (i.e., comprised of the three Ni_1 atoms) or through one of the side (1,1,2) nickel triangles (i.e., comprised of two Ni_1 and one Ni_2). Although Table IV indicates that the holes in these two different types of triangular nickel faces are of similar size, the fact that the least steric hindrance of a Lewis base with the carbonyl ligands will occur with its approach along the anion's threefold axis toward a top (1,1,1) nickel triangle makes it likely that preferential abstraction of the proton by Lewis bases such as the *tert*-butoxide anion will occur through a (1,1,1) triangular nickel face. Figure 9 is a qualitative

Table IV. Calculated Ni-to-Centroid Distances[a] for Triangular Nickel Faces in $[Ni_{12}(CO)_{21}H]^{3-}$ Trianion

Nickel Triangles[b]	Distance (Å)	H Migration
1,1,1 (top)[c]	1.40	out of oct. hole
1,1,2 (side)[d]	1.45	out of oct. hole
2,2,2	1.55	oct. to oct. hole
1,2,2	1.57	oct. to teh. hole
2,2,3	1.29	teh. to teh. hole
1,2,3	1.43	out of teh. hole

[a]The distances are requisite minimum Ni–H contacts encountered during possible migrations of the hydrogen from its octahedral interstice (i.e., based upon the assumption that the nickel atoms remain fixed while the hydrogen migrates through a given nickel face).

[b]The numbers refer to the equivalent nickel atoms as labeled in Figure 4.

[c]1,1,1 (top) denotes the top nickel triangle comprised of the three Ni_1 atoms (Figure 4) in the upper half of the trianion containing the interstitial hydrogen.

[d]1,1,2 (side) denotes any of the three equivalent side nickel triangles comprised of two Ni_1 and one Ni_2 atoms (Figure 4) in the upper half of the trianion containing the interstitial hydrogen.

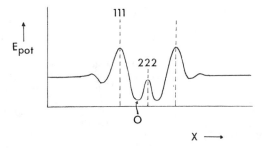

Figure 9. Qualitative potential energy diagram for the diffusion of a hydrogen along the principal C_3 axis of the $[Ni_{12}(CO)_{21}H]^{3-}$ trianion, as estimated from closest Ni–H contacts

potential energy diagram which incorporates these barrier considerations. It shows that, in visualizing the protonation of the trianion to give the dianion (or analogously of the tetra-anion to give the trianion), the proton, on approaching the (1,1,1) nickel face of the trianion along the principal threefold axis, first encounters a small potential barrier to absorption near the surface of the anion and then a larger potential barrier as it passes through the (1,1,1) triangular nickel face into the empty octahedral hole. The relatively smaller potential barrier at the central (2,2,2) nickel triangle points to the expected greater ease of the proton in the trianion to migrate between the two octahedral sites rather than outside through a (1,1,1) nickel triangle.

Acknowledgment

The synthesis and characterization of these compounds by G.L. and P.C. and their x-ray diffraction analyses by R.W.B. and L.F.D. were supported partially by a joint NATO grant to P.C. and to L.F.D. and by a National Science Foundation grant (No. GP-19175X) to L.F.D. The neutron diffraction measurements carried out at Argonne National Laboratory by R.W.B., A.J.S., and J.M.W. and the subsequent structural refinements by R.W.B. at the University of Wisconsin—Madison were performed under the auspices of the Division of Basic Energy Sciences of the U.S. Department of Energy. The support of the National Science Foundation under Grant NSF CHE 76-07409 to J.M.W. for this part of the research is gratefully acknowledged.

Literature Cited

1. Longoni, G., Ph.D. thesis, Instituto di Chimica, Generale ed Inorganica, Universita degli Studi, Milan, Italy (1971).
2. Calabrese, J. C., Dahl, L. F., Chini, P., Longoni, G., Martinengo, S., *J. Am. Chem. Soc.* (1974) **96**, 2614.

3. Calabrese, J. C., Dahl, L. F., Cavalieri, A., Chini, P., Longoni, G., Martinengo, S., *J. Am. Chem. Soc.* (1974) **96**, 2616.
4. Calabrese, J. C., Lower, L. D., Longoni, G., Chini, P., Dahl, L. F., unpublished data.
5. Lower, L. D., Ph.D. thesis, University of Wisconsin–Madison (1976).
6. Chini, P., Longoni, G., Albano, V. G., *Adv. Organomet. Chem.* (1976) **14**, 285.
7. Lower, L. D., Longoni, G., Chini, P., Dahl, L. F., unpublished data.
8. Longoni, G., Chini, P., *J. Am. Chem. Soc.* (1976) **98**, 7225.
9. Longoni, G., Chini, P., Cavalieri, A., *Inorg. Chem.* (1976) **15**, 3025.
10. Longoni, G., Chini, P., *Inorg. Chem.* (1976) **15**, 3029.
11. Chini, P., ADV. CHEM. SER. (1978) **167**, 1.
12. Bateman, L. R., Blount, J. F., Dahl, L. F., *J. Am. Chem. Soc.* (1966) **88**, 1082.
13. Simon, A., v. Schnering, H.-G., Schäfer, H., *Z. Anorg. Allg. Chem.* (1967) **355**, 295.
14. Schäfer, H., v. Schnering, H.-G., Tillack, J., Kuhnen, F., Wöhrle, W., Baumann, H., *Z. Anorg. Allg. Chem.* (1967) **353**, 281.
15. Simon, A., *Z. Anorg. Allg. Chem.* (1967) **355**, 311.
16. Albano, V. G., Ceriotti, A., Chini, P., Ciani, G., Martinengo, S., Anker, W. M., *J. Chem. Soc., Chem. Commun.* (1975) 859.
17. Martinengo, S., Heaton, B. T., Goodfellow, R. J., Chini, P., *J. Chem. Soc., Chem. Commun.* (1977) 39.
18. "Hydrogen in Metals" in *J. Less-Common Met.* (1976) **49**, 1–508.
19. Bennett, L. H., McAlister, A. J., Watson, R. E., *Physics Today* (1977) **30**, 34.
20. Wollan, E. O., Cable, J. W., Koehler, W. C., *J. Phys. Chem. Solids* (1963) **24**, 1141.
21. Donahue, J., "The Structures of the Elements," p. 210 and references cited therein, J. Wiley & Sons, New York, 1974.
22. Broach, R. W., Ph.D. thesis, University of Wisconsin–Madison (1977).
23. Germain, G., Main, P., Woolfson, M. M., *Acta Crystallogr.* (1971) **A27**, 368.
24. Longoni, G., Chini, P., Lower, L. D., Dahl, L. F., *J. Am. Chem. Soc.* (1975) **97**, 5034.
25. Ruff, J. K., White, R. P., Jr., Dahl, L. F., *J. Am. Chem. Soc.* (1971) **93**, 2159.
26. Lower, L. D., Dahl, L. F., *J. Am. Chem. Soc.* (1976) **98**, 5046.
27. Müller, J., Dorner, H., Huttner, G., Lorenz, H., *Angew. Chem. Int. Ed. Eng.* (1973) **12**, 1005.
28. Huttner, G., Lorenz, H., *Chem. Ber.* (1974) **107**, 996.
29. Koetzle, T. F., McMullan, R. K., Bau, R., Teller, R. G., Tipton, D. L., Wilson, R. D., "Abstracts of Papers," 2nd CIC-ACS Joint Conference, May 29–June 2, 1977.

RECEIVED January 30, 1978

A New Class of Hydrido-Bridged Platinum Complex with Application in Catalysis

MIGUEL CIRIANO, MICHAEL GREEN, JUDITH A. K. HOWARD, MARTIN MURRAY, JOHN L. SPENCER, F. GORDON A. STONE, and CONSTANTINOS A. TSIPIS

Department of Inorganic Chemistry, The University, Bristol BS8 1TS, England

The $Pt(O)$ monophosphine complex $[Pt(C_2H_4)_2\{(C_6H_{11})_3P\}]$ reacts with tri-organosilanes to give diplatinum compounds $[PtSiR_3'(\mu\text{-}H)\{(C_6H_{11})_3P\}]_2$ $[R_3'Si = EtMe_2Si, Me_2PhSi, Et_3Si, Me_2PhCH_2Si, ClMe_2Si, Cl_3Si, and (EtO)_3Si]$. An x-ray diffraction study of $[PtSiEt_3(\mu\text{-}H)\{(C_6H_{11})_3P\}]_2$ revealed a dihedral angle of $21°$ between the planes defined by the two SiPtP units, but failed, expectedly, to locate the bridging hydride ligands. An unusual feature of the diplatinum complexes is the absence in their ^1H NMR spectra of a high-field resonance above 12 τ. Detailed NMR studies on $[PtSi(OEt)_3(\mu\text{-}H)(Bu_2{}^t\text{-}MeP)]_2$ showed that these compounds are dynamic, and the signal corresponding to the hydrido bridge occurs at about 7 τ. Evidence is presented for asymmetric $Pt(\mu\text{-}H)_2Pt$ bridges leading to concomitant, multicenter $Pt\text{-}H\cdots Si$ bonding. The complexes are efficient hydrosilylation catalysts. $[PtSiMe_2Ph(\mu\text{-}H)\{(C_6H_{11})_3P\}]_2$ decomposes thermally to give, in low yield, a bis(dimethylsilyl)diplatinum compound, $[PtH(SiMe_2)\{(C_6H_{11})_3P\}]_2$.

The discovery (1) of pure-olefin complexes of Pt(O) is prompting rapid expansion of organo platinum chemistry. The olefin groups in compounds such as [Pt(cod)$_2$] (cod = cyclo-octa-1,5-diene) or [Pt(C$_2$H$_4$)$_3$] are displaced easily in a variety of reactions (2,3,4,5,6). Particularly important complexes that are derived from these olefin–platinum species are the bis(ethylene)tertiaryphosphineplatinum derivatives [Pt(C$_2$H$_4$)$_2$(R$_3$P)] [R$_3$P = Me$_3$P, Me$_2$PhP, MePh$_2$P, Ph$_3$P, and (cyclo-C$_6$H$_{11}$)$_3$P] (7), which effectively can be sources of the R$_3$PPt fragment in synthesis since they lose ethylene readily. In this chapter,

0-8412-0390-3/78/33-167-111/$05.00/0
© American Chemical Society

we report reactions with tri-organosilanes that give diplatinum complexes $[PtSiR_3'(\mu\text{-}H)(R_3P)]_2$ (8).

Most of the work so far has been carried out with the tricyclohexylphosphine complex $[Pt(C_2H_4)_2\{(C_6H_{11})_3P\}]$, which on treatment with the silanes in toluene or petroleum ether gives in high yield (60–90%) yellow, air-stable crystalline compounds $[PtSiR_3'(\mu\text{-}H)\{(C_6H_{11})_3P\}]_2$ [$R_3'Si$ = $EtMe_2Si$, Me_2PhSi, Et_3Si,

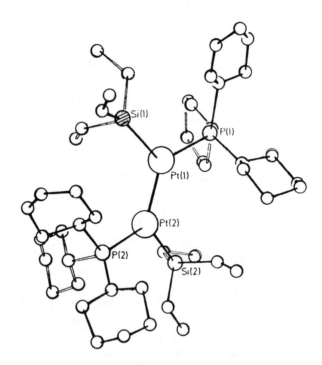

Journal of the Chemical Society, Chemical Communication

Figure 1. Molecular structure of $[PtSiEt_3(\mu\text{-}H)\text{-}$ $(C_6H_{11})_3P]_2$. $Pt(1)\text{-}P(1)$, $2.282(9)$; $Pt(2)\text{-}P(2)$, $2.271(9)$; $Pt(1)\text{-}Si(1)$, $2.33(1)$; $Pt(2)\text{-}Si(2)$, $2.34(1)$ Å; $< Si(1)\text{-}$ $Pt(1)\text{-}Pt(2)$, $124.9(3)$; $Si(2)\text{-}Pt(2)\text{-}Pt(1)$, $123.5(3)$; $P(1)\text{-}Pt(1)\text{-}Pt(2)$, $131.4(3)$; $P(2)\text{-}Pt(2)\text{-}Pt(1)$, 133.9- $(3)^\circ$

Me_2PhCH_2Si, $ClMe_2Si$, Cl_3Si, and $(EtO)_3Si$]. P-31 (^1H decoupled, chemical shift relative to H_3PO_4) NMR studies on these complexes gave singlet signals that had two sets of ^{195}Pt satellites arising from molecules with one and two active platinum nuclei, the latter being second order (AA'XX' system), e.g. for $[PtSi(OEt)_3(\mu\text{-}$ $H)\{(C_6H_{11})_3P\}]_2$, $\delta = -66.3$ ppm [^1J(PPt) 3127 Hz and ^2J(PPt) 82 Hz]. The dinuclear nature of the complexes was confirmed by a single crystal x-ray diffraction study on $[PtSiEt_3(\mu\text{-}H)\{(C_6H_{11})_3P\}]_2$ (Figure 1) (8). The six atoms, Si(1),

Pt(1), P(1); Si(2), Pt(2), P(2) comprising the main skeleton of the molecule, are significantly non-co-planar, with a dihedral angle of 21° between the Si(1)Pt(1)P(1) and Si(2)Pt(2)P(2) planes. The Pt–Pt distance (2.692 Å) is not unusual (2). No bridging hydrogen atoms were located.

Attempts to detect the hydrido ligands by ir and NMR spectroscopy were initially inconclusive. In the ir, broad bands were observed in the range 1695–1545 cm^{-1} which could be attributed to Pt(μ-H)$_2$Pt groups. The ^{1}H NMR spectra above 12 τ revealed no resonance attributable to a platinum hydride. Nevertheless, in spite of what was considered at first to be the absence of a NMR signal caused by a metal hydride, there was chemical evidence for Pt(μ-H)Pt bridges. Thus, treating [PtSiMe$_2$Ph(μ-H){(C$_6$H$_{11}$)$_3$P}]$_2$ with the ligands Bu^{t}NC, Me$_3$P, or Me$_2$PhAs gave the terminal hydrido complexes, *cis*-[PtH(SiMe$_2$Ph){(C$_6$H$_{11}$)$_3$P}L] [L = Bu^{t}NC, Me$_3$P, or Me$_2$PhAs], which were evidently formed by cleavage of the Pt(μ-H)Pt bridges. The spectroscopic properties of these products were as expected, and there was no difficulty in observing either the ir bands or the ^{1}H NMR resonances caused by PtH, e.g. for *cis*-[PtH(SiMe$_2$Ph){(C$_6$H$_{11}$)$_3$P}(Bu^{t}NC)], ν_{max}(PtH), 2100 cm^{-1}; 13.95 τ [*d*, *J*(PH) 20, *J*(PtH) 960 Hz. Moreover, the dinuclear complexes reacted with ethylene to give [Pt(C$_2$H$_4$)$_2${(C$_6$H$_{11}$)$_3$P}], with triphenylphosphite to give [Pt{(PhO)$_3$P}$_4$], and with carbon monoxide to give [Pt(μ-CO){(C$_6$H$_{11}$)$_3$P}]$_3$; the silane Me$_2$PhSiH also was formed in the reaction with (PhO)$_3$P.

Deuteration studies also provided evidence for hydrido bridges in the di-platinum compounds. It was expected that reaction of [Pt(C$_2$H$_4$)$_2${(C$_6$H$_{11}$)$_3$P}] with a deuteride R$_3'$SiD would give complexes [PtSiR$_3'$(μ-D){(C$_6$H$_{11}$)$_3$P}]$_2$. In practice, syntheses of the latter compounds were not clean; [Pt(μ-H)Pt]-containing species were produced simultaneously because of D–H exchange. This exchange could occur by reversing some of the steps in the hydrosilylation mechanism, discussed later.

An initial experiment involved treating [Pt(C$_2$H$_4$)$_2${(C$_6$H$_{11}$)$_3$P}] with Et$_3$SiD. The ^{1}H-decoupled ^{31}P NMR spectrum showed a resonance at −63.2 ppm [^{1}J(PPt), 3229; ^{2}J(PPt), 60 Hz, as expected for a [P·Pt(μ-X)Pt·P] (X = H or D) system, but the signal was broad because of ^{31}P–^{2}D coupling.

In other experiments, [Pt(C$_2$H$_4$)$_2${(C$_6$H$_{11}$)$_3$P}] reacted with Me$_2$PhSiD. The ir spectrum of the product showed a band at 1163 cm^{-1} with a much weaker band at 1610 cm^{-1}. Because of the relation between these bands, it is reasonable to assign the former to [Pt(μ-D)Pt] and the latter to [Pt(μ-H)Pt]. No band at 1163 cm^{-1} is observed in the spectrum of [PtSiMe$_2$Ph(μ-H){(C$_6$H$_{11}$)$_3$P}]$_2$. Reaction of the partially deuterated product with trimethylphosphine gave a mixture of the mononuclear deuterated and nondeuterated hydrides *cis*-[PtD(SiMe$_2$Ph){(C$_6$H$_{11}$)$_3$P}(Me$_3$P)] [ν_{max}(PtD), 1470 cm^{-1}; ^{2}D NMR: τ, 12.9 *d* of *d*, *J*(PD) 23.8 and 3.2, *J*(PtD) 122 Hz] and *cis*-[PtH(SiMe$_2$Ph){(C$_6$H$_{11}$)$_3$P}(Me$_3$P)] [ν_{max}(PtH), 2054 cm^{-1}; ^{1}H NMR τ, 12.85 *d* of *d*, *J*(PH) 162 and 22 Hz]. The relative intensities of the ir bands and NMR signals showed that the deuterated species predominated in this mixture of mononuclear hydrides.

From these results, it appears that the complexes $[PtSiR_3'(\mu\text{-}H)(R_3P)]_2$ undergo reactions in which they give either the fragment $PtH(SiR_3')(R_3P)$, subsequently captured by a σ-donor molecule L, or they reform Pt(O) complexes, e.g. $[Pt\{(PhO)_3P\}_4]$ and R_3' SiH. Moreover, in spite of the absence of a high-field ^1H NMR signal, the ir spectra provide good evidence for the $[Pt(\mu\text{-}H)Pt]$ group.

The enigma of the apparently absent PtH NMR signal in the ^1H NMR spectra of the diplatinum compounds at first appeared to be compounded by the results of Puddephatt et al. (9). These workers recently have characterized dinuclear bis(diphenylphosphino)methane cationic complexes of platinum containing both bridging and terminal hydrido ligands. In the compound $[Pt_2H_2(\mu\text{-}H)(dppm)_2]^+Cl^-$ (dppm = $Ph_2PCH_2PPh_2$), for example, the NMR resonances for terminal PtH and bridging $Pt(\mu\text{-}H)Pt$ appear at 16.8 τ [^1J(PtH) 1162, ^2J(PtH) 116 Hz] and 15.8 τ [^1J(PtH) 540 Hz], respectively. Thus, it appears that in a compound containing a symmetrical $[Pt(\mu\text{-}H)Pt]$ group, a high field NMR signal near 16 τ is expected. The absence of a signal in this region in complexes $[PtSiR_3'(\mu\text{-}H)(R_3P)]_2$ indicates that either the resonance is not seen because the molecules are undergoing dynamic behavior in solution and the measurement (at $+30°C$) is being made near the coalescence temperature, or the $Pt(\mu\text{-}H)Pt$ bridge bonding is of a nonclassical kind, giving rise to a hydride resonance with an unusual chemical shift. In view of these considerations detailed NMR experiments were carried out on the compound $[PtSi(OEt)_3(\mu\text{-}H)\text{-}(Bu_2{}^tMeP)]_2$.

This complex was prepared by reaction of $(EtO)_3SiH$ with $[Pt(C_2H_4)_2(Bu_2{}^t\text{-}MeP)]$, and was studied to simplify the 5–10 τ region of the NMR spectrum. Complexes $[PtSiR_3'(\mu\text{-}H)\{(C_6H_{11})_3P\}]_2$, because of the cyclohexyl groups and alkyl substituents on silicon, show NMR bands that tend to obscure signals below 10 τ. Moreover, the chosen complex is considerably more soluble than the cyclohexylphosphine derivatives. The resonances observed in the ^1H spectrum of $[PtSi(OEt)_3(\mu\text{-}H)(Bu_2{}^tMeP)]_2$, recorded at $+75°$, $+25°$, and $-60°C$, are listed in Table I. Resonances caused by the protons in the $Bu_2{}^tMeP$ and $Si(OEt)_3$ groups were assigned readily. Of considerable significance was the observation of a signal in the region $6.6–7\tau$ that had a chemical shift which varied with temperature. The nature of the platinum satellites of this resonance clearly showed that it was caused by a proton bonded to platinum which, in the limiting low temperature spectrum, was associated more closely with one metal atom than with the other. It was undergoing dynamic behavior at elevated temperatures so that there was a time-averaged J(PtH) value of 681 Hz at 75°C. Moreover, $^1H\{^{195}Pt\}$ INDOR measurements at 30°C gave a platinum resonance at -44 ppm (measured relative to 21.4 MHz $(Me_4Si = 100$ MHz)) with J(PtP) = 3122 Hz and average J(PtH) = 678 Hz. This $^{195}Pt–^1H$ coupling value agrees well with that measured from the ^1H spectrum at 75°C. The coupling constants ^1J(PtH) 894 Hz and ^2J(PtH) 456 Hz obtained from the spectrum of $[PtSi(OEt)_3(\mu\text{-}H)(Bu_2{}^t\text{-}MeP)]_2$ at $-60°C$ bracket the value observed [J(PtH, 540 Hz) (9)] for the group $[Pt(\mu\text{-}H)Pt]$ in $[Pt_2H_3(dppm)_2]^+Cl^-$.

Table I. ^{1}H NMR Data for the Diplatinum Complex [PtSi(OEt)$_3$ (μ-H)(Bu$_2^t$MeP)]$_2$ [a]

Resonance	+75°C	+25°C	−60°C	
Bu^{t}P	8.66	8.70	8.69	τ
	13	13	14	J(PH)
MeP	8.19	8.21	8.23	τ
	9	9	9	J(PH)
	45	45	47	J(PtH)
*Me*CH$_2$O	—[b]	8.69	—[b]	τ
CH$_2$O	6.0	5.97	5.94	τ
	7	7	7	J(HH)
PtH	7.05	6.87	6.65	τ
	681	—[c]	456,894	J(PtH)
	52	51	104, 0	J(HP)

[a] Measured in C$_6$D$_5$CD$_3$ on JEOL-PFT 100, operating in the Fourier Transform mode.
[b] Signal obscured by Bu^t resonance.
[c] Not observed because of fluxional behavior.

The discovery that the ^{1}H signal for the hydrido ligands in [PtSi(OEt)$_3$(μ-H)(Bu$_2^t$MeP)]$_2$ occurs at <10 τ, with two distinct J(PtH) values, leads us to propose that the dinuclear complexes [PtSiR$_3'$(μ-H)(R$_3$P)]$_2$ have unsymmetrical [Pt(μ-H)Pt] bridges, and, moreover, the low-field chemical shift indicates a multicenter interaction of the hydrogens, not only with two platinum atoms but also with the silicon atoms.   That is, the molecular structures are represented best by Structure 1 rather than by Structure 2.   The observed chemical shift 6.65 τ (see Table I) is similar but somewhat up field from that in the parent silane (EtO)$_3$SiH (5.8 τ).   As mentioned above, a chemical shift of 15.8 τ was observed (9) for the bridged hydride in [Pt$_2$H$_3$(dppm)$_2$]$^+$.   This is 9.2 τ higher than that for the hydrido-bridge ligand in [PtSi(OEt)$_3$(μ-H)(Bu$_2^t$MeP)]$_2$.   It is interesting to compare this 9.2 τ shift to low field with a change in relative chemical shift deduced from other work.   The signal for the protons in the tri-rhenium hydride [Re(μ-H)(CO)$_4$]$_3$ occurs at 27.1 τ (10).   Graham et al. (11) have studied extensively reactions of organosilanes with the carbonyls of manganese, rhenium, and tungsten, and one type of compound isolated was [Re$_2$H$_2$(μ-SiR$_2$)(CO)$_8$] (R = Me or Ph) (12).   NMR and crystallographic results (12,13) suggest the existence of Re–H–Si interactions in these compounds, that are formally derivatives of [Re(μ-H)(CO)$_4$]$_3$, in which a Re(CO)$_4$ group has been replaced by R$_2$Si.   The resonance of the hydride ligands in [Re$_2$H$_2$(μ-SiMe$_2$)(CO)$_8$] occurs at 20.6 τ, a

1   2

down-field shift of 6.5 τ from that found in [Re(μ-H)(CO)$_4$]$_3$. This compares with the 9.2 τ down-field shift between the resonances for the bridged hydrides [Pt$_2$H$_3$(dppm)$_2$]$^+$ and [PtSi(OEt)$_3$(μ-H)(Bu$_2^t$MeP)]$_2$.

An unsymmetrical, bridged hydrido structure (see Structure 1), where an interaction exists between the silicon atoms and the bridge system, correlates well with the chemical reactivity of these compounds (referred to above), where they either revert to [R$_3$PPt] and R$_3'$SiH or form mononuclear Pt(II) hydrido complexes [PtH(SiR$_3'$)(R$_3$P)L]. In Structure 1, the platinum atoms could be interacting with bonding pairs of electrons in H:SiR$_3'$, as proposed by Graham et al. (11) for the interaction of the manganese group [Mn(CO)$_2$(η^5-C$_5$H$_5$)] with Ph$_3$SiH in the complex [MnH(SiPh$_3$)(CO)$_2$(η^5-C$_5$H$_5$)]. In the case of Structure 1, the second platinum atom is assisting in rupturing the Si–H bond. Moreover, the interactions implied by Structure 1 could well be related to the mode of activation (14) by platinum of alkanes so that H–D exchange occurs.

The complexes [PtSiR$_3'$(μ-H)(R$_3$P)]$_2$ are remarkably efficient hydrosilylation catalysts for alkenes and alkynes (15). Adding silanes to double or triple carbon–carbon bonds usually proceeds exothermally immediately on mixing the reactants at room temperature. The catalyst:substrate ratio used is in the range 10^{-4} to 10^{-6}:1, and some representative results are summarized in Table II. The

Table II.   Hydrosilylation of Alkenes and Alkynes with [PtSiR$_3'$-
(μ-H){(C$_6$H$_{11}$)$_3$P}]$_2$ [a]

	Silane	Product	Yield (%)
Alkenes			
Pent-1-ene	Cl$_3$SiH	C$_5$H$_{11}$SiCl$_3$	70
	Me$_2$PhSiH	C$_5$H$_{11}$SiPhMe$_2$	91
Hex-1-ene	ClMe$_2$SiH	C$_6$H$_{13}$SiMe$_2$Cl	85
	Me$_3$SiH	C$_6$H$_{13}$SiMe$_3$	90
	(EtO)$_3$SiH	C$_6$H$_{13}$Si(OEt)$_3$	60
Styrene	Cl$_3$SiH	PhCH$_2$CH$_2$SiCl$_3$	94
	Me$_2$EtSiH	PhCH$_2$CH$_2$SiEtMe$_2$	92
Allyl chloride	HSiCl$_3$	ClCH$_2$CH$_2$CH$_2$SiCl$_3$	70
Dienes			
Hexa-1,5-diene	Me$_2$PhSiH	Me$_2$PhSi(CH$_2$)$_6$SiPhMe$_2$	100
Octa-1,7-diene	ClMe$_2$SiH	ClMe$_2$Si(CH$_2$)$_8$SiMe$_2$Cl	80
4-Vinylcyclohexene	Cl$_2$MeSiH	cyclo-C$_6$H$_9$-CH$_2$CH$_2$SiMeCl$_2$	82
Alkynes			
But-1-yne	Et$_3$SiH	*trans*-EtCH:CHSiEt$_3$	88
	Cl$_3$SiH	*trans*-EtCH:CHSiCl$_3$	88
Phenylacetylene	ClMe$_2$SiH	*trans*-PhCH:CHSiClMe$_2$	92
But-2-yne	Me$_2$PhSiH	*cis*-MeCH:C(Me)SiMe$_2$Ph	97
	Ph$_2$SiH$_2$	*cis*-MeCH:C(Me)SiPh$_2$H	92
Diphenylacetylene	Cl$_3$SiH	*cis*-PhCH:C(Ph)SiCl$_3$	76

[a] These and related results are discussed in detail in Ref. 15. Milligram quantities of catalysts are used.

Scheme 1.   (*i*) R$_3'$SiH; (*ii*) –2R CH:CH$_2$; (*iii*) RCH:CH$_2$; (*iv*) –RCH:CH$_2$; (*v*) –R$_3'$SiCH$_2$CH$_2$R

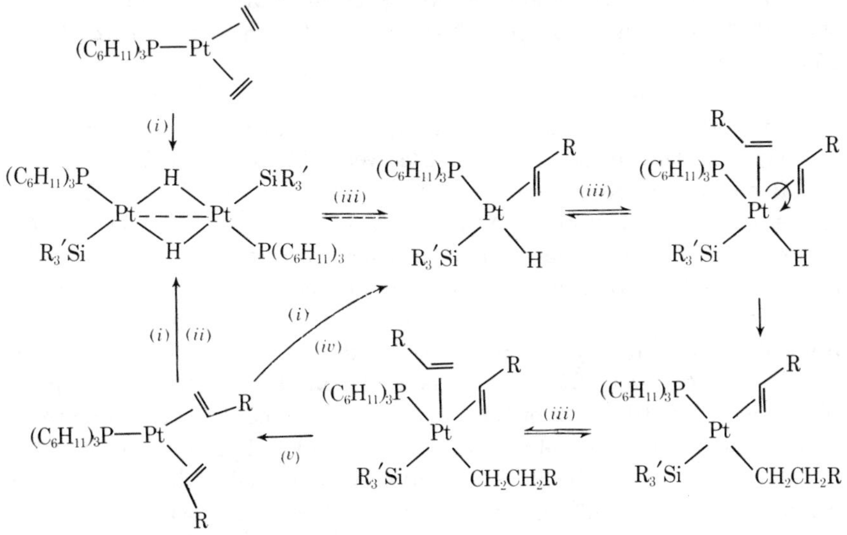

most commonly used catalyst in this field is H$_2$PtCl$_6$·6H$_2$O or Speier's catalyst (*16*) which only needs small amounts (circa 10^{-5} mol per mol silane) of the platinum compound.   However, when H$_2$PtCl$_6$·6H$_2$O is used, it is frequently necessary to heat the reactants, and there is often an induction period before hydrosilylation commences.   The latter effect undoubtedly is related to the fact that hexachloroplatinic acid is not the true catalyst.   An initial reduction of the Pt(IV) complex by the silane is probably necessary, giving a low-valent complex that can undergo a series of oxidative–addition and reductive–elimination steps with the reactants.   Diplatinum complexes [PtSiR$_3'$(μ-H){(C$_6$H$_{11}$)$_3$P}]$_2$ may function as catalysts as indicated in Scheme 1.   Bridge cleavage of the diplatinum complexes by olefin would give a square-planar Pt(II) species in which it is reasonable to assume that the olefin molecule would be trans to the R$_3'$ Si substituent because of the high trans directing effect of this group (*17*).   Further steps involve migration of the hydrogen onto the coordinated olefin, reductive elimination of an organosilane, and regeneration of the monophosphine-substituted Pt(0) complex.   A similar mechanism has been invoked for the hydrosilylation of alkynes (*15*).

During the research described in this chapter, another example of a molecular structure was discovered, that might involve incipient [Pt(μ-H)Si] bonding. When the diplatinum complex [PtSiMe$_2$Ph(μ-H){(C$_6$H$_{11}$)$_3$P}]$_2$ is heated at 60°–100°C in an inert organic solvent, a pale yellow, crystalline complex is formed in low yield.   The ir spectrum of the crystals showed a strong, broad band at 1650 cm^{-1}, but no bands caused by phenyl groups or by a terminal PtH stretch were detectable.   The ^{1}H NMR above 10 τ showed no resonance, but overlapping signals attributable to methyl and cyclohexyl groups were observed in the region 7.5–9.0 τ.   The ^{31}P (^{1}H decoupled) NMR spectrum clearly showed the presence of a P·Pt·Pt·P group with a singlet resonance at −52.6 ppm [^1J(PPt), 3982; ^2J(PPt), 253 and J(PtPt), 2832 Hz].

To establish the nature of this pyrolysis product, a single crystal x-ray diffraction study (18) was undertaken at 200°K on a Syntex P2$_1$ four-cycle diffractometer.   The following crystal data were obtained: $C_{40}H_{80}P_2Pt_2Si_2$, M 1071.5; monoclinic, space group P2$_1$/c, a = 9.158(2), b = 11.725(3), c = 20.252(9) Å, β = 92.370(31)°, U = 2172.7 Å^3, Z = 2, D$_m$ = 1.55, D$_c$ = 1.64, F(000) = 1072, μ(Mo–K$_\alpha$) = 69.2 cm^{-1}.   Out of 7123 reflections, 6024 measured to 2θ = 65° were retained {I $\geqslant$ 2.0 σ (I)}, and used to solve and refine the structure (R = 0.063).

The molecular structure is shown in Figure 2.   Two platinum atoms are bridged by two dimethylsilyl groups so that there are two distinct Pt–Si bond lengths [2.424(2) and 2.327(2) Å].   The Pt, Si, and P atoms are essentially coplanar, with a P·Pt·Pt angle of 160°.   the Pt–Pt distance of 2.708(1) Å is greater than the sum of the covalent radii (2.62 Å) but could reflect a degree of metal–metal bonding.   Thus, the Pt–Pt separations with (μ-CO) bridges in the cluster complex [Pt$_4$(CO)$_5$(Me$_2$PhP)$_4$] are 2.752–2.790 Å (19).   The environment of each platinum is essentially square planar with one phosphorus and two silicon atoms

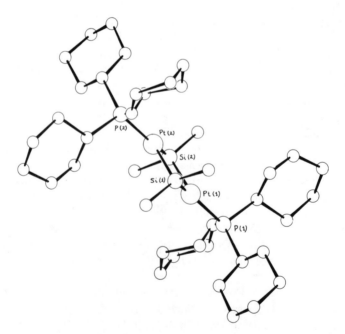

Figure 2.   Molecular structure of the product from the thermal decomposition of [Pt(SiMe$_2$Ph)(μ-H)- (C$_6$H$_{11}$)$_3$P]$_2$

occupying three sites and with a vacant site trans to the longer Pt(1)-Si(1) bond. This site is large enough [the P(1)Pt(1)Si(2) angle is 143.8°] to accommodate a terminal hydride.   Indeed, if there is no atom in this neighborhood, the coordination around platinum is highly unusual.   Accordingly, we searched for a terminal hydride since all the other hydrogen atoms in the molecule had been located.   The largest residual feature in a difference-density synthesis was found, in fact, in this vacant site 1.61(8) Å from platinum and 1.73(8) Å from silicon.   It is interesting to compare these parameters with the PtH distance (1.66 Å) in $[PtH(SB_9H_{10})(Et_3P)_2]$ (*20*) and the SiH distance (1.8 Å) in $[MnH(SiPh_3)$-$(CO)_2(\eta^5\text{-}C_5H_5)]$ (*21*).

Using the above results, we propose that Structure 3 represents the complex formed in the thermal decomposition of $[PtSiMe_2Ph(\mu\text{-}H)\{(C_6H_{11})_3P\}]_2$.   An interesting feature of 3 is that the $[Pt(\mu\text{-}H)Si]$ interactions proposed occur on the edges of the shorter of the Pt–Si bonds of the parallelogram.   This contrasts with the molecular structure of $[W_2H_2(SiEt_2)_2(CO)_8]$, where it has been argued (*22*) that the $[W(\mu\text{-}H)Si]$ bridges occur along the two longer sides of the W_2Si_2 parallelogram.   Nevertheless, if hydrogen atoms were placed along the longer sides of the Pt_2Si_2 fragment in $[PtH(SiMe_2)\{(C_6H_{11})_3P\}]_2$, this would result in the non-occupancy of a substantial site on platinum trans to the longer Pt–Si bond.

In Graham's (*11,12*) compounds, transition metal–hydrogen–silicon bridge bonding interactions have been invoked, but recent x-ray crystallographic studies (*23,24*) provide contrary evidence.   In a series of dirhenium complexes containing Re·Si·Re·Si units, the Re–Si distances are the same, irrespective of whether or not hydride ligands are present.   Therefore, it was concluded that when hydride ligands are present, they are bound terminally to rhenium with no attractive interaction with the silicon atoms.   If this conclusion is correct, it still leaves unanswered the problem of the absence in the ir spectra of some of these hydrido complexes of bands corresponding to terminal metal–hydrogen stretching frequencies.   Normally such absorptions are seen easily.

With platinum–silicon compounds described in this chapter, evidence supporting two-electron, three-center Pt–H–Si bonding seems to be strong, and the phenomenon is perhaps expected in view of the existence of numerous transition metal (M) compounds containing $M(\mu\text{-H})B$ linkages (25) and some evidence for incipient $M\cdots H\text{-}C$ interactions in palladium, molybdenum, and iridium complexes (26,27,28,29).

Acknowledgment

We thank the Science Research Council for support, R. J. Goodfellow and B. F. Taylor for INDOR and ^2D NMR measurements, and J. Proud for preliminary experiments. We also thank Martin Cowie for communicating to us the results given in Ref. 24 before their publication and for helpful discussion.

Literature Cited

1. Green, M., Howard, J. A. K., Spencer, J. L., Stone, F. G. A., *J. Chem. Soc., Dalton Trans.* (1977) 271.
2. Green, M., Howard, J. A. K., Laguna, A., Smart, L. E., Spencer, J. L., Stone, F. G. A., *J. Chem. Soc., Dalton Trans.* (1977) 278.
3. Forniés, J., Green, M., Spencer, J. L., Stone, F. G. A., *J. Chem. Soc., Dalton Trans.* (1977) 1006.
4. Green, M., Laguna, A., Spencer, J. L., Stone, F. G. A., *J. Chem. Soc., Dalton Trans.* (1977) 1010.
5. Green, M., Grove, D. M., Howard, J. A. K., Spencer, J. L., Stone, F. G. A., *J. Chem. Soc., Chem. Commun.* (1976) 759.
6. Barker, G. K., Green, M., Howard, J. A. K., Spencer, J. L., Stone, F. G. A., *J. Am. Chem. Soc.* (1976) **98**, 3373.
7. Green, M., Howard, J. A. K., Spencer, J. L., Stone, F. G. A., *J. Chem. Soc., Chem. Commun.* (1977) 449.
8. Green, M., Howard, J. A. K., Proud, J., Spencer, J. L., Stone, F. G. A., Tsipis, C. A., *J. Chem. Soc., Chem. Commun.* (1976) 671.
9. Brown, M. P., Puddephatt, R. J., Rashidi, M., Seddon, K. R., *Inorg. Chim. Acta* (1977) **23**, L27.
10. Kaesz, H. D., Saillant, R. B., *Chem. Rev.* (1972) **72**, 231.
11. Hart–Davis, A. J., Graham, W. A. G., *J. Am. Chem. Soc.* (1971) **94**, 4388 and references cited therein.
12. Hoyano, J. K., Elder, M., Graham, W. A. G., *J. Am. Chem. Soc.* (1969) **91**, 4568.
13. Elder, M., *Inorg. Chem.* (1970) **9**, 762.
14. Webster, D. E., *Adv. Organomet. Chem.* (1977) **15**, 147.
15. Green, M., Spencer, J. L., Stone, F. G. A., Tsipis, C. A., *J. Chem. Soc., Dalton Trans.* (1977) 1519, 1525.
16. Speier, J. L., Webster, J. A., Barnes, G. H., *J. Am. Chem. Soc.* (1957) **79**, 974.
17. Chatt, J., Eaborn, C., Ibekwe, S. D., Kapoor, P. N., *J. Chem. Soc. A* (1970) 1343.
18. Howard, J. A. K., Pugh, N., Woodward, P., unpublished data.
19. Vranka, R. G., Dahl, L. F., Chini, P. Chatt, J., *J. Am. Chem. Soc.* (1969) **91**, 1574.
20. Kane, A. R., Guggenberger, L. J., Muetterties, E. L., *J. Am. Chem. Soc.* (1970) **92**, 2571.
21. Hutcheon, W. L., Doctoral dissertation. University of Alberta, 1971. See also *Chem. Eng. News* (1970) **48**, 75.
22. Bennett, M. J., Simpson, K. A., *J. Am. Chem. Soc.* (1971) **93**, 7156.
23. Smith, R. A., Bennett, M. J., *Acta Crystallogr.* (1977) **B33**, 1113, 1118.
24. Cowie, M., Bennett, M. J:, *Inorg. Chem.* (1977) **16**, 2321, 2325.
25. Wegner, P. A., "Boron Hydride Chemistry," Chapter 12, E. L. Muetterties, Ed., Academic, 1975.
26. Roe, D. M., Bailey, P. M., Moseley, K., Maitlis, P. M., *J. Chem. Soc., Chem. Commun.* (1972) 1273.

27. Cotton, F. A., LaCour, T., Stanislowski, A. G., *J. Am. Chem. Soc.* (1974) **96**, 754.
28. Cotton, F. A., Stanislowski, A. G., *J. Am. Chem. Soc.* (1974) **96**, 5074.
29. Van Baar, J. F., Vrieze, K., Stufkens, D. J., *J. Organomet. Chem.* (1975) **97**, 461.

RECEIVED July 19, 1977

9

Ruthenium and Rhodium Hydrides Containing Chiral Phosphine or Chiral Sulfoxide Ligands, and Catalytic Asymmetric Hydrogenation

BRIAN R. JAMES, RODERICK S. MC MILLAN, ROBERT H. MORRIS, and DANIEL K. W. WANG

Department of Chemistry, University of British Columbia, British Columbia, Canada V6T 1W5

The Ru(II) complexes $Ru_2Cl_4(diop)_3$, $HRuCl(diop)_2$, and Ru-$Cl_2(diop)_2$, were synthesized and found to be effective in solution for catalytic asymmetric hydrogenation of some prochiral olefinic substrates. The systems, including some kinetic data, are compared with the related $HRh(diop)_2$-catalyzed systems; $(-)diop = 2R,3R$-$(-)$-2,3-0-isopropylidene-2,3-dihydroxy-1,4-bis(diphenylphosphino)butane. Two trimeric Ru(II) complexes, $[RuCl_2L_2]_3$ were made, where L is either the chiral sulfur-bonded R-$(+)$-methyl para-tolyl sulfoxide or (S,R;S,S)-$(+)$-2-methylbutyl methyl sulfoxide; the latter functions as an asymmetric hydrogenation catalyst, but optical yields of only 15% were attained. There is evidence for hydride intermediates. Ru(II) catalysts containing $(-)dios$, the 2R,3R-bis-(methyl sulfinyl)butane analog of $(-)diop$, give optical yields up to 25% with excess enantiomer opposite to that obtained for the corresponding $(-)diop$ system. The cationic rhodium complex $[(norbornadiene)Rh(PPh_3)(dios)]^+$ is a precursor to an active hydrogenation catalyst, but no asymmetry has been observed in products.

Asymmetric synthesis (1) has gained new momentum with the potential use of homogeneous catalysts. The use of a transition metal complex with chiral ligands to catalyze a synthesis asymmetrically from a prochiral substrate is beneficial in that resolution of a normally obtained racemate product may be avoided. In certain catalytic hydrogenations of olefinic bonds, optical purities approaching 100% have been attained (2, 3, 4, 5); hydrogenations of ketones (6,

0-8412-0390-3/78/33-167-122/$05.00/0

7) and imines (6) and hydroformylation experiments (8) have thus far achieved much lower enantioselectivities. Catalytic asymmetric syntheses using metal complexes have also been concerned with hydrosilylation (7, 9, 10), carbon–carbon bond formation as in oligomerization (11), and, more recently, oxidations involving peroxides (12, 13). The metal complexes have been used homogeneously in solution or on various supports (7, 14).

The use of chiral tertiary phosphine ligands has been studied most widely, but other chiral ligands such as carboxylic acids (15), imines (8, 16), amides (17), amines (18), alkoxides (19), and hydroxammates (13) have been investigated, and we reported recently on some sulfoxide systems (20, 21).

The detailed pathways of asymmetric induction during catalysis are not well understood even for the more widely studied hydrogenations (8, 22), and matching substrates with the most suitable transition-metal chiral catalyst remains very much an empirical art. We are not aware of kinetic studies except our own on catalyzed asymmetric hydrogenations, although they usually are assumed to follow well-studied nonchiral analogs; for example, $RhClP_3$* systems, where

R = PPh_2 gives 2R, 3R-(−)diop
R = $SOCH_3$ gives 2R, 3R-(−)dios

CH_3—CH_2—$\overset{*}{CH}$—CH_2—S

MBMSO

MPTSO

P* is an optically active tertiary phosphine, likely will resemble the $RhCl(PPh_3)_3$ system (23). However, even in this exhaustively studied system, both hydride and/or unsaturate routes are feasible (23, 24); by varying conditions, the choice of route could affect stereoselectivity. Most asymmetric hydrogenations have used prochiral olefinic acid substrates, and these systems have not been thoroughly studied even with nonchiral catalysts.

This chapter reports principally on studies with ruthenium chiral phosphine and chiral sulfoxide complexes and their use for catalytic hydrogenation. We have used the familiar diop ligand, [2R,3R-(−)-2,3-O-isopropylidene-2,3-dihydroxy-1,4-bis(diphenylphosphino) butane] (7); a related chiral chelating sulfoxide ligand dios, the bis(methyl sulfinyl)butane analog (21); (S,R;S,S)-(+)-2-methylbutyl methyl sulfoxide(MBMSO), chiral in the alkyl group; and R-(+)-methyl para-tolyl sulfoxide(MPTSO), chiral at sulfur. Preliminary data on some corresponding Rh(I) complexes are presented also.

We have aimed to characterize hydride species with the intent to study them as hydrogenation catalysts; mechanistic studies should aid eventually in understanding the enantioselective processes better. We have published brief reports (20, 25, 26) on some of the studies described in this chapter.

Results and Discussion

Complexes Containing Diop. A phosphine exchange reaction using $RuCl_2(PPh_3)_3$ and diop (~1:2) readily yields the neutral, green complex $Ru_2Cl_4(diop)_3$ (Complex 1) containing a bridging diop ligand. A single $\nu(Ru–Cl)$ at 310 cm^{-1} and the proton-decoupled ^{31}P NMR data (Figure 1) can be interpreted in terms of five-coordinate, square pyramidal geometry at each ruthenium. Such a structure generally gives an ABX pattern that is simplified if $J_{AX} = J_{BX}$, which seems reasonable here. The spectra are interpreted readily with $J_{AX} = J_{BX} = 30$ Hz and strong trans coupling, $J_{AB} = 310$ Hz. The electronic spectrum with maxima at 455 and 700 nm also resembles that of the five-coordinate $RuCl_2(PPh_3)_3$ with maxima at 480 and 750 nm (27, 28). Of interest, the expected octahedral $RuCl_2(diphosphine)_2$ complexes are obtained with $Ph_2P(CH_2)_nPPh_2$ ligands with $n = 1, 2, 3$ while the diphenylphosphinobutane(dpb) derivative with $n = 4$ (comparable to diop) again gives the bridged dimer product (25, 26, 29), presumably because of steric problems.

1

Consistent with its unsaturated character, Complex 1 in the solid state readily absorbs 2.0 mol CO at 1 atm to give $Ru_2Cl_4(diop)_3(CO)_2$, $\nu(CO)$ at 2010 cm^{-1}; the dpb analog behaves similarly (29). Interestingly, in toluene solution the $Ru_2Cl_4(diop)_3$ complex absorbs 3.0 mol CO and solution ir gives bands at 2100 and 1990 cm^{-1}. What must be a mixture of carbonyls has not been separated, but assuming that the $Ru–P_A$ bond is cleaved in solution, likely products are six-coordinate $RuCl_2(diop)(CO)_2$ and $RuCl_2(diop)(diop^*)(CO)$, where diop* refers to a monodentate, dangling diop.

Reactions of $Ru_2Cl_4(diop)_3$ with H_2 at ambient conditions in dimethylacetamide solution (dma) also indicated the existence of diop* complexes. Complex 1 absorbs 1.0 mol H_2 per dimer according to Reaction 1; no reaction occurs in toluene. The yellow hydride (Complex 2) was prepared independently by a phosphine exchange method from $HRuCl(PPh_3)_3$ and diop(1:2.5) and is characterized by an absorption maximum at 375 nm ($\epsilon = 2000$); subtracting this spectrum from that of the products of Reaction 1 leaves an absorption curve close to that measured for $RuCl_2(PPh_3)_2$, with a broad maximum around 500 nm (28, 30). This spectrum is attributed to $RuCl_2(diop)$. We have been unable to isolate this in a pure form, but it is almost certainly like $RuCl_2(PPh_3)_2$, a chloride-bridged

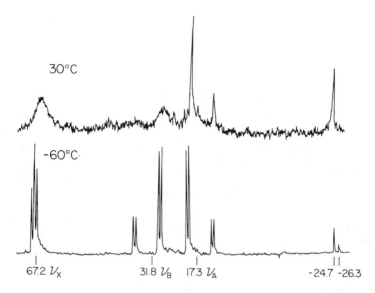

Figure 1.   Observed $^{31}P-\{H\}$ spectrum of $Ru_2Cl_4(diop)_3$ in $CDCl_3$ or toluene, relative to 85% H_3PO_4; downfield shifts are reported as positive

dimer (27). Solutions of $RuCl_2(PPh_3)_2$ also do not react with H_2 (30) but do react readily in the presence of 1 mol PPh_3 to yield $HRuCl(PPh_3)_3$. Similarly, in the presence of added diop, Complex 1 absorbs 2.0 mol H_2 according to Reaction 2, which gives a second method of synthesizing the hydride. Preliminary kinetic data on Reaction 1 showed a complex dependence on ruthenium, which was consistent with predissociation (such as Reaction 3) indicated by the CO uptake experiments. Such H_2 and CO gas-uptake studies encouraged us to attempt to synthesize the five-coordinate Complex 3, even though the six-coordinate $RuCl_2(diop)_2$ isomer did not exist. Complex 3 was prepared simply by treating Complex 1 with excess diop or by carrying out a $RuCl_2(PPh_3)_3$/diop (1:10) exchange reaction. The NMR spectrum of Complex 3 showed the same ABX pattern of Complex 1 plus a sharp line of correct intensity at a much higher field caused by the dangling phosphorus, close to that of uncoordinated diop. The extra lines at higher field in Figure 1 (-26.3 and -24.7 δ) are attributed to small amounts of Complex 3 and free diop, respectively. Complex 3 also reacts with H_2 (1:1) to give $HRuCl(diop)_2$ (cf. Reaction 2).

$$Ru_2Cl_4(diop)_3 + H_2 \rightarrow HRuCl(diop)_2 + RuCl_2(diop) + HCl \qquad (1)$$
$$\quad\;\; 1 \qquad\qquad\qquad\qquad 2$$

$$Ru_2Cl_4(diop)_3 + 2H_2 \xrightarrow{\text{diop}} 2HRuCl(diop)_2 + 2HCl \qquad (2)$$

$$Ru_2Cl_4(diop)_3 \rightleftharpoons RuCl_2(diop)(diop^*) + RuCl_2(diop) \qquad (3)$$
$$\mathbf{3}$$

The high field ^{1}H NMR of $HRuCl(diop)_2$ at 25°C in $CDCl_3$ or toluene-d_8 consists of seven equally spaced bands with intensities roughly in the ratios 1: 2:3:4:3:2:1 (Figure 2), and, based on conductivity measurements in nitromethane ($\Lambda = 19$ cm^2 ohm^{-1} M^{-1}; usually ~60 for a 1:1 electrolyte), this spectrum was interpreted as a trigonal-bipyramidal cation complex with an associated chloride anion (25, 26). The hydride, however, is nonconducting in dma ($\Lambda \sim 0.9$ cm^2 ohm^{-1} M^{-1}; we found a value of ~60 for a 1:1 electrolyte) and is almost certainly so in the NMR solvents used, chloroform and toluene. Thus the structure in these solvents and in the solid state is now considered to be the neutral six-coordinate cis Complex 2. The septet results if $J_{HP_a} = J_{HP_{a'}} = 14.2$ Hz and $J_{HP_e} = J_{HP_{e'}} = 28.8$ Hz. Such an assignment implies that $P_{e'}$ and P_e are exchanging rapidly at the temperature studied (25°C), and there is precedent in the literature for such behavior (31). The proton-decoupled ^{31}P NMR spectra (Figure 2) show a consistent symmetrical A_2X_2 type pattern ($J_{PP} = 40$ Hz), and the uncoupled spectrum shows the expected splitting pattern.

Further support for Complex 2 comes from reaction of the hydride in toluene with 1 mol CO to yield $HRuCl(CO)(diop)(diop^*)$; the room temperature ^{1}H NMR in toluene consists of two triplets centered at 16.4 τ (J_{PH}(trans) = 115 Hz, J_{PH}(cis) = 23 Hz), consistent with Structure 4. The dangling phosphorus produces an intense sharp high field line even at room temperature (cf. Figure 1). More detailed variable-temperature NMR studies on these interesting hydrides are in progress, but our present approximate analyses are consistent with Complex 2 and Structure 4. Presumably, steric factors again prevent a trans arrangement of two chelating diop moieties.

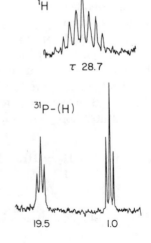

^{1}H

τ 28.7

^{31}P-(H)

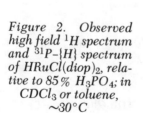

Figure 2. Observed high field ^{1}H spectrum and ^{31}P-{H} spectrum of $HRuCl(diop)_2$, relative to 85% H_3PO_4; in $CDCl_3$ or toluene, ~30°C

19.5   1.0

2

4

Catalytic Hydrogenation Using the Diop Complexes. The $HRuCl(diop)_2$ catalyst system is effective for asymmetric hydrogenation. Dma solutions of Complexes **1, 2,** or **3** readily catalyze the homogeneous hydrogenation of activated olefinic substrates at $30°–60°C$ and 1 atm H_2 (Table I). Complex **2** and its precursor, Complex **3**, give similar maximum hydrogenation rates for the same conditions and concentrations while Complex **1** only has one-half the ruthenium available as the potential hydride catalyst (*see* Reaction 1) and shows about one-half the activity in terms of total ruthenium. Optical yields of up to 60% were obtained for some prochiral α,β-unsaturated carboxylic acid substrates. We have not attempted to optimize such yields; as usual, use of (+)-diop gives one conformer in excess while use of (−)-diop gives the opposite conformer (*25*).

The diop system is the most effective of the Ru(II) chiral phosphine complexes that we have found for asymmetric hydrogenation (*25, 26*). The hydrogenation rates are about $\frac{1}{50}$ as large as those using $HRuCl(PPh_3)_3$ under corresponding conditions (*32*) but are reasonably efficient nevertheless. For example, 1M solutions of atropic acid are converted quantitatively to 2-phenylpropionic acid (40% enantiomeric excess (ee)) in one day with 10^{-2} M catalyst at 1 atm H_2.

Preliminary kinetic data on the catalyzed hydrogenation of acrylamide using $HRuCl(diop)_2$ generally show a first-order dependence on hydrogen, between a first- and a zero-order on both ruthenium and substrate, and an inverse dependence on added diop at lower substrate concentrations. These dependences are consistent with the mechanism outlined below (Reaction 4) and the corresponding rate law (Equation 5). The less than first-order dependence on ru-

(4)

$$\frac{-d[H_2]}{dt} = \frac{kK[Ru]_{total}[H_2][olefin]}{K[olefin] + [diop]}$$

(5)

thenium is reflected in the denominator term ([diop]) which contributes at lower substrate concentrations; at zero-order substrate, the alkyl is fully formed (K-[olefin] $\gg$ [diop]), and the rate is strictly first-order in ruthenium with no inverse diop dependence. The mechanism closely resembles that of $HRuCl(PPh_3)_3$ (*32,*

Table I. Hydrogenation of Unsaturated Organic Substrates Using Homogeneous Chiral Ruthenium and Rhodium Catalysts

Substrate	Catalyst	% Hydrogenation	Product % ee	Maximum Rate × 10^6 (M sec^{-1})
Itaconic acid	1[a,b]	100	R(+), 23	5.80[c]
	2 or 3[a,b]	70	R(+), 38	10.0[c]
	HRh[(+)diop]$_2$	100	R(+), 20	58.3[d]
	[RuCl$_2$-(MBMSO)$_2$]$_3$[a,e]	50	R(+), 15[f]	—
	RuCl$_2$[(−)dios][(−)ddios][g]	50	R(+), 25	—
	RuCl$_2$[(−)dios][(−)ddios][h]	100	R(+), 8	—
Atropic acid	1[a,b]	100	R(−), 40	8.73[c]
	2[a,b]	100	R(−), 27	—
	2[(−)diop][a,b]	100	S(+), 27	—
	HRh[(+)diop]$_2$	100	R(−), 37	0.28[d]
	RuCl$_2$[(−)dios][(−)ddios][g]	20	S(+), 4	—
2-Acetamido-acrylic acid	1[a,b]	100	S(−), 59	1.34[c]
	HRh[(+)diop]$_2$	100	S(−), 56	0.33[d]
	[RuCl$_2$-(MBMSO)$_2$]$_3$[i]	100	R(+), 1.5	—
	[RuCl$_2$-(MBMSO)$_2$]$_3$[j]	90	S(−), 0.5	—
	RuCl$_2$[(−)dios][(−)ddios][g]	60	S(−), 7	—
Citraconic acid	1[a,b]	100	0.0	3.25[c]
	HRh[(+)diop]$_2$	0.0	0.0	0.0
	[RuCl$_2$-(MBMSO)$_2$]$_3$[a,e]	0.0	0.0	0.0
Acrylamide	1[a,b]	100	—	42.7[c]
	2[a,b]	100	—	80.0[c]
	HRh[(+)diop]$_2$	100	—	2.8[d]
	[RuCl$_2$-(MBMSO)$_2$]$_3$	100	—	2.5[k]

[a] 0.7–1.3 M substrate, except for rate data (see c,d,k).
[b] Complex 1, Ru$_2$Cl$_4$[(+)diop]$_3$; Complex 2, HRuCl[(+)diop]$_2$; Complex 3, RuCl$_2$[(+)diop]-[(+)diop*]; 60°C in dma, 10^{-2} M Ru (monomer), 1 atm H$_2$.
[c] Measured using 4×10^{-3} M Ru (monomer). 0.2 M substrate at 60°C in dma, 1 atm H$_2$.
[d] 1.5×10^{-3} M Rh, 0.04 M substrate, in 1-butanol–toluene (2:1); 30°C except for atropic acid at 50°C (35, 36).
[e] 2×10^{-2} M Ru (monomer), 40°C in dma, ~4 atm H$_2$, 10 days.
[f] The ee is determined by optical rotation measurements (21) and by using a chiral shift reagent on the dimethyl ester of the α-methylsuccinic acid product.
[g] 1.50×10^{-2} M Ru, 0.5 M substrate, 55°C in dma, 3 atm H$_2$, 7 days (21).
[h] As for g but at 70°C.
[i] 2×10^{-2} M Ru (monomer), 0.4 M substrate, 60°C in dma, ~100 atm, 10 days.
[j] 1.5×10^{-2} M Ru (monomer), 0.4 M substrate, 40°C in dma, ~4 atm H$_2$, 7 days.
[k] Measured using 10^{-3} M Ru (monomer), 0.4 M substrate at 70°C in dma, 1 atm H$_2$.

33, 34). The corresponding diphos complex, *trans*-HRuCl[Ph₂P(CH₂)₂PPh₂]₂ (*25, 26*), is completely ineffective for hydrogenation under similar conditions presumably because of less labile phosphine ligands; the more hindered diop ligands of HRuCl(diop)₂ are replaced readily by olefinic substrates. A nonreduction of α- and β-methylcinnamic acids and mesoconic (methylfumaric) acids suggests that steric factors governing olefin coordination are important, and the selectivity pattern again resembles that of the HRuCl(PPh₃)₃ catalyst (*32, 33, 34*).

Also included in Table I åre data for hydrogenation of the same substrates catalyzed by the analogous rhodium complex, HRh[(+)-diop]₂ (*35, 36*). Considering the different conditions used (30°C in 2:1 1-butanol–toluene solution), the optical yields are remarkably similar and give the same conformer in excess. Internal olefins again were not reduced. Our suggested mechanism (*35, 36*) corresponds closely to that outlined in Reaction 4, except that in the rhodium system, kinetic data indicate that the diop ligand does not dissociate completely but becomes monodentate and dangles in solution to provide a coordination site for the substrate. The isolation of the RuCl₂(diop)(diop*) complex reported in this chapter, and of Rh(I) carbonyls containing.monocoordinated diphosphines (*37*) adds credence to our postulated mechanism.

$$\text{R}_1\!-\!\text{S}\!\!\begin{array}{c}\nearrow\!\!\text{O}\\\searrow\!\!\text{ONa}^+\end{array}\xrightarrow{\text{SOCl}_2}\text{R}_1\!-\!\text{S}\!\!\begin{array}{c}\nearrow\!\!\text{O}\\\searrow\!\!\text{Cl}\end{array}\xrightarrow{(-)\text{menthol}}\text{R}_1\!-\!\text{S}\!\!\begin{array}{c}\nearrow\!\!\text{O}\\\searrow\!\!\text{O-menthyl}\end{array}$$

(R₁ = *p*-tolyl)

$$\xrightarrow{\text{CH}_3\text{MgI}}\text{R}_1\!-\!\text{S}\!\!\begin{array}{c}\nearrow\!\!\text{O}\\\searrow\!\!\text{CH}_3\end{array}\qquad(6)$$

Chiral Sulfoxide Ligands.   The MBMSO ligand, a clear, colorless oil at 20°C, was synthesized starting from 95% S(−)-2-methylbutanol, with successive conversion of the -OH group to -Br, -SH, and -SCH₃ and finally to -SO(CH₃) using hydrogen peroxide in acetone [ν(SO) = 1025 cm⁻¹]. The S-chirality at the carbon center remains unchanged throughout the synthesis, and because the final oxidation step is nonstereospecific, the ligand is prepared as a mixture of two diastereomers, (S,R) and (S,S). ¹H NMR measurements on solutions of the ligand in the presence of a chiral lanthanide shift reagent (Kiralshift E7, Alfa Chemicals) showed equal amounts of each isomer. For example, the S-CH₃ singlet is split into two equal-area singlets separated by 0.30 ppm (*38*).

The synthetic route to R(+)-MPTSO is outlined in Reaction 6. The experimental detail followed that of Boucher and Bosnich (*39*) and is a modification of literature methods (*40, 41*). Chlorination of the *para*-toluene sulfinate gives the sulfinyl chloride that was converted to the (−)menthyl sulfinates. This diastereomer mixture was resolved by fractional crystallization, and the subsequent Grignard reaction, that is known to proceed by direct inversion at the sulfur (*42*), gave the solid, white chiral sulfoxide in an overall yield of 25%, with optical purity of 96% $R(+)$; $\nu(SO)$ at 1055 cm^{-1}.

We reported recently on the $(-)$dios ligand synthesis from L-$(+)$-tartaric acid and a related ddios ligand, the dihydroxy derivative that resulted from acid cleavage of the isopropylidene acetal group (21).

Complexes Containing Chiral Sulfoxides. The blue solutions obtained by refluxing RuCl$_3$·3H$_2$O in polar solvents under H$_2$ have provided a useful route to Ru(II) complexes (43, 44). As described in the experimental section, treatment of such methanolic solutions with monodentate sulfoxides (Ru:ligand = 1:2) yielded the trimeric [RuCl$_2$L$_2$]$_3$ species (where L = chiral ligands MBMSO and MPTSO) and a polymeric complex [RuCl$_2$L$_2$]$_n$ (where L is racemic methyl phenyl sulfoxide).

The [RuCl$_2$L$_2$]$_3$ complexes are yellow-gold, neutral species, and the molecular weights in benzene correspond to a trimer. Gouy magnetic measurements show moments of about 0.6 BM per trimer at 20°C which, in view of the air sensitivity of the compounds, is consistent with diamagnetic Ru(II) complexes. Strong ir bands at 1105 and 1110 cm^{-1}, respectively, for the MBMSO and MPTSO

5   6

complexes are attributed to $\nu(SO)$ of S-bonded sulfoxide (45, 46); medium broad bands at 330 and 325 cm^{-1}, respectively, are probably terminal Ru–Cl stretches. The presence of ir bands caused by O-bonded sulfoxide expected in the 900–980 cm^{-1} region (45, 46) is equivocal since this region is complicated by other bands present in the free ligand.

^{1}H NMR studies are useful for detecting the presence of S- or O-bonded sulfoxide. In dmso complexes, for example, the methyl protons shift downfield from the free ligand value by $\sim$1 ppm in S-bonded dmso and by $\sim$0.1 ppm in O-bonded complexes (45). For [RuCl$_2$(MBMSO)$_2$]$_3$ in CDCl$_3$, broad peaks caused by the protons of the γ- and δ-carbons are found in the 0.7–1.75 δ region; the β-carbon protons are at 1.85–2.50 δ, and the α-carbon protons are at 2.8–4.2 δ. The peak areas show that mainly S-bonded sulfoxides are present, but the possibility of some O-bonded species can not be ruled out entirely for the system because of the close proximity of the peaks. The NMR spectrum of [RuCl$_2$(MPTSO)$_2$]$_3$ in CDCl$_3$, however, consists of three broad resonance regions at 2.04–2.58 δ (protons of the *para*-tolyl methyl group), 3.34–3.96 δ (S-bonded sulfoxide methyl protons), and 6.44–7.90 δ (aromatic protons). There are no

resonances attributable to O-bonded sulfoxide; thus, both trimers are considered
to contain only S-bonded sulfoxide.   A linear structure such as Structure 5 seems
plausible, but triangular clusters of R(II) are known (47, 48).   The existence (48)
of $[Ru(CO_2Me)_2(H_2O)]_3$ with bridging acetates ($\mu \sim 0.4$ BM per Ru_3) suggests
that Structure 6 would also be a reasonable formulation.

The gold solid $[RuCl_2(CH_3SOC_6H_5)_2]_n$, isolated from the methyl phenyl
sulfoxide reaction, is diamagnetic or feebly paramagnetic and shows a S-bonded
$\nu(SO)$ at 1130 cm^{-1} and $\nu(Ru-Cl)$ at 330 cm^{-1}.   The limited solubility suggests
a polymeric structure.

We reported recently (21) the synthesis and characterization of ruthenium
complexes containing chelating chiral sulfoxides; $RuCl_2(dios)(ddios)$, $RuCl_2(d$-
dios$)_2$, and $RuCl_2(ddios)(dmso)(CH_3OH)$ were made either via the blue solutions
or via sulfoxide exchange with *cis*-$RuCl_2(dmso)_4$ (49, 50).   Cationic Rh(I) complexes of the type $[Rh(diene)(PPh_3)(sulfoxide)]^+$, with a range of sulfoxides including chiral ones, have also been synthesized (46).

Catalytic Hydrogenation Using the Chiral Sulfoxide Complexes.   The
$[RuCl_2(MBMSO)_2]_3$ trimer did effect asymmetric hydrogenation of the terminal
olefins, 2-acetamidoacrylic and itaconic acids, under homogeneous conditions
in dma (see Table I).   A 15% enantiomeric excess was obtained for hydrogenation
of itaconic to *R*-methylsuccinic acid using ~ 4 atm H_2 at 40°C while the ami-
doacrylic acid gave *N*-acetylalanine in only $\sim 1.0\%$ ee.   Trisubstituted olefins
such as α- and β-methylcinnamic, citraconic, and mesaconic acids were not reduced at all.   The $[RuCl_2(MPTSO)_2]_3$ system was more interesting because it
contained chiral sulfur centers.   Activated olefins, including prochiral ones, were
hydrogenated readily in dma at 60°C even at 1 atm H_2.   Unfortunately, the
reductions occur concomitant with ruthenium metal formation and probably
are catalyzed heterogeneously; no stereoselectivity was observed.   The MPTSO
complex itself is stable under H_2 and produces no metal; thus, reduction to metal
likely occurs via an unstable hydrido–olefin complex.

The MBMSO complex readily hydrogenates acrylamide in dma at 70°C
and 1 atm H_2, at rates convenient for kinetic studies.   Such an investigation revealed a kinetic dependence on Ru of one third (51); this is consistent with the
trimer complex dissociating to a small extent to an active monomeric catalyst.
Dma yellow solutions of this trimer ($\sim 10^{-2}$ M) did absorb small amounts of H_2
in the absence of substrate ($\sim 5\%$ based on a 1 H_2 per Ru reaction).   This absorption increased to $\sim 20\%$ to produce an orange solution in the presence of
proton sponge [the strong base 1,8-bisdimethylamino(naphthalene)], presumably
by promoting formation of intermediate hydrides via a reaction such as 7 (30).
We have been unable to detect any high field ^{1}H NMR signals in these solutions,
and we have experienced similar difficulties even for concentrated solutions of
$HRuCl(PPh_3)_3$ in dma.   Similar studies on base-promoted hydride formation
from *cis*-$RuCl_2(dmso)_4$ have given metal hydride species detectable at 28 τ in
dmso-d_6; yellow to orange color changes were observed again (51).   Under similar
conditions, the $RuCl_2(dios)(ddios)$ complex leads to higher optical yields than

the [RuCl$_2$(MBMSO)$_2$]$_3$ trimer (*see* Table I), presumably because of the rigidity of the chelate ligands. The data indicate decreasing asymmetric induction with increasing temperature that could result from greater ligand dissociation, less restriction on motion of coordinated substrate, and hence less stereoselective catalytic species. Interestingly, the (−)dios system, that has the same *R* chirality as (−)diop at carbons atoms 2 and 3, gives the enantiomer in excess opposite to the HRuCl(diop)$_2$ system for hydrogenation of itaconic and 2-acetamidoacrylic acids. Differences in substrate binding could be an important factor; in addition to π-bonding and carboxylate coordination, H-bonding is a possibility in the sulfoxide systems (-CO$_2$H or —NH with sulfoxide oxygen or -OH of ddios).

The cationic rhodium complexes [Rh(diene)(PPh$_3$)(sulfoxide)]$^+$, in which all the monodentate sulfoxides are O-bonded (*46*), tend to give metal on treatment with H$_2$ in solution, even in the presence of coordinating olefinic substrates; the O-bonded sulfoxides presumably are not strong enough π-acceptors. Interestingly, a [(NBD)Rh(PPh$_3$)(dios)]$^+$ species formed in situ from the [(NBD)-Rh(PPh$_3$)(acetone)]$^+$ cation (*46*) in dma absorbs 3 mol H$_2$ at ambient conditions, producing norbornane. Reaction 8 seems likely, although the metal hydride has not been detected by NMR. The ^{1}H NMR indicates that the in situ reactant only has O-bonded dios (*21*) while the product shows resonances typical of S-

$$Ru^{II} + H_2 \rightleftharpoons Ru^{II}H + H^+ \tag{7}$$

$$[(NBD)Rh(PPh_3)(dios)]^+ \xrightarrow{3H_2} [H_2Rh(PPh_3)(dios)]^+ + \text{norbornane} \tag{8}$$

bonded dios (*21*); such bonding reasonably appears to stabilize the hydride. The hydrogenated solutions are active homogeneous hydrogenation catalysts under mild conditions, but the use of prochiral substrates has yielded products with zero enantiomeric excess. The substrates may displace the chiral ligand.

Experimental

General. Dried, degassed reagent grade solvents were used throughout, and dma was purified as described previously (*52*). Complexes were made under inert atmospheres using Schlenk-tube techniques. Micro-analyses were done by P. Borda of this department. The hydrogenation procedures and the work-up of products for determining optical yields are reported elsewhere (*21*).

Itaconic acid (Eastman) and 2-acetamidoacrylic acid (Fluka) were CP grade; atropic acid was prepared according to the literature (*53*). Ruthenium and rhodium trichlorides were obtained as trihydrates from Johnson Matthey Limited. (+)Diop was obtained from Strem Chemicals; a literature synthesis provided (−)diop (*7*). Racemic methyl phenyl sulfoxide was a K & K product.

Sulfoxide Ligands. (*S*,*R*;*S*,*S*)-(+)-2-METHYLBUTYL METHYL SULFOXIDE (MBMSO). 80 g 95% (*S*)-(−)-2-methylbutan-1-ol (K & K) was titrated with Br$_2$ in dry DMF in the presence of triphenylphosphine according to the method of Wiley et al. (*54*). The chiral bromide was converted to the mercaptan via an isothiourea hydrobromide salt that was then decomposed by aqueous NaOH (55). Treatment of the mercaptan in aqueous NaOH with iodomethane (56) gave (S)-2-methylbutyl methyl sulfide that was separated as a clear liquid by steam distillation; $\delta_{TMS}^{CDCl_3}$ 0.90–1.70 (multiplet, 9H, CH_3-CH_2-CH-CH_3), 2.1 (singlet, 3H, S-CH_3), 2.60–2.78 (multiplet, 2H,-CH_2-S). This sulfide reacted overnight at room temperature with H_2O_2 in acetone; the vacuum distillate fraction (65–70°C, 0.2 mm) was dried over BaO and was vacuum distilled to give the clear oily sulfoxide in 36% overall yield; bp 63°C, 0.1 mm; $[\alpha]_D^{25}$ = +20.3; neat. Found: C, 53.90; H, 10.64. $C_6H_{14}SO$ requires C, 53.68; H, 10.51. $\delta_{TMS}^{CDCl_3}$ 0.90–1.20 (multiplet, 8H, CH_3-CH_2-C-CH_3), 1.25–1.60 (multiplet, 1H, -CH-), 2.40–2.80 (triplet, 2H, -CH_2-S), 2.55 (singlet, 3H, S-CH_3).

(R)-(+)-METHYL *PARA*-TOLYL SULFOXIDE (MPTSO). Sodium *para*-toluene sulfinate (Eastman) (96 g) was added slowly to 200 mL thionyl chloride under N_2, and the mixture was stirred overnight. Excess $SOCl_2$ was flash evaporated, and the *para*-toluenesulfinyl chloride, an orange, viscous, air- and water-sensitive liquid, was distilled from the reaction mixture [bp 87°C, 0.01 mm, 78 g (99%)]. This chloride, 71 g in 100 mL anhydrous ether, was added to 64 g (−)menthol in 100 mL anhydrous ether/35 mL dry pyridine under N_2 at dry ice-acetone temperature; the mixture was stirred and warmed to 20°C over 1 hr. Then, 100 mL of 1 M aqueous HCl was added; ether extractions ($MgSO_4$ dried) yielded an oily residue that was dissolved in 30 mL petroleum ether and stored at 0°C. Menthyl *para*-toluenesulfinate (88 g) was collected and recrystallized from acetone as colorless rods; $[\alpha]_D^{25}$ = −197 (C 1, acetone), which agrees with the literature (57). Methyl magnesium iodide solution (5.3 g Mg, 35 g CH_3I, 100 mL ether) was added under N_2 to 30 g menthyl sulfinate in 250 mL ether at 0°C; after stirring at 20°C for 0.5 hr, 150 mL of saturated NH_4Cl solution were added, and the aqueous phase was made just basic with aqueous ammonia. The dried ether layer was evaporated, and the residue was diluted with 50 mL hexane and cooled to 0°C to give 6.4 g of sulfoxide; further ether extractions yielded another 5 g. Recrystallization from cyclohexane gave white flakes; mp 74°C, $[\alpha]_D^{25}$ = +143° (C 1, acetone), which agrees with the literature (58). Found: C, 62.52; H, 6.64; O, 10.53; S, 20.82. $C_8H_{10}SO$ requires C, 62.34; H, 6.49; O, 10.39; S, 20.78. $\delta_{TMS}^{CDCl_3}$ 2.45 (singlet, 3H, CH_3-), 2.74 (singlet, 3H, CH_3-S), 7.46 (quartet, 4H, Ar).

Ruthenium Complexes. μ-DIOPBIS[DICHLOROMONO(DIOP)RUTHENIUM(II)], $RU_2CL_4(DIOP)_3$. One gram of $RuCl_2(PPh_3)_3$ (59) and 0.9 g diop (1:1.7 mole ratio) were refluxed in 100 mL hexane under N_2 for 16 hr; the green precipitate was washed with hexane and vacuum dried (80%). Found: C, 60.6; H, 5.3; Cl, 7.8. $C_{93}H_{96}O_6Cl_4P_6Ru_2$ requires C, 60.7; H, 5.2; Cl, 7.7.

CHLOROHYDRIDOBIS(DIOP)RUTHENIUM(II), HRUCL(DIOP)$_2$. One gram of HRuCl(PPh$_3$)$_3$ dma (60) and 1 diop were refluxed in 100 mL hexane under Ar for 24 hr; the yellow precipitate was washed with hexane and vacuum dried (60%). Found: C, 65.2; H, 6.0. $C_{62}H_{65}O_4ClP_4Ru$ requires C, 65.6; H, 5.8.

DICHLOROBIS(DIOP)RUTHENIUM(II), $RUCL_2(DIOP)(DIOP^*)$. One gram of $Ru_2Cl_4(diop)_3$ and 1 g diop were stirred in 25 mL benzene under Ar at 20°C for 10 hr; the residue that was obtained by evaporation was washed thoroughly with hexane to yield a light green powder (70%). Found: C, 63.3; H, 5.7. $C_{62}H_{64}O_4Cl_2P_4Ru$ requires C, 63.7; H, 5.5.

TRIMERIC DICHLOROBIS[(S,R;S,S)-(+)-2-METHYLBUTYL METHYL SULFOXIDE]RUTHENIUM(II), $[RUCL_2(MBMSO)_2]_3$. A total of 1.1 mL of the sulfoxide was added to the blue solution formed by refluxing 1 g of $RuCl_3\cdot3H_2O$ in 40 mL CH_3OH under H_2, and the refluxing continued under H_2 for two days; filtering to remove metal and the removal of the CH_3OH left a brown oil.

Benzene, 40 mL, was added and the solution was freeze dried; this gave 1.75 g (97%) of the complex. Found: C, 33.1; H, 6.6; Cl, 16.2; S, 14.7. $C_{36}H_{84}O_6S_6Cl_6Ru_3$ requires C, 32.7; H, 6.4; Cl, 16.1; S, 14.6. Mol wt 1275 g/mol (benzene). $\delta_{TMS}^{CCl_4}$ 0.7–1.75 (multiplet, 16H, CH_3-CH_2-C-CH_3), 1.85–2.5 (multiplet, 2H, -CH-), 2.8–4.2 (multiplet, 10H, CH_3-S-CH_2-).

TRIMERIC DICHLOROBIS[(R)-(+)-METHYL PARA-TOLYL SULFOXIDE]-RUTHENIUM(II), [$RuCL_2(MPTSO)_2$]$_3$. A total of 1.3 g of the sulfoxide was added to a methanolic blue solution formed from 1.0 g $RuCl_3\cdot3H_2O$, and the refluxing continued under H_2 for four days. The brown solution was pumped to dryness, and 15 mL $CHCl_3$ was added; the solution was filtered, and when ether was added, the complex precipitated (1.2 g, 60%). Found: C, 40.4; H, 4.4; Cl, 14.5. $C_{48}H_{60}O_6S_6Cl_6Ru_3$ requires C, 40.0; H, 4.2; Cl, 14.7. Mol wt 1360 g/mol (benzene). $\delta_{TMS}^{CDCl_3}$ 2.04–2.58 (multiplet, 6H, CH_3-Ar), 3.34–3.96 (multiplet, 6H, S-CH_3), 6.44–7.90 (multiplet, 8H, Ar).

POLYMERIC DICHLOROBIS(METHYL PHENYL SULFOXIDE)RUTHENIUM(II), [RU-$CL_2(CH_3SOC_6H_5)_2$]$_n$. A total of 1.3 g of the sulfoxide was added to the methanolic blue solution formed from 0.75 g $RuCl_3\cdot3H_2O$, and the refluxing under H_2 continued overnight. The gold solid that separated was filtered, washed with ethanol and acetone, and vacuum dried (0.9 g, 65%). Found: C, 36.9; H, 3.7; Cl, 15.9. $C_{14}H_{16}O_2S_2Cl_2Ru$ requires C, 37.2; H, 3.6; Cl, 15.7.

Acknowledgment

We thank the National Research Council of Canada for financial support, including a scholarship (R.H.M.), and Johnson Matthey Ltd. for loan of the ruthenium and rhodium salts.

Literature Cited

1. Morrison, J. D., Mosher, H. S., "Asymmetric Organic Reactions," American Chemical Society, Washington, D. C., 1976.
2. Knowles, W. S., Sabacky, M. J., Vineyard, B. D., Weinkauff, D. J., J. Am. Chem. Soc. (1975) 97, 2567.
3. Tanaka, M., Ogata, I., J. Am. Chem. Soc., Chem. Commun. (1975) 735.
4. Achiwa, K., J. Am. Chem. Soc. (1976) 98, 8265.
5. Fryzuk, M. D., Bosnich, B., J. Am. Chem. Soc. (1977) 99, 6262.
6. Levi, A., Modena, G., Scorrano, G., J. Am. Chem. Soc., Chem. Commun. (1975) 6.
7. Dumont, W., Poulin, J-C., Dang, T-P., Kagan, H. B., J. Am. Chem. Soc. (1973) 95, 8295.
8. Pino, P., Consiglio, G., Botteghi, C., Salomon, C., ADV. CHEM. SER. (1974) 132, 295.
9. Yamamoto, K., Hayashi, T., Zembayashi, M., Kumada, M., J. Organomet. Chem. (1976) 118, 161.
10. Corriu, R. J. P., Moreau, J. J. E., J. Organomet. Chem. (1974) 64, C51.
11. Bogdanovic, B., Angew. Chem. Int. Ed. (1973), 12, 954.
12. Yamada, S., Mashiko, T., Terashima, S., J. Am. Chem. Soc. (1977) 99, 1988.
13. Michaelson, R. C., Palermo, R. E., Sharpless, K. B., J. Am. Chem. Soc. (1977) 99, 1990.
14. Takaishi, N., Imai, H., Bertelo, C. A., Stille, J. K., J. Am. Chem. Soc. (1976) 98, 5400.
15. Sbrana, G., Braca, G., Giannetti, E., Intern. Conf. Organomet. Chem., 7th, Venice, 1975, p. 188.
16. Hirai, H., Furuta, T., J. Polym. Sci., Part B (1971) 9, 729.

17. McQuillin, F. J., "Homogeneous Hydrogenation in Organic Chemistry," p. 98, Reidel, Dordrecht, 1976.
18. Ohgo, Y., Kobayashi, K., Takeuchi, S., Yoshimura, J., *Bull. Chem. Soc. Jpn.* (1972) **45**, 933.
19. Carlini, C., Politi, D., Ciardelli, F., *J. Am. Chem. Soc., Chem. Commun.* (1970) 1260.
20. James, B. R., McMillan, R. S., Reimer, K. J., *J. Mol. Catal.* (1976) **1**, 439.
21. James, B. R., McMillan, R. S., *Can. J. Chem.* (1977) **55**, 3927.
22. Knowles, W. S., Sabacky, M. J., Vineyard, B. D., ADV. CHEM. SER. (1974) **132**, 274.
23. Tolman, C. A., Meakin, P. Z., Lindner, D. L., Jesson, J. P., *J. Am. Chem. Soc.* (1974) **96**, 2762.
24. James, B. R., "Homogeneous Hydrogenation," p. 400, Wiley, New York, 1973.
25. James, B. R., Wang, D. K. W., Voigt, R. F., *J. Am. Chem. Soc., Chem. Commun.* (1975) 574.
26. James, B. R., Wang, D. K. W., *Inorg. Chim. Acta* (1976) **19**, L17.
27. Hoffman, P. R., Caulton, K. G., *J. Am. Chem. Soc.* (1975) **97**, 4221.
28. James, B. R., Markham, L. D., *Inorg. Chem.* (1974) **13**, 97.
29. Bressan, M., Rigo, P., *Inorg. Chem.* (1975) **14**, 2286.
30. James, B. R., Rattray, A. D., Wang, D. K. W., *Chem. Commun.* (1976) 792.
31. Miller, J. S., Caulton, K. G., *J. Am. Chem. Soc.* (1975) **97**, 1067.
32. Markham, L. D., Ph.D. dissertation, University of British Columbia (1973).
33. James, B. R., "Homogenous Hydrogenation," p. 83, Wiley, New York, 1973.
34. Hallman, P. S., McGarvey, B. R., Wilkinson, G., *J. Chem. Soc. A* (1968) 3143.
35. Cullen, W. R., Fenster, A., James, B. R., *Inorg. Nucl. Chem. Lett.* (1974) **10**, 167.
36. James, B. R., Mahajan, D., *Isr. J. Chem.* (1977), in press.
37. Sanger, A., *J. Chem. Soc., Dalton Trans.* (1977) 120.
38. Whitesides, G. M., Lewis, D. W., *J. Am. Chem. Soc.* (1971) **93**, 5914.
39. Boucher, H., Bosnich, B., *J. Am. Chem. Soc.*, (1977) **99**, 6253.
40. Holloway, J., Kenyon, J., Phillips, H., *J. Chem. Soc.* (1928) 3000.
41. Axelrod, M., Bickart, P., Jacobus, J., Green, M. M., Mislow, K., *J. Amer. Chem. Soc.* (1968) **90**, 4835.
42. Jacobus, J., Mislow, K., *J. Am. Chem. Soc.* (1967) **89**, 5228.
43. Gilbert, J. D., Rose, D., Wilkinson, G., *J. Chem. Soc. A* (1970) 2765.
44. James, B. R., McMillan, R. S., *Inorg. Nucl. Chem. Lett.* (1975) **11**, 837.
45. McMillan, R. S., Mercer, A., James, B. R., Trotter, J., *J. Chem. Soc., Dalton Trans.* (1975) 1006.
46. James, B. R., Morris, R. H., Reimer, K. J., *Can. J. Chem.* (1977) **55**, 2353.
47. Rose, D., Wilkinson, G., *J. Chem. Soc. A* (1970) 1791.
48. Spencer, A., Wilkinson, G., *J. Chem. Soc., Dalton Trans.* (1972) 1570.
49. Mercer, A., Trotter, J., *J. Chem. Soc., Dalton Trans.* (1975) 2480.
50. Evans, I. P., Spencer, A., Wilkinson, G., *J. Chem. Soc., Dalton Trans.* (1973) 204.
51. McMillan, R. S., Ph.D. dissertation, University of British Columbia (1976).
52. James, B. R., Rempel, G. L., *Discuss. Faraday Soc.* (1968) **46**, 48.
53. Ames, G. R., Davey, W., *J. Chem. Soc.* (1958) 1794.
54. Wiley, G. A., Hershkowitz, R. L., Rein, B. M., Chung, B. C., *J. Am. Chem. Soc.* (1964) **86**, 964.
55. Rabjohn, N., Ed., *Org. Synth.* (1963) **4**, 937.
56. McAllen, D. T., Cullum, T. V., Dean, R. A., Fidler, F. A., *J. Am. Chem. Soc.* (1951) **73**, 3627.
57. Herbrandson, H. F., Dickerson, R. T., Jr., Weinstein, J., *J. Am. Chem. Soc.* (1956) **78**, 2576.
58. Mislow, K., Green, M. M., Laur, P., Melillo, J. T., Simmons, T., Ternay, A. L., Jr., *J. Am. Chem. Soc.* (1965) **87**, 1958.
59. Stephenson, T. A., Wilkinson, G., *J. Inorg. Nucl. Chem.* (1966) **28**, 945.
60. James, B. R., Markham, L. D., *J. Catal.* (1972) **27**, 442.

RECEIVED July 19, 1977.

10

Chemistry of Bis(pentamethylcyclopentadienyl)-zirconium Dihydride

JOHN E. BERCAW

Arthur Amos Noyes Laboratory of Chemical Physics, California Institute of Technology, Pasadena, CA 91125

The title compound, $(\eta^5\text{-}C_5Me_5)_2ZrH_2$ (7), is prepared by treating $\{(\eta^5\text{-}C_5Me_5)_2ZrN_2\}_2N_2$ with dihydrogen. Structure 7 is monomeric, in contrast to polymeric $\{(\eta^5\text{-}C_5H_5)_2ZrH_2\}_x$, and is very soluble in hydrocarbons and in ethers. Its ^1H NMR spectrum exhibits a resonance attributable to the two equivalent hydrides (singlet, δ 7.46) at an unusually low field, and the Zr–H stretching frequency ($1555\ cm^{-1}$) is much lower than other monomeric transition metal hydrides. The chemical reactivity of the Zr–H moiety is distinctly hydridic in contrast to the acidic character of group VI–VIII transition metal hydrides. In the presence of D_2, the hydride ligands are exchanged readily, even at $-80°C$, to yield H_2 and HD, and at $+70°C$, all 30 methyl hydrogens exchange with a D_2 atmosphere. Structure 7 promotes the reduction of CO and olefins. Several intermediates in these reactions have been identified, and possible mechanisms for these transformations are discussed.

Most of the interest in transition metal hydrides has focused on complexes of group VI to group VIII metals. These hydrides of the latter transition metals are understood now fairly well and have been identified as intermediates in important catalytic reactions such as the hydrogenation, hydroformylation, hydrosilylation, and isomerization of olefins. On the other hand, relatively little is known about the properties of group IV and V transition metal hydrides. Recent work by Schwartz and co-workers (1) with $\{(C_5H_5)_2Zr(H)Cl\}_x$ demonstrated that zirconium hydrides promote certain difficult transformations with high regioselectivity; therefore, hydrozirconation has gained recognition as a useful method in organic synthesis. Furthermore, there are indications that

0-8412-0390-3/78/33-167-136/$05.00/0

early transition metal hydrides can be used as catalysts for CH bond activation (*2, 3, 4*) and for CO reduction (*5*).

The best-defined zirconium hydrides are derivatives of bis(cyclopentadienyl)-zirconium. The first were prepared in 1966 by treating $(\eta^5\text{-}C_5H_5)_2$-$Zr(BH_4)_2$ with trimethylamine (Reactions 1 and 2) (*6, 7*). The compound $\{(\eta\text{-}$

$$(\eta^5\text{-}C_5H_5)_2Zr(BH_4)_2 + N(CH_3)_3 \rightarrow (\eta^5\text{-}C_5H_5)_2Zr(H)(BH_4) + (CH_3)_3NBH_3 \tag{1}$$

$$(\eta^5\text{-}C_5H_5)_2Zr(BH_4)_2 + 2N(CH_3)_3 \rightarrow \{(\eta^5\text{-}C_5H_5)_2ZrH_2\}_x + 2(CH_3)_3NBH_3 \tag{2}$$

$$2(\eta^5\text{-}C_9H_7)_2Zr(CH_3)_2 \xrightarrow[140^\circ C]{H_2(80\ atm)} [(\eta^5\text{-}C_9H_{11})_2ZrH_2]_2 + 4CH_4 \tag{3}$$

$C_5H_5)_2ZrH_2\}_x$ can also be obtained by treating $\{(\eta^5\text{-}C_5H_5)_2ZrCl\}_2O$ with $LiAlH_4$, whereas treating $(\eta^5\text{-}C_5H_5)_2ZrCl_2$ with $LiAlH(O\text{-}tert\text{-}C_4H_9)_3$ yields $\{(\eta^5\text{-}C_5H_5)_2Zr(H)Cl\}_x$ (*1, 9*). Investigations of the chemical and physical properties of these compounds are severely limited by their solubility, however. Only $(\eta^5\text{-}C_5H_5)_2Zr(H)(BH)_4$ appears to be monomeric; both $\{(\eta^5\text{-}C_5H_5)_2Zr(H)(Cl)\}_x$ and $\{(\eta^5\text{-}C_5H_5)_2ZrH_2\}_x$ are insoluble, apparently as a result of polymeric structures with Zr–H–Zr bridging groups. More recently, a related zirconium hydride species was prepared by hydrogenation of bis(η^5-indenyl)dimethylzirconium(IV) (Reaction 3) (*10*). This compound is dimeric, and on the basis of ^{1}H NMR and ir data, its structure is probably that shown below:

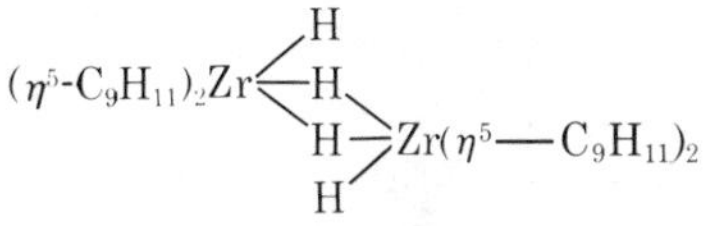

Our own interest in zirconium hydrides stems from work with dinitrogen and carbonyl derivatives of bis(pentamethylcyclopentadienyl)zirconium(II) (*11, 12, 13*). We find that these compounds are much more amenable to study than the parent $(\eta^5\text{-}C_5H_5)_2Zr$ derivatives because of their greater stability, higher solubility, and enhanced crystallizability. During these studies, we devised synthetic routes to the title compound $(\eta^5\text{-}C_5Me_5)_2ZrH_2$ and began to investigate its unusual chemical properties.

Synthesis and Characterization of $(\eta^5\text{-}C_5Me_5)_2ZrH_2$

The development of the chemistry of organometallic compounds containing the pentamethylcyclopentadienyl ligand has been hampered by the lack of a simple, high yield synthesis of 1,2,3,4,5-pentamethylcyclopentadiene (*14*). We have recently expanded the method first reported by Sorensen et al. (*15*) to a large scale preparation which follows Scheme 1 (*16*). This procedure is applicable

Scheme 1

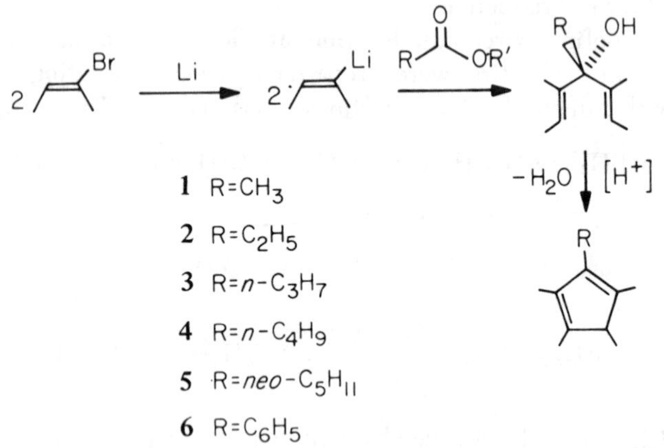

1 $R = CH_3$

2 $R = C_2H_5$

3 $R = n\text{-}C_3H_7$

4 $R = n\text{-}C_4H_9$

5 $R = neo\text{-}C_5H_{11}$

6 $R = C_6H_5$

to the synthesis of various alkyl- and aryltetramethylcyclopentadienes (1–6) by using the appropriate ester in the first step. Yields are generally good (50–75%).

Bis(pentamethylcyclopentadienyl)zirconium dichloride is prepared readily in 50–60% yield following Reactions 4 and 5. Although $(\eta^5\text{-}C_5Me_5)_2ZrH_2$ (7) can be obtained by treating $(\eta^5\text{-}C_5Me_5)_2ZrCl_2$ with $Li^+[BH(C_2H_5)_3]^-$ or with $Na^+[AlH_2(OCH_2CH_2OCH_3)_2]^-$ in THF or benzene, respectively, its isolation from these reaction mixtures is extremely difficult. We therefore used an indirect synthesis (Reactions 6 and 7). Compound $\{(\eta^5\text{-}C_5Me_5)_2ZrN_2\}_2N_2$ (8) is obtained as a pure crystalline material in 60% yield; its structure is reported elsewhere (13). Structure 8 reacts quantitatively (NMR) with H_2 at $0°C$ according to Reaction 7.

$$C_5Me_5H + n\text{-}C_4H_9Li \xrightarrow[\text{DME}]{} Li^+(C_5Me_5)^- + C_4H_{10} \tag{4}$$

$$2 Li^+(C_5Me_5)^- + ZrCl_4 \xrightarrow[\text{DME}]{70°C} (\eta^5\text{-}C_5Me_5)_2ZrCl_2 + 2 LiCl \tag{5}$$

$$(\text{DME} = 1,2\text{-dimethoxyethane})$$

$$2(\eta^5\text{-}C_5Me_5)_2ZrCl_2 \xrightarrow[N_2]{Na(Hg)} \{(\eta^5\text{-}C_5Me_5)_2ZrN_2\}_2N_2 + 4NaCl \tag{6}$$

$$\{(\eta^5\text{-}C_5Me_5)_2ZrN_2\}_2N_2 + 4H_2 \rightarrow 2(\eta^5\text{-}C_5Me_5)_2ZrH_2 + 3N_2 \tag{7}$$

Unlike polymeric $\{(\eta^5\text{-}C_5H_5)_2ZrH_2\}_x$, pale yellow $(\eta^5\text{-}C_5Me_5)_2ZrH_2$, (7) is very soluble in hydrocarbons and ethers. Its molecular weight (cryoscopically determined for 112 mg of 7 per g of benzene as 369; calculated as 364), analytical data, NMR spectrum (Table I), and ir spectrum (ν(Zr–H) 1555 cm^{-1} (ms, br);

$\nu(\text{Zr–D})$ 1100 cm^{-1} (ms, br)) are entirely in accord with a monomeric, pseudo-tetrahedral structure analogous to $(\eta^5\text{-}C_5Me_5)_2ZrCl_2$. The low field chemical shift observed for the hydride hydrogen atoms of 7 (7.46 δ) is in direct contrast to the characteristic high field resonances for group V–VIII transition metal hydrides but intermediate to the hydride resonances for $(\eta^5\text{-}C_5Me_5)_2TiH_2$ (0.28 δ) (17) and $(\eta^5\text{-}C_5Me_5)_2HfH_2$ (15.6 δ) (18). Although the factors responsible for these downfield shifts in group IV hydrides are not clear, it appears that the paramagnetic contribution to the total chemical shift is negative for Ti, Zr, and Hf. The Zr–H stretching frequency is found at a lower energy than $\nu(\text{M–H})$ for group VI–VIII transition metal hydrides (1800–2200 cm^{-1}), and is closer to that observed for $(\eta^5\text{-}C_5H_5)_2NbH_3$ (1710 cm^{-1}) (4) and for $(\eta^5\text{-}C_5H_5)_2TaH_3$ (1735 cm^{-1}) (19). The observed Zr–H stretching frequencies in the 1500–1600 cm^{-1} region were taken previously to indicate a bridging·hydride ligand (8, 10); however, these new data for monomeric 7 make such assignments suspect.

The Zr–H bonds for 7 are clearly hydridic, again in contrast to the group VIII transition metal hydrides that behave chemically more like protonated metal complex anions. Thus 7 readily reduces HCl, CH$_3$I, and CH$_2$O (Reactions 8–10). The Zr$^+$–H$^-$ polarization of the zirconium hydride bonds for 7 is not altogether unexpected in light of the position of Zr in the periodic table.

$$(\eta^5\text{-}C_5Me_5)_2ZrH_2 + 2\ HCl \xrightarrow{-80^\circ C} (\eta^5\text{-}C_5Me_5)_2ZrCl_2 + 2\ H_2 \qquad (8)$$

$$(\eta^5\text{-}C_5Me_5)_2ZrH_2 + 2CH_3I \xrightarrow{-80^\circ C} (\eta^5\text{-}C_5Me_5)_2ZrI_2 + 2CH_4 \qquad (9)$$

$$(\eta^5\text{-}C_5Me_5)_2ZrH_2 + 2\ CH_2O \xrightarrow{25^\circ C} (\eta^5\text{-}C_5Me_5)_2Zr(OCH_3)_2 \qquad (10)$$

Deuterium Exchange Studies

The two hydride positions for $(\eta^5\text{-}C_5Me_5)_2ZrH_2$ (7) exchange readily in a D$_2$ atmosphere. After 15 min at -80°C, the hydrogen atmosphere above a toluene solution of 7 consists of a statistical mixture of H$_2$, HD, and D$_2$, ignoring thermodynamic isotope effects. At higher temperatures (>ca. 50°C), all 30 methyl hydrogen positions of the two $[\eta^5\text{-}C_5(CH_3)_5]$ rings also exchange with a D$_2$ atmosphere. Kinetic measurements of the methyl hydrogen resonance in the ^{1}H NMR spectrum of 7 (in benzene-h_6 relative to a ferrocene standard with $\geq$13 g-atom excess of D) indicate that the disappearance rate is first order in 7 and is independent of the $\bar{p}(D_2)$ over the range 0.6–2.0 atm. Thus the rate of deuterium incorporation into the $(\eta^5\text{-}C_5Me_5)_2ZrH_2$ methyl groups follows the relationship: rate = k[Zr], $k = 5 \times 10^{-4}$ sec^{-1} at 84°C.

Since the product isolated after N$_2$ is removed from 8 is not $(\eta^5\text{-}C_5Me_5)_2Zr$, but rather its tautomer with Structure 9 (*see* Scheme 2), it is reasonable to propose the mechanism shown in Scheme 2 to account for deuterium incorporation into

Scheme 2

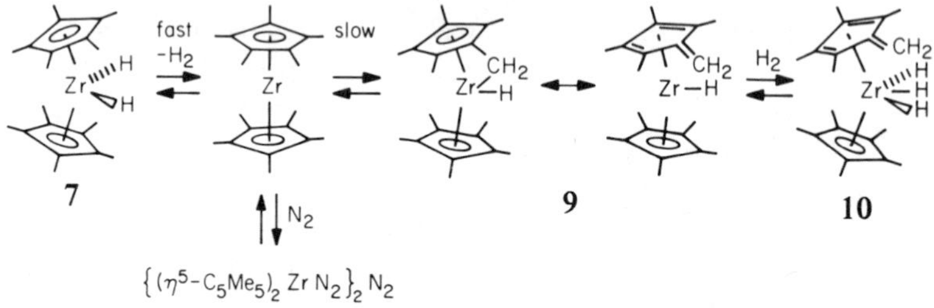

$$\{(\eta^5\text{-}C_5Me_5)_2\,Zr\,N_2\}_2\,N_2$$
8

the hydride and methyl groups of **7**. However, two features of this mechanism
require closer scrutiny. First, the observation of HD in the faster exchange of
D_2 with the two hydride positions of **7** requires a rapid bimolecular exchange
of hydride ligands (Reaction 11) in addition to the reductive elimination–oxidative
addition sequence (shown as the first step in Scheme 1). Second, the high thermal
stability of **7** (>120°C, vacuum) casts considerable doubt that reductive elimination of H_2 could proceed as rapidly as required at −80°C.

$$(\eta^5\text{-}C_5Me_5)_2ZrH_2 + (\eta^5\text{-}C_5Me_5)_2ZrD_2 \rightleftarrows 2(\eta^5\text{-}C_5Me_5)_2ZrHD \qquad (11)$$

Reversible addition of H_2 to **9**, forming **10**, requires conjugation of the
Zr–CH_2 moiety with the cyclopentadienyl ring; otherwise, an oxidative addition
of H_2 to a formal Zr(IV) species must be invoked. Therefore, we studied systematically a series of bis(η^5-alkyltetramethylcyclopentadienyl)zirconium(IV)
dihydrides to assess the regioselectivity of deuterium incorporation. Scheme
3 summarizes some preliminary results. As can be seen from these data, conjugation of a CH bond with the cyclopentadienyl ring is not required; rather, for
11, deuterium incorporation occurs readily into the methyl C–H bonds of the
ethyl groups and preferentially for the *tert*-butyl group of **12**. The stability of
the product of zirconium addition to a *tert*-butyl C–H bond in **12** is demonstrated
by Reaction 12. Thus, unlike thermally stable **7**, **12** reversibly evolves H_2 at 25°C
to yield **13**, which was characterized readily by its ^{1}H NMR spectrum.

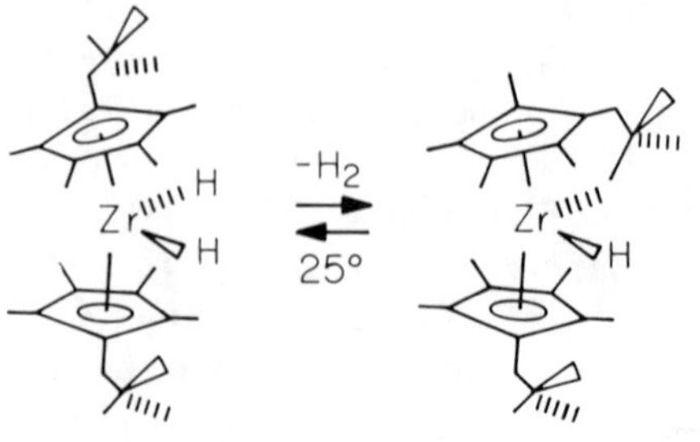

(12)

12   **13**

Scheme 3

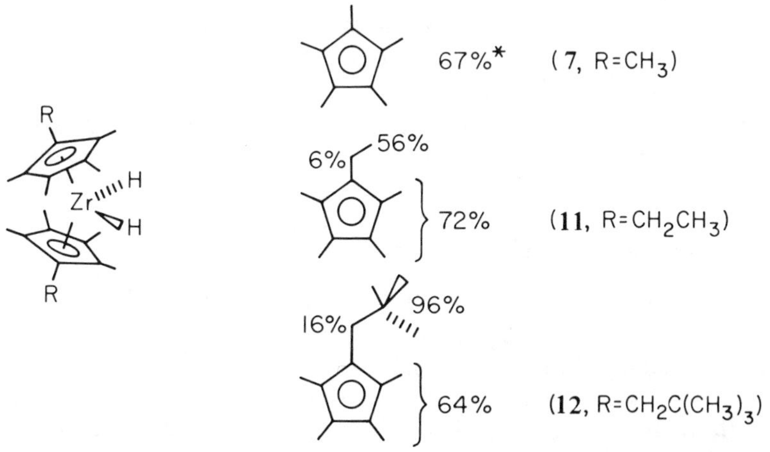

67%* (**7**, R=CH$_3$)

6% 56%
72% (**11**, R=CH$_2$CH$_3$)

16% 96%
64% (**12**, R=CH$_2$C(CH$_3$)$_3$)

*percentage of predicted statistical deuterium incorporation after 1 hr at 70° in C$_6$H$_6$ (1 atm D$_2$)

In light of these new data and the difficulties with Scheme 3 discussed above, we propose the mechanism outlined in Scheme 4 (shown for **7** only). Its key feature is a facile and reversible metal-to-ring hydride transfer giving (pentamethylcyclopentadienyl)(1 - *exo* - methyltetramethylcyclopentadiene)hydridozirconium(II), i.e., **7** ⇌ **14**. Benfield and Green (*19*) reported that an analogous rearrangement of (η^5-C$_5$H$_5$)$_2$Mo(C$_2$H$_5$)Cl to (η^5-C$_5$H$_5$)(1-*endo*-ethyl C$_5$H$_5$)-Mo(PR$_3$)Cl is promoted by addition of phosphines to the former. A reversible hydride and/or alkyl transfer to a (η^5-C$_5$H$_5$) ring has been proposed also to ac-

Scheme 4

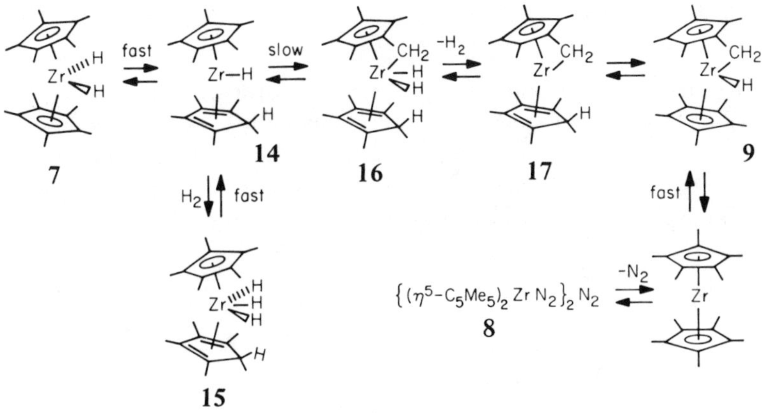

count for the formation of a bis-mesityl derivative of tungstenocene when $(\eta^5\text{-}C_5H_5)_2WH_2$ is irradiated in mesitylene (20, 21). Although not proved, Scheme 4 does account for two essential features of ZrH_2 and $[\eta^5\text{-}C_5(CH_3)_5]$ hydrogen exchange with D_2: facile HD formation at $-80°C$, and the lack of dependence of the $(CH_3)\text{-}D_2$ exchange rate on the $\bar{p}(D_2)$ (providing that **16** $\rightleftarrows$ **17** is much faster than **14** → **16**). In addition, the mechanism provides an associative pathway for $ZrH_2\text{-}D_2$ exchange, avoiding the need to invoke a rapid reductive elimination of H_2 from **7**.

Reduction of CO and Olefins Promoted by $(\eta^5\text{-}C_5Me_5)_2ZrH_2$

Compound $(\eta^5\text{-}C_5Me_5)_2ZrH_2$ (**7**) is formally a 16–electron complex and as such can be expected to add donors such as tertiary phosphines and CO. Accordingly, **7** absorbs PF_3 (0.89 mol/mol of **7**) at $-80°C$ in toluene to yield the unstable, 18-electron complex $(\eta^5\text{-}C_5Me_5)_2Zr(H)_2(PF_3)$(**18**). On the basis of its ^1H NMR spectrum at $-50°C$ (Table I), the structure of **8** appears to be analogous to $(\eta^5\text{-}C_5H_5)_2TaH_3$, with PF_3 occupying the central equatorial position mutually cis to both hydride ligands, i.e.,

$$(\eta^5\text{-}C_5Me_5)_2Zr\overset{H}{\underset{H}{<}}PF_3$$

18

Similarly, **7** absorbs CO (0.97 mol/mol of **7**) in toluene at $-80°C$ to generate the carbonyl hydride $(\eta^5\text{-}C_5Me_5)_2Zr(H)_2(CO)$ (**19**). Although **19** is not stable enough to be isolated (vide infra), it has been characterized in solution at low temperature. Thus, **19** reacts with excess HCl at $-80°C$ to yield $(\eta^5\text{-}C_5Me_5)_2ZrCl_2$, H_2(1.78 mol/mol of **19**), and CO (0.85 mol/mol of **19**), close to the stoichiometry for Reaction 13. The ^1H NMR spectrum for **19** at $-64°C$ (toluene-d_8) consists of

$$(\eta^5\text{-}C_5Me_5)_2Zr(H)_2(CO) + 2HCl \rightarrow (\eta^5\text{-}C_5Me_5)_2ZrCl_2 + 2H_2 + CO \quad (13)$$

a singlet at 1.84 δ, attributable to the methyl hydrogens of the two equivalent $(\eta^5\text{-}C_5Me_5)$ rings, and a singlet at 1.07 δ, attributable to two equivalent hydride ligands. The spectrum of $(\eta^5\text{-}C_5Me_5)_2Zr(H)_2(^{13}CO)$ exhibits the same singlet at 1.84 δ, but the hydride resonance is now the expected doublet ($^2J_{^1H-^{13}C} = 25.1$ Hz). Thus the structure of **19** appears to be analogous to **18**, i.e.,

$$(\eta^5\text{-}C_5Me_5)_2Zr\overset{H}{\underset{H}{<}}CO$$

19

When solutions of **19** are warmed above ca. $-50°$C, the appearance of a new set of ^{1}H NMR signals consisting of a (η^5-C$_5$Me$_5$) resonance at 1.94 δ (30H), a singlet at 5.73 δ (1H), and singlet at 6.55 δ (1H), accompanies the disappearance of the spectrum of **19**.   Independent experiments indicate that no H$_2$ or CO evolution is associated with the conversion of **19** to this new compound (**20**), so both must have the same empirical formula.   While our initial interpretation of these NMR data was that **20** was a formyl hydride species, further experiments revealed that it is a dimer with an unusual enne-dioxy bridge of the structure shown in **20**.   This structure is supported by its ir spectrum (ν (Zr–H) 1580 cm^{-1},

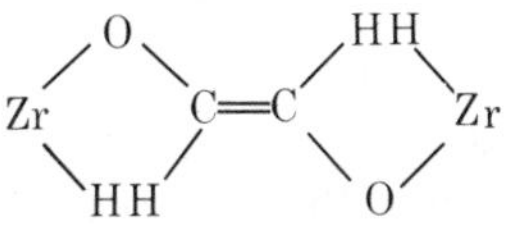

20

ν(Zr–D) 1130 cm^{-1}, ν(C–O) 1205 cm^{-1}, and ν(^{13}C–O) 1180 cm^{-1}) and most convincingly by ^{1}H and ^{13}C NMR data for {(η^5-C$_5$Me$_5$)$_2$ZrH}$_2$(O^{13}CH = ^{13}CHO), prepared from **7** and ^{13}CO, a characteristic *AA'XX'* pattern (**22**) for the protons of the ($-$O^{13}CH = ^{13}CHO$-$) bridge (Table I).   The non-{^{1}H}^{13}C NMR shows the same *AA'XX'* pattern.   Furthermore, **20** reacts smoothly with methyl iodide to yield methane and the orange crystalline complex {(η^5-C$_5$Me$_5$)$_2$ZrI}$_2$ (OCH = CHO) (**21**) (Reaction 14).   On the basis of its ir spectrum (ν(CO) 1195 cm^{-1}

$$\{(\eta^5\text{-C}_5\text{Me}_5)_2\text{ZrH}\}_2 \text{ (OCH}=\text{CHO)} + 2\text{ CH}_3\text{I}$$
$$\rightarrow \{(\eta^5\text{-C}_5\text{Me}_5)_2\text{ZrI}\}_2 \text{ (OCH}=\text{CHO)} + 2\text{CH}_4 \quad (14)$$

(vs); ν(^{13}C–O) 1175 cm^{-1} (vs)) and its ^{1}H NMR spectrum (Table I), the structure of **21** appears to be analogous to that for **20**.   Although we have encountered some difficulties in final refinement, an x-ray structure determination for **21** confirmed this tentative assignment (*23*).

When warmed under H$_2$ atmosphere in the presence of (η^5-C$_5$Me$_5$)$_2$ZrH$_2$ (**7**), (η^5-C$_5$Me$_5$)$_2$Zr(H)$_2$(CO) (**19**) is reduced to (η^5-C$_5$Me$_5$)$_2$Zr(H)(OCH$_3$) (**22**) competitively with its dimerization to **20**, and **7** is recovered quantitatively.   Concurrent with our studies, a report suggested a possible origin of the unusual reactivity for **19**.   Floriani and co-workers found that (η^5-C$_5$H$_5$)$_2$Zr(CH$_3$)$_2$ reacts reversibly with CO at room temperature to produce a "π-acyl" derivative (Reaction 15) (*24*).

$$(\eta^5\text{-C}_5\text{H}_5)_2\text{Zr(CH}_3)_2 + \text{CO} \rightleftarrows (\eta^5\text{-C}_5\text{H}_5)_2\text{Zr(CH}_3)(\text{CH}_3\text{CO}) \quad (15)$$

Their x-ray structure determination revealed that the acyl group was bonded to zirconium via both acyl carbon and acyl oxygen atoms.   While they consider the acyl group as a three-electron ligand bonded to zirconium by both σ(C) and

π(C=O) interactions, an alternative description can be invoked in simple valence bond formalism:

$$(a) \qquad (b)$$

The oxycarbene representation (b) is consistent with the reduced C–O bond order (ν(CO) = 1545 cm^{-1}) and furthermore offers an explanation for the reactivity of 19 (*see* Scheme 5). Thus, 19 may be expected to rearrange to a formyl hydride intermediate analogous to (η^5-C$_5$H$_5$)$_2$Zr(CH$_3$)(CH$_3$CO), which is reduced by 7 to the methoxy hydride species 22 in competition with its dimerization to 20. Both pathways, (a) dimerization via coupling of carbene centers, and (b) insertion of carbene into a Zr–H bond of 7, are thus reconciled. The intermediacy of bis(pentamethylcyclopentadienyl)zirconium(II) in the reduction of 19 to 22 is supported by an experiment in which CO was slowly diffused into an N$_2$-blanked solution of 7. A transient dark red solution, characteristic of $\{(\eta^5$-C$_5$Me$_5)_2$ZrN$_2\}_2$N$_2$ and/or $\{(\eta^5$-C$_5$Me$_5)_2$Zr(CO)$\}_2$N$_2$, was noted prior to the

Table I. Proton Nuclear Magnetic Resonance Data

Compound	Solvent			
(η^5-C$_5$Me$_5$)$_2$ZrH$_2$ (7)	benzene-d_6	[C$_5$(CH$_3$)$_5$]	s	2.02 δ
		ZrH$_2$	s	7.46 δ
(η^5-C$_5$Me$_4$Et)$_2$ZrH$_2$ (11)	benzene-d_6	[C$_5$(CH$_3$)$_4$(CH$_2$CH$_3$)]	s, s	2.03, 2.05 δ
		[C$_5$(CH$_3$)$_4$(CH$_2$CH$_3$)]	t	1.02 δ ($^3J_{HH}$ = 7 Hz)
		[C$_5$(CH$_3$)$_4$(CH$_2$CH$_3$)]	q	2.60 δ ($^3J_{HH}$ = 7 Hz)
		ZrH$_2$	s	7.45 δ
(η^5-C$_5$Me$_4$CH$_2$CMe$_3$)$_2$-ZrH$_2$ (12)	benzene-d_6	[C$_5$(CH$_3$)$_4$(CH$_2$C-(CH$_3$)$_3$)]	s, s	2.02, 2.08 δ
		[C$_5$(CH$_3$)$_4$(CH$_2$C-(CH$_3$)$_3$)]	s	2.65 δ
		[C$_5$(CH$_3$)$_4$(CH$_2$C-(CH$_3$)$_3$)]	s	0.95 δ
		ZrH$_2$	s	7.43 δ
(η^5-C$_5$Me$_5$)$_2$ZrH$_2$(PF$_3$) (18)	toluene-d_8 -50 °C	[C$_5$(CH$_3$)$_5$]	s	1.77 δ
		ZrH$_2$(PF$_3$)	dq	0.55 δ ($^2J_{H^{31}P}$ = 1.08 Hz) $^3J_{H^{19}F}$ = = 21.5 Hz)

Table I. Continued

Compound	Solvent			
$(\eta^5\text{-}C_5Me_5)_2ZrH_2(CO)$ **(19)**	toluene-d_8 −64 °C	$[C_5(CH_3)_5]$	s	1.84 δ
		$ZrH_2(CO)$	s	1.07 δ
		$ZrH_2(^{13}CO)$	d	1.07 δ
				$(^2J_{H^{13}C} = 25.1$ Hz$)$
$\{(\eta^5\text{-}C_5Me_5)_2ZrH\}_2\text{-}$ (OCHCHO) **(20)**	toluene-d_8	$[C_5(CH_3)_5]$	s	1.94 δ
		$OCH{=}CHO$	s	6.55 δ
		ZrH	s	5.73 δ
		$O^{13}CH{=}^{13}CHO$	10 line $AA'XX'$ pattern	$(^1J_{H^{13}C} = 176.5$ Hz, $^1J_{^{13}C^{13}C} = 99$ Hz, $^2J_{H^{13}C} = 7.5$ Hz, $^3J_{HH} = 9$ Hz$)$
$\{(\eta^5\text{-}C_5Me_5)_2ZrI\}_2\text{-}$ (OCHCHO) **(21)**	benzene-d_6	$[C_5(CH_3)_5]$	s	1.94 δ
		$OCH{=}CHO$	s	6.83 δ
		$O^{13}CH{=}^{13}CHO$	10 line $AA'XX'$ pattern	$(^1J_{H^{13}C} = 180.3$ Hz, $^1J_{^{13}C^{13}C} = 100.3$ Hz, $^2J_{H^{13}C} = 6.7$ Hz, $^3J_{HH} = 10.4$ Hz$)$
$(\eta^5\text{-}C_5Me_5)_2Zr(H)\text{-}$ (CH_2CHMe_2) **(23)**	benzene-d_6	$[C_5(CH_3)_5]$	s	1.93 δ
		ZrH	s	6.43 δ
		$ZrCH_2CH(CH_3)_2$	d	−0.04 δ $(^3J_{HH} = 7.0$ Hz$)$
		$ZrCH_2CH(CH_3)_2$	d	0.99 δ $(^3J_{HH} = 6.5$ Hz$)$
		$ZrCH_2CD(CH_3)_2$	s	−0.04 δ
		$ZrCH_2CD(CH_3)_2$	s	1.04 δ
$(\eta^5\text{-}C_5Me_5)_2Zr(H)\text{-}$ (OCH=CHCH- Me$_2$) **(24)**	benzene-d_6	$[C_5(CH_3)_5]$	s	1.94 δ
		ZrH	s	6.04 δ
		$ZrOCH{=}CHCH\text{-}(CH_3)_2$	d	6.63 δ $(^3J_{HH} = 12$ Hz$)$
		$ZrOCH{=}CHCH\text{-}(CH_3)_2$	dd	4.61 δ $(^3J_{HH} = 12$ Hz, $^3J_{HH} = 8$ Hz$)$
		$ZrOCH{=}CHCH\text{-}(CH_3)_2$	d	1.07 δ $(^3J_{HH} = 6.5$ Hz$)$
		$ZrOCD{=}CHCD\text{-}(CH_3)_2$	s	4.58 δ
		$ZrOCD{=}CHCD\text{-}(CH_3)_2$	s	1.10 δ

Scheme 5

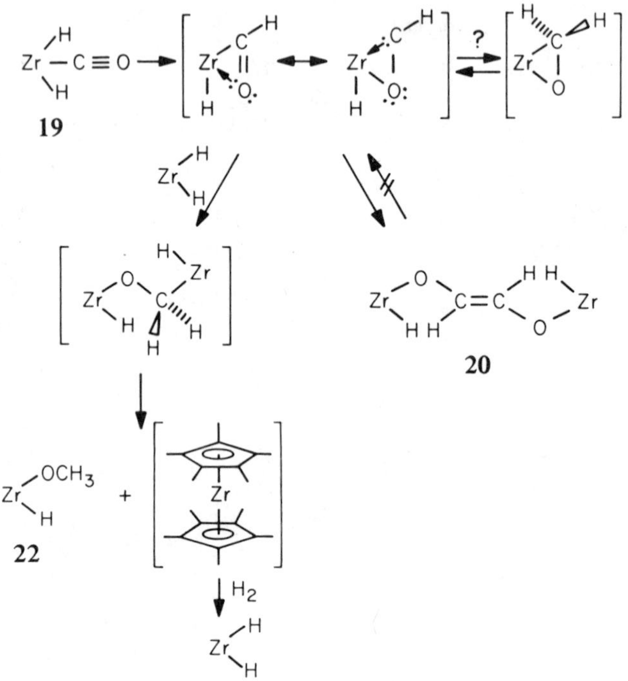

eventual conversion of the solution to a 1:1 mixture of **22** and $(\eta^5\text{-}C_5Me_5)_2Zr(CO)_2$ (Reaction 16):

$$2(\eta^5\text{-}C_5Me_5)_2ZrH_2 \xrightarrow{\text{CO}} (\eta^5\text{-}C_5Me_5)_2Zr(H)(OCH_3) + (\eta^5\text{-}C_5Me_5)_2Zr(CO)_2$$

$$(16)$$

The migratory insertion of CO into a Zr–H bond postulated in Scheme 5 is unprecedented for other transition metal carbonyl hydrides and thus is of considerable significance. It was of particular interest to compare the propensity of CO to insert into a Zr–H bond relative to a Zr–C bond; hence, an alkyl hydride derivative of bis(pentamethylcyclopentadienyl)zirconium was synthesized. Compound $(\eta^5\text{-}C_5Me_5)_2ZrH_2$ (**7**) reacts smoothly with isobutylene at room temperature to produce the isobutyl hydride complex $(\eta^5\text{-}C_5Me_5)_2Zr(H)$-$(CH_2CHMe_2)$ (**23**) in quantitative yield (Reaction 17). When treated with CO (1 atm, 25°C), **23** is converted rapidly to $(\eta^5\text{-}C_5Me_5)_2Zr(H)(OCH=CHCHMe_2)$ (**24**) (Reaction 18). By carefully monitoring the changes accompanying this reaction by ^1H NMR spectrometry at −50°C, we could identify a transient intermediate with a spectrum (toluene-d_8; [$C_5(CH_3)_5$], s (30H), 1.82 δ; ZrH, s (1H), 3.66 δ; ZrOCCH_2CH(CH$_3$)$_2$, d (2H), 2.53 δ ($^3J_{HH}$ = 8 Hz); ZrOCCH$_2$CH(CH$_3$)$_2$, d (6H), 1.20 δ ($^3J_{HH}$ = 7 Hz); ZrO^{13}CCH_2CH(CH$_3$)$_2$, dd (2H), 2.50 δ ($^3J_{HH}$ = 7 Hz, $^2J_{H^{13}C}$ = 5 Hz); ZrH(O^{13}CCH$_2$CH(CH$_3$)$_2$), d (1H), 3.61 δ ($^2J_{H^{13}C}$ = 9 Hz)) consistent with that expected for (η^5-C$_5$Me$_5$)$_2$Zr(H)(OCCH$_2$CHMe$_2$). Furthermore, the results of a deuterium labeling study, outlined in Reactions 19, 20, and 21, established conclusively that alkyl and hydride move to CO during the conversion of **23** to **24** (*see* Table I).

$$(\eta^5\text{-C}_5\text{Me}_5)_2\text{ZrH}_2 + \text{CH}_2{=}\text{CMe}_2 \rightarrow (\eta^5\text{-C}_5\text{Me}_5)_2\text{Zr(H)(CH}_2\text{CHMe}_2) \quad (17)$$

$$(\eta^5\text{-C}_5\text{Me}_5)_2\text{Zr(H)(CH}_2\text{CHMe}_2) + \text{CO}$$
$$\rightarrow (\eta^5\text{-C}_5\text{Me}_5)_2\text{Zr(H)(OCH}{=}\text{CHCHMe}_2) \quad (18)$$

$$\{(\eta^5\text{-C}_5\text{Me}_5)_2\text{ZrN}_2\}_2\text{N}_2 + 2\text{D}_2 \rightarrow 2(\eta^5\text{-C}_5\text{Me}_5)_2\text{ZrD}_2 + 3\text{N}_2 \quad (19)$$

$$(\eta^5\text{-C}_5\text{Me}_5)_2\text{ZrD}_2 + \text{CH}_2 = \text{CMe}_2 \rightarrow (\eta^5\text{-C}_5\text{Me}_5)_2\text{Zr(D)(CH}_2\text{CDMe}_2) \quad (20)$$

$$(\eta^5\text{-C}_5\text{Me}_5)_2\text{Zr(D)(CH}_2\text{CDMe}_2) + \text{CO}$$
$$\rightarrow (\eta^5\text{-C}_5\text{Me}_5)_2\text{Zr(H)(OCD} = \text{CHCDMe}_2) \quad (21)$$

The results implicate the mechanism shown in Scheme 6. Thus initial migratory insertion of CO into the Zr–C bond is followed by intramolecular insertion of the oxycarbene into the Zr–H bond to produce what could be considered an aldehyde adduct of bis(pentamethylcyclopentadienyl)zirconium(II). The final step involves simple β–hydride abstraction to give the enolate hydride, **24**.

Scheme 6

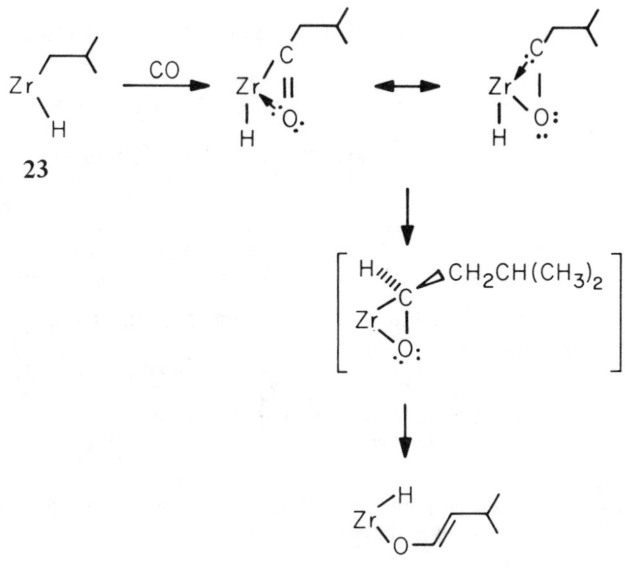

23

24

Zirconium hydride reactivity with carbon monoxide demonstrates the strong driving force toward products with a Zr–O bond. Indeed, the facility of the CO migratory insertion into Zr–C and especially Zr–H bonds may be from a carbonyl oxygen–zirconium interaction that stabilizes the transition state to the acyl and formyl complexes.

Acknowledgment

The author thanks his graduate students, Juan M. Manriquez, Donald R. McAlister, Robert D. Sanner, David K. Erwin, and Richard S. Threlkel for their contributions to this work.

Literature Cited

1. Schwartz, J., Labinger, J. A., *Angew. Chem., Int. Ed. Engl.* (1976) **15**, 33.
2. Barefield, E. K., Parshall, G. W., Tebbe, F. N., *J. Am. Chem. Soc.* (1970) **92**, 5234.
3. Guggenberger, L. J., *Inorg. Chem.* (1973) **12**, 294.
4. Tebbe, F. N., Parshall, G. W., *J. Am. Chem. Soc.* (1971) **93**, 3793.
5. Manriquez, J. M., McAlister, D. R., Sanner, R. D., Bercaw, J. E., *J. Am. Chem. Soc.* (1976) **98**, 6733.
6. James, B. D., Nanda, R. K., Wallbridge, M. G. H., *J. Chem. Soc. D.* (1966) 849.
7. James, B. D., Nanda, R. K., Wallbridge, M. G. H., *Inorg. Chem.* (1967) **6**, 1979.
8. Wailes, P. C., Coutts, R. S. P., Weigold, H., "Organometallic Chemistry of Titanium, Zirconium and Hafnium," Academic, New York, 1974.
9. Wailes P. C., Weigold, H., *J. Organomet. Chem.* (1970) **24**, 405.
10. Weigold, H., Bell, A. P., Willing, R. I., *J. Organomet. Chem.* (1974) **73**, C23.
11. Manriquez, J. M., Bercaw, J. E., *J. Am. Chem. Soc.* (1974) **96**, 6229.
12. Manriquez, J. M., Sanner, R. D., Marsh, R. E., Bercaw, J. E., *J. Am. Chem. Soc.* (1976) **98**, 3042.
13. Sanner, R. D., Manriquez, J. M., Marsh, R. E., Bercaw, J. E., *J. Am. Chem. Soc.* (1976) **98**, 8351.
14. Feitler, D., Whitesides, G. M., *Inorg. Chem.* (1976) **15**, 466.
15. Campbell, P., Chiu, N. W. K., Deugan, K., Miller, I. J., Sorensen, T. S., *J. Am. Chem. Soc.* (1969) **91**, 6404.
16. Threlkel, R. S., Bercaw, J. E., *J. Organomet. Chem.* (1977) **136**, 1.
17. Bercaw, J. E., Marvich, R. H., Bell, L. G., Brintzinger, H. H., *J. Am. Chem. Soc.* (1972) **94**, 1219.
18. Manriquez, J. M., Gay, R., Bercaw, J. E., unpublished data.
19. Benfield, F. W. S., Green, M. L. H., *J. Chem. Soc., Dalton Trans.* (1974) 1324.
20. Green, M. L. H., McCleverty, J. A., Pratt, L., Wilkinson, G., *J. Am. Chem. Soc.* (1961) 4854.
21. Elmit, K., Green, M. L. H., Forder, R. A., Jefferson, I., Prout, K., *J. Chem. Soc., Chem. Commun.* (1974) 747.
22. Emsley, J. W., Feeney, J., Sutcliffe, L. H., "High Resolution Magnetic Resonance Spectroscopy," Vol. 1, p. 396, Pergamon, Oxford, 1965.
23. Sanner, R. D., Marsh, R. E., Bercaw, J. E., unpublished data.
24. Fachinetti, G., Floriani, C., Marchetti, F., Merlino, S., *J. Chem. Soc., Chem. Commun.* (1976) 522.

RECEIVED July 25, 1977.   Contribution No. 5637 from the Arthur Amos Noyes Laboratory of Chemical Physics.   Work supported by the National Science Foundation.

Chemistry of Hydridoniobium Complexes: Approaches to Homogeneous Catalysis of the Carbon Monoxide–Hydrogen Reaction

J. A. LABINGER

Department of Chemistry, University of Notre Dame, Notre Dame, IN 46556

Based on the chemistry of transition metal hydride and carbonyl complexes, a mechanism for the homogeneously catalyzed reduction of CO by H_2 that involves inserting CO into a metal–hydrogen bond might not be feasible. An alternate route for formation of the first C–H bond, nucleophilic attack by a hydridic metal hydride on CO, suggests several criteria for the type of complex that might be a successful catalyst. Several hydridoniobium complexes possess the desired hydridic character and hence are potential catalysts (although CO reduction could not be effected by the hydridocarbonyl complex $(\eta^5\text{-}C_5H_5)NbH(CO)$). The mechanism of the formation of $(\eta^5\text{-}C_5H_5)_2NbH_3$ by reducing $(\eta^5\text{-}C_5H_5)_2NbCl_2$ is complex and appears to involve disproportionation of a hydridoniobium(IV) intermediate.

Catalysis of reactions of carbon monoxide and hydrogen (synthesis gas) is a subject of current interest because of the growing importance of effective coal utilization, from which synthesis gas is obtained. In the near future, it will be necessary to extensively substitute coal for oil not only as an energy source but also as a source for petrochemicals (*1*). Presently, there are several processes for transforming synthesis gas into organic chemicals that are based upon heterogeneous catalysts: methanol synthesis, methanation, and the Fischer–Tropsch process. Except for two recent reports (*see below*), no homogeneous system capable of effecting any reaction of CO with H_2 has been reported.

A homogeneous system for catalysis of such reactions could have both fundamental and practical significance. Such a system would be more amenable

0-8412-0390-3/78/33-167-149/$05.00/0

to mechanistic studies than the heterogeneous catalysts and hence might be useful as a model for the latter, although it is not clear whether there should be any extensive mechanistic similarity between reactions of these different types of catalysts. More importantly, a homogeneously catalyzed process offers greater possibilities for product selectivity and control, which would be of considerable value in improving the economic feasibility of obtaining energy and especially petrochemicals and precursors of it from coal.

In trying to construct a homogeneous system capable of effecting a heterogeneously catalyzed process, two approaches appear attractive. One approach is to try to incorporate any features unique to a heterogeneous catalyst into a soluble species—specifically, the potential involvement of more than one metal atom in the active site. This approach suggests studying metal cluster complexes as potential catalysts; the merits of such systems were outlined by Muetterties (2). In fact, some preliminary successes have been achieved along these lines. Rhodium carbonyl clusters catalyze the formation of alcohols, especially ethylene glycol, from hydrogen and carbon monoxide (3) while $Ir_4(CO)_{12}$ in molten $NaCl\cdot2AlCl_3$ leads to alkane formation with selectivity for ethane (4). However, neither of these appears to be of immediate practical use; the first requires excessively high pressures while in the second, water that is formed as a reduction byproduct will react with the medium ($AlCl_3$).

The other approach, that is followed here, involves proposing a hypothetical mechanism for the homogeneously catalyzed process using only steps for which there is good precedent in organo-transition metal chemistry and involves choosing a system to optimize those steps that appear to be potentially the most troublesome. A reasonable pathway for reducing CO to methanol is shown in Scheme 1. Steps (a) and (d) show oxidative addition of H_2 to a coordinatively unsaturated metal center, a well-known process (5). Step (b) is the insertion of

Scheme 1

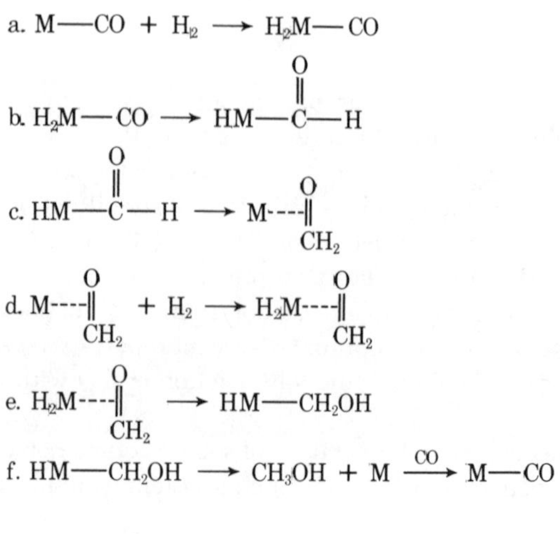

a. M—CO + H_2 $\longrightarrow$ H_2M—CO

b. H_2M—CO $\longrightarrow$ HM—$\overset{\displaystyle O}{\overset{\displaystyle \|}{C}}$—$H$

c. HM—$\overset{\displaystyle O}{\overset{\displaystyle \|}{C}}$—$H$ $\longrightarrow$ M----$\overset{\displaystyle O}{\underset{\displaystyle CH_2}{\|}}$

d. M----$\overset{\displaystyle O}{\underset{\displaystyle CH_2}{\|}}$ + H_2 $\longrightarrow$ H_2M----$\overset{\displaystyle O}{\underset{\displaystyle CH_2}{\|}}$

e. H_2M----$\overset{\displaystyle O}{\underset{\displaystyle CH_2}{\|}}$ $\longrightarrow$ HM—CH_2OH

f. HM—CH_2OH $\longrightarrow$ CH_3OH + M $\xrightarrow{CO}$ M—CO

CO into a metal–hydrogen bond, apparently analogous to the common insertion of CO into a metal–alkyl bond (*6*). Step (c) is the reductive elimination of an acyl group and a hydride, observed in catalytic decarbonylation of aldehydes (*7,8*). Steps (d–f) correspond to catalytic hydrogenation of an organic carbonyl compound to an alcohol that can be achieved by several mononuclear complexes (*9,10*). Schemes similar to this one have been proposed for the mechanism of CO reduction by heterogeneous catalysts, the latter considered to consist of effectively separate, one-metal atom centers (*11,12*). As noted earlier, however, this may not be a reasonable model.

By examining the chemistry of systems relevant to these various steps, in spite of the apparent close similarity to the alkyl analog, one finds that no example of the insertion of CO into an M–H bond has ever been demonstrated. This suggests that the formation of the first C–H bond in homogeneous CO reduction by such an insertion step might be a highly unfavorable process and that some other mode of bond formation must be sought. One possibility would be via nucleophilic attack by H^- on coordinated CO. Indeed, several examples have been reported wherein treatment of a metal carbonyl with a boron or aluminum hydride reduces the CO to a formyl, hydroxymethyl, or methyl group, all attractive intermediates for a catalytic CO reduction process ($Cp = \eta^5\text{-}C_5H_5$):

$$Fe(CO)_5 + LiAlH_4 \rightarrow HC(O)Fe(CO)_4^- \qquad (13)$$

$$Fe(CO)_4(P(OPh)_3) + KHB(OR)_3 \rightarrow HC(O)Fe(CO)_3(P(OPh)_3)^- \qquad (14)$$

$$CpRe(CO)_2(NO)^+ + NaBH_4 \xrightarrow{\text{aq}} CpRe(CO)(NO)(CH_2OH) \qquad (15)$$

$$CpRe(CO)_2(NO)^+ + NaBH_4 \xrightarrow{\text{anhyd}} CpRe(CO)(NO)(CH_3) \qquad (16)$$

$$CpMo(CO)_3(PPh_3)^+ + NaBH_4 \rightarrow CpMo(CO)_2(PPh_3)(CH_3) \qquad (17)$$

Of course, main-group metal hydrides such as these cannot be incorporated into a useful catalytic scheme since they do not form readily from H_2. The following, apparent criteria for a suitable catalyst are based on the above discussion.

(1) Hydride complexes must be formed directly from H_2 under catalytic conditions. This probably requires a transition metal hydride.

(2) The complex must have substantial hydridic character; that is, it must be able to react as $M^+\text{-}H^-$.

Most transition metal hydrides do not exhibit such behavior since they tend to be covalent or acidic in nature, but several examples of hydridic behavior have been reported:

$$(CpTiH)_2(C_{10}H_8) + 2H_2O \rightarrow (CpTi(OH))_2(C_{10}H_8) + 2H_2 \qquad (18)$$

$$Cp_2ZrHCl + ROH \rightarrow Cp_2ZrCl(OR) + H_2 \qquad (19)$$

$$Cp_2ZrH_2 + CH_3COCH_3 \rightarrow Cp_2Zr(OCH(CH_3)_2)_2 \qquad (20)$$

These examples all involve group IVa metal complexes, suggesting that the desired behavior might be possibly an attribute of early transition metal hydride complexes, an area that has been studied considerably less than complexes of metals that are further to the right of the early transition metals on the periodic table. Indeed, as discussed in this volume (see Chapter 10), a zirconium hydride complex recently has been found to reduce CO to methanol among other products.

However, for catalytic reduction of CO, there is one additional problem: the products of CO reduction (alcohols or water) react readily with these strongly hydridic group IVa complexes. This would lead to self-destruction of the catalyst after a single cycle. Hence a third criterion must be added:

(3) The hydride must not react irreversibly with alcohols or water under catalytic conditions.

Ideally one would like reactivity similar to that of $NaBH_4$, that readily reduces organic carbonyl groups at room temperature but that reacts very slowly with alcohols and water.

We investigated niobium hydride complexes as possible catalysts by assuming (possibly naively) that a property exhibited strongly by group IVa metals might be present to a lesser degree in group Va. Several such complexes are known and apparently satisfy criteria (1) and (3): they are stable to water and alcohols, as seen during their preparation (see below), and they were active in hydrogen-exchange reactions (21), suggesting that they can activate H_2 directly. We have observed the desired hydridic behavior, but we have not demonstrated that these species can in fact reduce CO. Some results bearing on the mechanism of formation of these hydrides are also presented below.

Experimental

All manipulations were carried out in an inert atmosphere, using Schlenk techniques (22) or a nitrogen-filled dry box (Vacuum Atmospheres). Solvents were purified by distillation from sodium benzophenone ketyl under argon. $NbCl_5$ was obtained from Alfa Products, Ventron Corp., and was used without further purification. NMR spectra were measured on Varian A-60 and XL-100 spectrometers.

Preparation of Cp_2NbH_3 and Derivatives. Cp_2NbCl_2 prepared by the method of Lucas and Green (23,24) was suspended in benzene and treated with an excess of $NaAlH_2(OCH_2CH_2OCH_3)_2$ (Red-al, available as a 70% benzene solution from Aldrich Chemicals). The resulting dark brown solution was cooled in ice and hydrolyzed with degassed water. The benzene layer was decanted, filtered through anhydrous sodium sulfate, and evaporated, giving a brown solid product in 70–90% yield. Spectral properties were identical to those previously reported (25).

This trihydride was converted to $Cp_2NbH(CO)$ by heating it in benzene solution for 1 hr at 80°C, under 1–2 atm CO, followed by removal of the solvent.

This gave a brown, solid hydridocarbonyl in 90–95% yield. Refluxing the trihydride in benzene with excess PMePh$_2$ gave previously unreported Cp$_2$NbH(PMePh$_2$) in low yield. Its spectroscopic properties (NMR peaks at 5.58 τ, double doublet (J_{PH} = 2.0 Hz; J_{HH} = 0.4 Hz), C$_5$H$_5$; 8.33 τ, doublet (J_{PH} = 6.0 Hz), P–CH$_3$; and 17.70 τ, broad doublet (J_{PH} = 29 Hz), Nb-H) agree well with values reported for the PMe$_2$Ph (*24*) and PEt$_3$ (*25*) analogs.

Reactions with Acetone. Reagent-grade acetone was dried further by refluxing with P$_2$O$_5$ under argon, and distilled in vacuo directly onto a dry sample of the hydridoniobium complex. A weighed amount of toluene was added as a reference for quantitative NMR determinations, and the resulting solution was transferred by syringe to a serum-capped, argon-flushed NMR tube. The reactions were followed by NMR (*see* "Results and Discussion"). After the starting hydride complex disappeared completely, the solution was hydrolyzed with a small amount of degassed water, and the 2-propanol yield was determined by NMR.

Reaction of Cp$_2$NbH(CO) with H$_2$. A solution of Cp$_2$NbH(CO) (prepared as described above) in benzene that contained a weighed amount of toluene was heated under 1–2 atm of H$_2$ at 100°–130°C in a Fischer–Porter bottle. Progress of the reaction was monitored by periodically cooling, venting, removing a small aliquot, and recording its NMR spectrum. The signals for Cp$_2$NbH(CO) decreased steadily, with a half-life of about one day at 130°C; only traces of new soluble products could be observed by NMR. Accompanying this decrease of starting material was the formation of a brown precipitate in the reaction vessel. After decanting supernatant solution, treatment of this precipitate (that was highly air-sensitive) with water resulted in a change in appearance (lighter in color, more granular) and formation of methanol, established by comparison of NMR (in D$_2$O), VPC, and mass spectral properties to an authentic sample. The yield of methanol (determined by NMR in D$_2$O, with dioxane added as reference) varied from 3 to 15% based on Cp$_2$NbH(CO) consumed.

Preparation of Cp$_2$NbH(^{13}CO). A break-seal ampoule containing 100 mL of ^{13}CO (90.5% ^{13}C, obtained from Merck) was sealed to a small flask equipped with a high-vacuum O-ring stopcock. The flask was charged with 650 mg Cp$_2$NbCl$_2$, 2 mL benzene, and 1.5 mL Red-al, was frozen and evacuated, and the stopcock was closed off. The break-seal was broken, and the flask was heated at 80°C for several hours, and the resulting brown solution worked up as above. The product had spectral properties similar to the unlabeled carbonyl, with CO appearing at 1860 cm^{-1} (vs. 1900 cm^{-1} for ^{12}CO) and the NMR showing coupling of both the Cp (a doublet, with J_{HH} = 0.6 Hz, for ^{12}CO and a triplet, with J_{HH} $\approx$ J_{CH} $\approx$ 0.6 Hz for ^{13}CO) and the Nb–H (broad singlet, with half-width of ca. 18 Hz for ^{12}CO and 24 Hz for ^{13}CO) protons with the ^{13}C nucleus. On heating the labeled compound under H$_2$, exactly the same behavior as above was observed. However, the methanol produced was found to be entirely ^{12}CH$_3$OH by NMR and mass spectral analyses.

Reaction of Cp$_2$NbCl$_2$ with LiAlH$_4$. A suspension of Cp$_2$NbCl$_2$ in THF was treated with an excess of solid LiAlH$_4$, resulting in an orange solution with suspended gray solid. The NMR of the supernatant solution (*see* Table I) showed an equimolar mixture of Cp$_2$NbH$_3$ and a new product, identified by NMR as Cp$_2$NbH$_2$AlH$_2$ (*see* "Results and Discussion"); on standing, the former was gradually transformed into the latter. Evaporation of THF and extraction with benzene gave orange Cp$_2$NbH$_2$AlH$_2$ in ca. 65% yield. (This would be the only product even if the evaporation and extraction were carried out immediately after reaction.) Hydrolysis of a benzene solution of the latter gives virtually quantitative conversion to Cp_2NbH_3 that was obtained, after drying and evaporation, as a white crystalline solid. Aside from the difference in color, this product appears identical to that prepared by the method described above. Conversion of this sample to $Cp_2NbH(CO)$ gives a very pale, violet product, which upon heating under conditions described above gives no methanol.

When the same procedure was carried out with $LiAlD_4$, the THF solution showed the same NMR spectrum in the Cp region but no peaks that were attributable to Nb–H, as expected. However, on standing, the peak assigned to $Cp_2NbH_2AlH_2$ began to grow very slowly. Evaporation and benzene extraction gave a product with a substantial peak at 19.8 τ, that continued to grow until the ratio of its integral to that for the Cp signal was 2:10. The intensity of the Cp peak (relative to added toluene) decreased concurrently, roughly one-fifth as rapidly. From the intensity of the Nb–H peak 15 min after extraction into benzene, the rate of the exchange process was estimated (neglecting any back-exchange or possible isotope effects) at ca. 6×10^{-4} sec^{-1}, giving $\Delta G^{\ddagger} \approx 22$ kcal/mol at room temperature.

Reduction by $NaAlH_2Et_2$ was studied similarly, using an excess of the reagent (toluene solution, from Alfa) added to a benzene suspension of Cp_2NbCl_2. The niobium-containing product could not be separated from excess $NaAlH_2Et_2$, as they both appear to have very similar solubility properties.

Trapping Experiments. Cp_2NbCl_2 was suspended in a solution of benzene containing a weighed amount of $PMePh_2$, with the ratio of phosphine to niobium varying from less than one to a large excess. The solution was treated with excess Red-al, and the composition of the resulting solution was determined by NMR, with yields obtained by adding a weighed amount of toluene as reference. The product distribution was essentially independent of whether or not excess Red-al was hydrolyzed before NMR analysis. A similar trapping experiment was carried out using $LiAlH_4$ as reductant in THF. Also, trapping by CO or C_2H_4 was achieved by reducing the niobium complex under 1 atm of the appropriate gas. In all cases, the product was identified by comparing NMR signals for the Cp and Nb–H protons as well as ligand signals for $PMePh_2$ and C_2H_4 with those of authentic samples (25).

Results and Discussion

Hydridic Nature of Niobium Hydrides. The hydridoniobium complexes, Cp_2NbH_3 and Cp_2NbHL, are stable to water and alcohols at room temperature and hence do not exhibit the strongly hydridic behavior of group IVa hydrides that was discussed earlier. However, both Cp_2NbH_3 and $Cp_2NbH(CO)$ react slowly with acetone at room temperature and thus are somewhat hydridic, although the reactions are rather complex. Cp_2NbH_3 reacts within about one day; disappearance of the NMR signals for the starting material is accompanied by formation of 1 mol of free cyclopentadiene and smaller amounts of 2-propanol. No new peaks corresponding to any product niobium complex are observed, although the dark brown color of the solution indicates such species are present. After hydrolysis the NMR shows only 1 mol of cyclopentadiene and ca. 2 mol of 2-propanol per niobium. With $Cp_2NbH(CO)$, basically similar behavior is observed, but an intermediate that appears to be $Cp_2Nb(OCH(CH_3)_2)(CO)$ can be seen by NMR (C_5H_5, 4.77 τ, singlet; $OCH(CH_3)_2$, 6.75 τ, multiplet; $CH(CH_3)_2$,

9.40 τ, doublet, $J = 6.0$ Hz).   This product increases up to about 25% of the total niobium (this amount varies somewhat) and then decreases.   After complete consumption of starting material, the only NMR-detectable products are cyclopentadiene and 2-propanol (1 mol each per niobium).

Thus, the niobium hydrides in fact do exhibit the desired hydridic character; they can reduce acetone but cannot react with CO reduction products.   From the color, the absence of any NMR, and the formation of 1 mol of free cyclopentadiene, the niobium-containing product appears to be a monocyclopentadienyl Nb(IV) complex.   The rather facile removal of one of the Cp groups appears somewhat surprising but has been observed in related systems.   For example, alcoholysis of Cp_2TiCl_2 gives $CpTiCl_2(OEt)$ plus cyclopentadiene, even at room temperature (26).

Source of Methanol in the Reaction of $Cp_2NbH(CO)$ with H_2.   The hydridocarbonyl complex is thermally quite stable, having a half-life of about one day at 130°C.   When heated at this temperature under H_2, no soluble products

Table I.   NMR Parameters of Hydridoniobium Complexes

Complex	$C_5H_5{}^a$	$Nb{-}H^a$
$Cp_2NbH_3{}^b$	5.17	12.6(1), 13.5(2)
$Cp_2NbH(CO)^b$	5.42	16.5(1)
$Cp_2NbH(PMePh_2)$	5.58	17.7(1)
$Cp_2NbH_2AlH_2$	5.15 [5.47]^c	19.8(2) [21.1]^c
$Cp_2NbH_3 \cdot AlHEt_2$	5.12	14.5(2), 15.0(1)
$Cp_2NbH_2AlEt_2{}^b$	5.37	21.4(2)

^{a} Shifts measured in benzene solution on the τ scale; numbers in parentheses are approximate integral of Nb–H signal relative to Cp = 10.
^{b} Also reported in Ref 22.
^{c} In THF solution.

are formed.   The decrease in starting material is accompanied by formation of a brown precipitate.   When this extremely air-sensitive solid reacts with water, it changes in appearance and liberates small amounts of methanol.   The methanol yield (based on consumed niobium complex) ranged from 3 to 15%.   This is small enough that the possibility of the methanol arising from an impurity needs to be considered, even though both the Cp_2NbH_3 and $Cp_2NbH(CO)$ were free of NMR-detectable impurities.   By preparing the ^{13}CO-labeled complex, it was shown readily that the methanol produced contained no ^{13}C; therefore, it is not produced by CO reduction.   The remaining possible sources are the presence of an impurity or degradation of the Cp rings.   It has been shown that degradation products such as ethylene can be obtained from the Cp rings of $Cp_2Ti(CH_3)_2$ at temperatures comparable with those used here (27).

Since the most likely impurity leading to methanol would come from the reducing agent Red-al that contains methoxy groups, a sample of Cp_2NbH_3 was prepared using $LiAlH_4$ as reductant (*see* below).   When this sample was converted to $Cp_2NbH(CO)$, heated under H_2, and the precipitate treated with water, no methanol was observed. This suggests strongly that the methanol is an artifact resulting from some Red-al derived impurity that was not removed from the hydridoniobium complexes. The Cp_2NbH_3 and $Cp_2NbH(CO)$, prepared using $LiAlH_4$, were white and pale violet, respectively, while both products prepared using Red-al were brown. The latter preparations probably contained some Nb(IV) complex that has alkoxide groups from Red-al. Such a species would probably be brown (as is Cp_2NbCl_2) but would not show any NMR signal because of its paramagnetism.

Mechanism of the Reduction of Cp_2NbCl_2. The reaction of Cp_2NbCl_2 with Red-al results in a net increase in the formal oxidation state of niobium, from Nb(IV) to Nb(V). This result is frequently observed when preparing hydrido complexes from halides—reaction of a halo complex L_nMCl_x with a reducing agent such as $LiAlH_4$ or $NaBH_4$ yields L_nNH_y where $y > x$. (Several examples can be found elsewhere in this volume.) Little is known about how such transformations take place. The only previously reported reduction of Cp_2NbCl_2 is the reaction with $NaBH_4$ that yields a Nb(III) complex $Cp_2NbH_2BH_2$ (23).

Because of the virtual insolubility of Cp_2NbCl_2 in organic solvents, kinetic studies cannot be used. However, NMR monitoring of the course of the reactions with several reductants showed intermediate products (*see* Table I) that suggest some interesting mechanistic possibilities:

(1) When $LiAlH_4$ is used as reducing agent, the reaction mixture (in THF) initially contains an equimolar mixture of Cp_2NbH_3 and a second product that is characterized by a high-field NMR signal at 21.1 τ with intensity one-fifth the Cp signal. A similar high-field signal has been observed by Tebbe for complexes wherein the hydride bridges between niobium and aluminum, specifically the complex $Cp_2NbH_2AlEt_2$, shows a similar signal at 22.8 τ (28). Based on this analogy, the new product is assigned structure $Cp_2NbH_2AlH_2$. This structure is similar to that of the borohydride complex mentioned above. Also, the related Ti(III) complex $Cp_2TiH_2AlH_2$ has been observed by EPR (29). After allowing the THF solution to stand, complete conversion to $Cp_2NbH_2AlH_2$ gradually occurs; this is also observed when the THF is evaporated, and the product is extracted into benzene.

Analogous behavior is observed when $NaAlH_2Et_2$ is used as reductant, except that instead of the high-field signals typical of Cp_2NbH_3, peaks are observed at 14.5 and 15.0 τ in a 2:1 ratio. The position of the Cp peak is the same as that for Cp_2NbH_3. This pattern resembles that observed for $Cp_2NbH_3\cdot AlEt_3$, in which the triethylaluminum group is coordinated to the central hydrogen on niobium and shifts the hydride resonances to 13.6 and 15.3 τ while the Cp peak is unchanged (28,30). Accordingly, these peaks are assigned as $Cp_2NbH_3\cdot AlHEt_2$. Again, on standing, complete conversion to $Cp_2NbH_2AlEt_2$ is observed.

(2) When Cp_2NbCl_2 is reduced with Red-al in the presence of potential ligands such as $PMePh_3$, ethylene, or CO, a substantial amount of the product is Cp_2NbHL, in addition to the expected Cp_2NbH_3. These Nb(III) complexes are not formed by subsequent reaction with the trihydride with L: such reactions are insignificant at room temperature, taking place only around 80°C. Hence, a species that is trapped by L to give Cp_2NbHL must be an intermediate in the

reaction.   By using varying concentrations of PMePh$_2$ up to 50%, but no more, of the total product is Cp$_2$NbH(PMePh$_2$), even if a large excess of trapping ligand is used.

(3) When LiAlH$_4$ is used as reductant in the presence of PMePh$_2$, all three products, Cp$_2$NbH$_3$, Cp$_2$NbH(PMePh$_2$), and Cp$_2$NbH$_2$AlH$_2$, are formed.   In contrast to Cp$_2$NbH$_2$BH$_2$, which reacts readily with phosphines to give Cp$_2$NbHL (23), Cp$_2$NbH$_2$AlH$_2$ does not react with PMePh$_2$, so again an intermediate is being trapped.

Reduction of a Nb(IV) complex initially leads to an equimolar mixture of Nb(III) and Nb(V), which suggests a disproportionation step.   Scheme 2 presents a possible mechanism to account for these observations.   In this scheme, a key intermediate is proposed to be Cp$_2$NbH$_2$.   This species has been observed by EPR at low temperature but is unstable, decomposing rapidly even below 0°C (31,32). Disproportionation would yield equimolar quantities of Cp$_2$NbH$_3$ and the coordinatively unsaturated Cp$_2$NbH, which should be trapped rapidly by an

Scheme 2

IV   IV
Cp$_2$NbCl$_2$ + 2MH$_2$AlR$_2$ $\longrightarrow$ Cp$_2$NbH$_2$ + 2HAlR$_2$ + 2MCl

$\downarrow$

V   III
Cp$_2$NbH$_3$ + [Cp$_2$NbH]

[Cp$_2$NbH] $\xrightarrow{\text{HAlR}_2}$ Cp$_2$NbH$_2$AlR$_2$ (R = H, Et)
$\xrightarrow{?}$ Cp$_2$NbH$_3$ (R = OCH$_2$CH$_2$OCH$_3$)
$\xrightarrow{L}$ Cp$_2$NbHL

Overall stoichiometry.

2Cp$_2$NbCl$_2$ + 4MH$_2$AlR$_2$ $\longrightarrow$ Cp$_2$NbH$_3$ + Cp$_2$NbH$_2$AlR$_2$ + 4MCl + 3HAlR$_2$

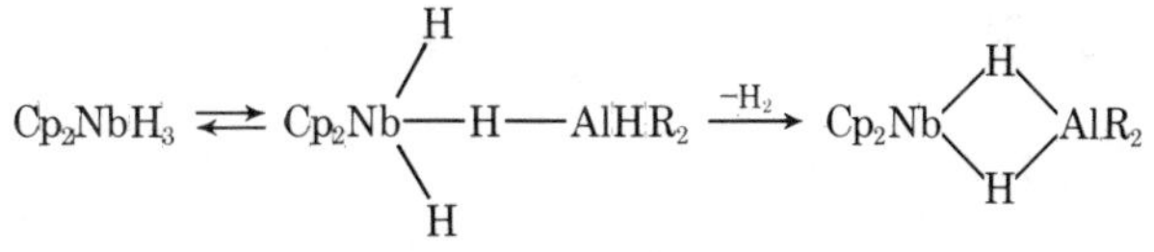

added ligand.   In the absence of an added ligand, it would be trapped by the aluminum hydride by-product AlHR$_2$ to give the bridged species Cp$_2$NbH$_2$AlR$_2$. Subsequent reaction of AlHR$_2$ with Cp$_2$NbH$_3$ and the elimination of hydrogen would account for the gradual conversion to the bridged product.   With AlHEt$_2$, this reaction is quite slow; the decomposition of Cp$_2$NbH$_3$·AlEt$_3$ to Cp$_2$NbH$_2$AlEt$_2$ is rapid above −40°C (28).   Steric factors possibly are responsible for the difference in stability.

Scheme 2 fails to explain how Red-al gives Cp$_2$NbH$_3$ as the sole product in good yield.   Assuming the same basic mechanism is followed, as is indicated by the trapping experiments, there must be some facile route for converting the Nb(III) intermediate Cp$_2$NbH$_2$Al(OCH$_2$CH$_2$OCH$_3$)$_2$ to the trihydride.   Possibly, the availability of relatively reactive hydrogen atoms on the alkoxide groups might be involved; labeling studies are being designed to try to resolve this question.

Hydrogen Atom Exchange in $Cp_2NbH_2AlH_2$. The Nb(III) complex $Cp_2NbH_2AlH_2$ upon hydrolysis is converted cleanly and quantitatively to Cp_2NbH_3. We attempted to prepare the deuterated analog, $Cp_2NbD_2AlD_2$, in hopes that the isotopic substitution in the trihydride would provide some information on the course of this reaction. Surprisingly however, the product isolated from the reaction of Cp_2NbCl_2 with $LiAlD_4$ had an NMR spectrum identical to the undeuterated analog: the high-field hydride signal was present with ca. one-fifth the intensity of the Cp signal. Clearly, some exchange process must be taking place.

Upon examining the NMR of the THF solution immediately after reaction, no high-field signals were present; the peak characteristic of $Cp_2NbH_2AlH_2$ grew very slowly over several hours. Evaporation of THF and extraction into benzene gave a sample whose spectrum showed a much larger high-field peak, that continued to grow until it reached its intensity of one-fifth the Cp peak. The latter decreased over this period in comparison with an added standard; the rate of decrease was approximately one-fifth the rate of increase of the high-field signal. These results suggest that the two bridging hydrides are exchanging with the Cp hydrogens. A similar process has been observed in the related species $Cp_2M(BH_4)_2$ (M = Zr, Hf), although in these compounds all four BH_4 hydrogens exchange (33). In the present case, the hydrogens bonded only to aluminum cannot be observed, but the relative rates of peak integral changes suggest that only the bridging hydrogens are involved.

The exchange process appears to be much more facile in benzene than in THF; presumably, THF coordination blocks a vacant site needed for exchange. From the extent of exchange over the first few minutes after extraction into benzene, a rough rate constant of $6 \times 10^{-4} \ sec^{-1}$ can be estimated, leading to a $\Delta G^{\ddagger}$ value ≈ 22 kcal/mol at room temperature. This can be compared with the value of 21 kcal/mol at 395 °K that was found by NMR line broadening for $Cp_2Zr(BH_4)_2$ (33).

Acknowledgment

We thank Fred Tebbe for helpful discussions and communication of results prior to publication. We also acknowledge the donors of the Petroleum Research Fund, administered by the ACS, for partial support of this research.

Literature Cited

1. Wender, I., *Catal. Rev.* (1976) **14**, 97.
2. Muetterties, E. L., *Bull. Soc. Chim. Belg.* (1975) **84**, 959.
3. Walker, W. E., Pruett, R. L., Belgian Patent 793, 086.
4. Demitras, G. C., Muetterties, E. L., *J. Am. Chem. Soc.* (1977) **99**, 2796.
5. Deeming, A. J., MTP Int. Rev. of Science, *Inorg. Chem. Ser. 1* (1972) **9**, 118.
6. Wojcicki, A., *Adv. Organomet. Chem.* (1973) **11**, 87.
7. Osborn, J. A., Jardine, F. H., Young, J. F., Wilkinson, G., *J. Chem. Soc. A* (1966) 1711.

8. Ohno, K., Tsuji, J., *J. Am. Chem. Soc.* (1968) **90**, 99.
9. Coffey, R. S., *J. Chem. Soc., Chem. Commun.* (1967) 923.
10. Schrock, R. R., Osborn, J. A., *J. Chem. Soc., Chem. Commun.* (1970) 567
11. Henrici-Olive, G., Olive, S., *Angew. Chem. Int. Ed. Engl.* (1976) **15**, 136.
12. Pichler, H., Schulz, H., *Chem. Ing. Tech.* (1970) **42**, 1162.
13. Siegl, W., *J. Organomet. Chem.* (1975) **92**, 321.
14. Casey, C. P., Neumann, S. M., *J. Am. Chem. Soc.* (1976) **98**, 5395.
15. Nesmeyanov, A. N., Anisimov, K. N., Kokobova, N. E., Krasnoslobodskaya, L. L., *Izv. Akad. Nauk SSSR, Ser. Khim.* (1970) 860.
16. Stewart, R. P., Okamoto, N., Graham, W. A. G., *J. Organomet. Chem.* (1972) **42**, C32.
17. Treichel, P. M., Shubkin, R. L., *Inorg. Chem.* (1967) **6**, 1329.
18. Guggenberger, L. J., Tebbe, F. N., *J. Am. Chem. Soc.* (1976) **98**, 4137.
19. Schwartz, J, Labinger, J. A., *Angew. Chem. Int. Ed. Engl.* (1976) **15**, 333.
20. Wailes, P. C., Weigold, H., *J. Organomet. Chem.* (1970) **24**, 413.
21. Parshall, G. W., *Acc. Chem. Res.* (1970) **3**, 139.
22. Shriver, D. F., "The Manipulation of Air-Sensitive Compounds," McGraw-Hill, New York, 1969.
23. Lucas, C. R., Green, M. L. H., *J. Chem. Soc., Chem. Commun.* (1972) 1005.
24. Lucas, C. R., *Inorg. Synth.* (1976) **16**, 107.
25. Tebbe, F. N., Parshall, G. W., *J. Am. Chem. Soc.* (1973) **95**, 3793.
26. Bharara, P. C., *J. Organomet. Chem.* (1976) **121**, 199.
27. Alt, H. G., di Sanzo, F. P., Rausch, M. D., Uden, R. C., *J. Organomet. Chem.* (1976) **107**, 257.
28. Tebbe, F. N., *J. Am. Chem. Soc.* (1973) **95**, 5412.
29. Henrici-Olive, G., Olive, S., *J. Organomet. Chem.* (1970) **23**, 155.
30. Tebbe, F. N., personal communication.
31. Elson, I. H., Kochi, J. K., Klabunde, U., Manzer, L. E., Parshall, G. W., Tebbe, F. N., *J. Am. Chem. Soc.* (1974) **96**, 7374.
32. Elson, I. H., Kochi, J. K., *J. Am. Chem. Soc.* (1975) **97**, 1262.
33. Marks, T. J., Kolb, J. R., *J. Am. Chem. Soc.* (1975) **97**, 3397.

RECEIVED July 19, 1977.

12

Reactivity Patterns in the Formation of Platinum (II) Hydrides by Protonation Reactions

D. MAX ROUNDHILL

Department of Chemistry, Washington State University, Pullman, WA 99164

Protonation reactions of Pt(PPh$_3$)$_3$ with acids HX give a wide product range. When the conjugate base X$^-$ only poorly coordinates to Pt(II), compounds of type [PtH(PPh$_3$)$_3$]X are formed. The second-order high field ^{1}H NMR spectrum and crystal structure of [PtH(PPh$_3$)$_3$](CF$_3$CO$_2$)$_2$H show the compound to be a planar Pt(II) complex. When the conjugate base strongly coordinates to Pt(II), the products are of the type trans-PtHX(PPh$_3$)$_2$. With 1-ethynylcyclohexanol the product is the Pt(IV) dihydride PtH$_2$(C≡CC$_6$H$_{10}$(OH)$_2$)(PPh$_3$)$_2$. Compounds HSCH$_2$CH$_2$SMe and HSCH$_2$CH$_2$SCH$_2$CH$_2$-CH$_2$SMe react with Pt(PPh$_3$)$_3$ to give hydrides PtH(SCH$_2$-CH$_2$SMe)PPh$_3$ and PtH(SCH$_2$CH$_2$SCH$_2$CH$_2$CH$_2$SMe)PPh$_3$, respectively. In the latter compounds, the terminal thioether is uncoordinated. Complexes PtH(OPPh$_2$)(HOPPh$_2$)L (L = PPh$_3$, PMePh$_2$) result from the oxidative addition of diphenylphosphine oxide to PtL$_3$. Data are presented on the ^{1}H and ^{31}P NMR spectra of PtH(OPPh$_2$)(HOPPh$_2$)PMePh$_2$ and PtH(OPPh$_2$)(F$_2$BOPPh$_2$)PMePh$_2$.

Hydrides of Pt(II) are the most numerous of any transition metal hydride group. In addition to the presence of the hydride ligand, the complexes invariably have a coordinated phosphine, and synthetic routes to these compounds using both hydridic and protonic reagents have been reported (1). The pure complexes are usually both air stable and kinetically inert. The purpose of this chapter is to show the diversity of hydrides that can be obtained from protonation reactions on zero-valent and di-valent triphenylphosphine platinum compounds, and to rationalize the type and nature of the product formed from the character of the acid HX.

0-8412-0390-3/78/33-167-160/$05.00/0

The zero-valent compound Pt(PPh₃)₃ undergoes a wide range of chemistry with protonic acids. The initial product in the protonation with acids HX is the ionic compound [PtH(PPh₃)₃]X. Under suitable conditions, this complex is the favored product and can be isolated in high yield. Such a situation exists when the acid HX is strong and the conjugate base X⁻ is a sufficiently poor ligand for Pt(II); it will not undergo subsequent substitution reactions with triphenylphosphine. Examples of such conjugate bases are trifluoroacetate (2), sulfate and fluoroborate (3), and picrate (4). However, when an acid is used that is weak and has a conjugate base that only poorly coordinates to Pt(II), such as acetic acid, the equilibrium in (Reaction 1) can be shifted only to the hydride product in the presence of an excess of acid, and the hydride product is not isolable from solution. For HCl, both an ionic, [PtH(PPh₃)₃]Cl, and a covalent, *trans*-PtHCl(PPh₃)₂ hydride have been isolated. The former type is obtained in donor solvents of high dielectric constant, and the latter formed in solvents such as benzene (3).

$$Pt(PPh_3)_3 + HX \rightleftharpoons [PtH(PPh_3)_3] \overset{-PPh_3}{\rightleftharpoons} trans\text{-}PtHX(PPh_3)_2 \qquad (1)$$

The ¹H NMR spectra of complexes [PtH(PPh₃)₃]X have been interesting (2, 5) because of the apparent difference in the spin–spin coupling patterns between the center line and the ¹⁹⁵Pt-satellite portion of the high field resonance. This spectrum shows the hydride resonance centered 5.83 ppm upfield of Me₄Si with ²J(H–P(trans)) of 164 Hz and coupled to two apparently nonequivalent cis-triphenylphosphines with ²J(H–P(cis)) of 13 and 17 Hz. The ¹⁹⁵Pt satellite portion of the spectrum appears as a doublet of triplets with ²J(H–P(trans)) of 160 Hz and ²J(H–P(cis)) of 13 Hz (Figure 1). At that time, we considered that this additional multiplicity was caused by second-order effects involving ¹J(P–Pt) and ²J(P–P) and not by strong ion pairing between the cation and anion in the complex leading to chemically inequivalent phosphorus nuclei in pyramidal geometry (2). To verify this premise, we have determined the structure of the

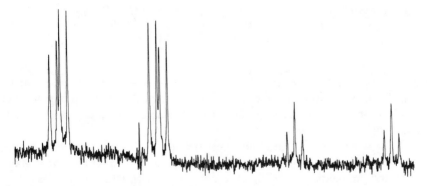

Figure 1. Center multiplet and upfield satellite of the ¹H NMR spectrum of [PtH(PPh₃)₃](CF₃CO₂)₂H

compound [PtH(PPh$_3$)$_3$](CF$_3$CO$_2$)$_2$H in the solid state. The molecule is strictly planar around platinum, and the anion is too far away from the metal center to be involved in coordination (6). The bond angles for P–Pt–P are 99.6(2)° and 100.6(2)°, with distances of 2.315(7) and 2.309(7)Å for the Pt–P distances of the mutually trans-triphenylphosphines (Figure 2). The Pt–P distance of the triphenylphosphine trans to the hydride is significantly longer at 2.363(7)Å, reflecting the high trans influence of the hydride ligand. This structure confirms earlier ideas that the multiplicity of the high-field ^{1}H NMR pattern is a consequence of second-order effects (2, 7). The planar geometry for the ion

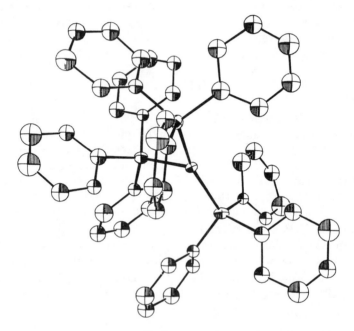

Figure 2. Structure of the cation in [PtH(PPh$_3$)$_3$]-(CF$_3$CO$_2$)$_2$H showing the arrangement of the three triphenylphosphines about Pt(II)

[PtH(PPh$_3$)$_3$]$^+$ also shows that the species is designated correctly as a complex of Pt(II) since, if a formal two-electron transfer from Pt(0) to the proton had not occurred, the geometry of a protonated Pt(0) complex should be a distorted tetrahedron.

The second step in Reaction 1 occurs when X$^-$ is a strongly coordinating ligand for Pt(II). Consequently, complexes of type *trans*-PtHX(PPh$_3$)$_2$ are formed when the protonation of Pt(PPh$_3$)$_3$ is carried out with HCN(3), imides (8), or thioacids (9). An alternate route to these complexes *trans*-PtHX(PPh$_3$)$_2$, however, is to treat the compound Pt(PPh$_3$)$_2$(C$_2$H$_4$) with the appropriate acid HX.

$$Pt(PPh_3)_3 \ + \ 2HC\equiv C-\overset{HO}{\underset{}{}} \xrightarrow{-PPh_3} \qquad\qquad\qquad\qquad \tag{2}$$

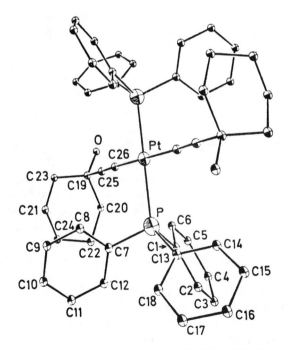

This latter procedure is favored for hydrides where X only poorly coordinates to Pt(II). The hydride complexes from $[PtH(PPh_3)_3]X$ have been investigated further with respect to both the addition of a second molecule of HX to give Pt(IV) dihydride and the use of ligand X that will undergo further phosphine substitution. The Pt(IV) dihydride complex $PtH_2Cl_2(PEt_3)_2$ has been prepared by treating *trans*-$PtHCl(PEt_3)_2$ with HCl (*10*), but it appears that the similar compound $PtH_2Cl_2(PPh_3)_2$ does not arise from HCl addition to *trans*-$PtHCl(PPh_3)_2$ (*11*). An unusual Pt(IV) dihydride nevertheless does result from 1-ethynylcyclohexanol addition to $Pt(PPh_3)_3$ (Reaction 2). This complex initially was characterized spectroscopically (*12*) but more recently has been confirmed by crystal structure determination (Figure 3) (*13*). The hydride resonance is

Figure 3. Structure of $PtH_2(C\equiv CC_6H_{10}(OH))_2(PPh_3)_2$

12.9 ppm upfield to Me_4Si, and it is inferred from the crystal structure that the hydride ligands are mutually trans. Complexes such as these, formed by oxidative addition of alkyne C–H bonds to low-valent metal centers, resemble proposed intermediates in alkyne polymerization reactions.

Generally, it is expected that HX addition to $Pt(PPh_3)_3$ will occur readily if HX is strong, or if the conjugate base X^- is a sufficiently good ligand for Pt(II), that a reaction between X^- and the small equilibrium concentration of $[PtH(PPh_3)_3]X$ would give $PtHX(PPh_3)_2$.

Except for 1-ethynylcyclohexanol, it appears that the addition of protonic acid to triphenylphosphine platinum hydrides is unfavorable. Nevertheless, the existence of such complexes with triethylphosphine ligands is proved sufficiently since, in addition to the isolation of complexes with hydrochloric acid (10, 14), good evidence is presented for the intermediacy of triethylphosphine Pt(IV) hydrides with silanes and phosphines (15, 16).

A somewhat different type of product is obtained when the conjugate base is a multidentate ligand incorporating an anion that will strongly coordinate to Pt(II). An example of such a ligand is 1,4-dithiapentane $MeSCH_2CH_2SH$. When the compound $Pt(PPh_3)_3$ is treated with this mixed thioether-thiol, the compound $PtH(SCH_2CH_2SMe)PPh_3$ is formed where both the thiolate and thioether are coordinated (Reaction 3). This reaction likely proceeds through initial formation

$$Pt(PPh_3)_3 \ + \ HSCH_2CH_2SMe \ \longrightarrow \ \underset{Ph_3P}{\overset{H}{>}}Pt\underset{S}{\overset{S}{<}} \ + \ 2PPh_3 \qquad (3)$$

of an intermediate $PtH(SCH_2CH_2SMe)(PPh_3)_2$ having an uncoordinated thioether group. Such a compound then must undergo rapid substitution of one of the triphenylphosphines by the thioether to form the chelate complex (17). Interestingly, the presence of the chelating thioether group renders the compound isolable. The stronger acid H_2S leads to a complex $PtH(SH)(PPh_3)_2$ from $Pt(PPh_3)_3$ (18), but we have been unable to isolate characterizable products from the more weakly acidic thiols. Solutions containing excess thiol exhibit high-field resonances characteristic of platinum hydride complexes, but attempted isolation in the absence of excess thiol causes reversion back to $Pt(PPh_3)_3$.

We have tried to extend this concept of using the thermodynamic advantage of the chelate effect (19) with the compound 1,4,8-trithianone, $MeSCH_2CH_2CH_2SCH_2CH_2SH$. This new ligand can be prepared by first preparing the intermediate aldehyde $MeSCH_2CH_2CHO$ from the cupric ion-catalyzed addition of methanethiol to acrolein. This intermediate compound is condensed with 1,2-ethanedithiol in the presence of boron trifluoride etherate to give, after addition of base, the cyclic trithia compound $MeSCH_2CH_2CH(SCH_2)_2$, that cleaves with calcium in ammonia to form 1,4,8-trithanonane (20) (Reaction 4). This compound reacts with $Pt(PPh_3)_3$ to

$$\text{CH}_2\text{=CH–CHO} + \text{MeSH} \xrightarrow{\text{Cu(OAc)}_2} \text{Me–S–CH}_2\text{CH}_2\text{–CHO} \xrightarrow{\text{HS–SH}}$$

$$\xrightarrow[\text{(ii) KOH}]{\text{(i) BF}_3\cdot\text{Et}_2\text{O}} \quad \text{Me–S} \quad \text{(dithiolane)} \quad \xrightarrow{\text{Ca/NH}_3} \quad \text{Me–S–CH}_2\text{CH}_2\text{CH}_2\text{–S–CH}_2\text{CH}_2\text{–SH} \tag{4}$$

$$\text{Pt(PPh}_3)_3 + \text{HSCH}_2\text{CH}_2\text{SCH}_2\text{CH}_2\text{CH}_2\text{SMe} \longrightarrow \quad \text{[PtH(PPh}_3)\text{ complex]} \quad + \ 2\text{PPh}_3 \tag{5}$$

give the hydride complex $\text{PtH(SCH}_2\text{CH}_2\text{SCH}_2\text{CH}_2\text{CH}_2\text{SMe)(PPh}_3)$ (Reaction 5). The ^1H NMR spectrum shows a high-field doublet ($^2J(\text{PH}) = 18$ Hz) centered 10.6 ppm upfield of Me_4Si. The remainder of the spectrum shows singlets 1.90 ppm downfield of Me_4Si for the methylenes of the ethane moiety, and resonances at 2.83, 1.42, and 2.56 ppm downfield of Me_4Si for the methylenes of the propane group. Interestingly, this latter triplet ($^2J(\text{HH}) = 7$ Hz) is unshifted from the free ligand, as is the methylene group at position 7 along the chain. Since the terminal methyl group at 2.10 ppm downfield of Me_4Si also is unshifted from the free ligand position, the structure must be four-coordinate, with the terminal thioether group uncoordinated.

Platinum hydrides having a single, unsubstituted triphenylphosphine also have been prepared by the addition of diphenylphosphine oxide to $\text{Pt(PPh}_3)_3$. Diphenylphosphine oxide is a weak protonic acid that exists predominantly as this tautomer in equilibrium with diphenylphosphinous acid (Reaction 6). The compound probably is not monomeric but likely is aggregated by hydrogen bonding (21). The reaction with $\text{Pt(PPh}_3)_3$ readily occurs to give $\text{PtH(OPPh}_2)\text{-}$ $(\text{HOPPh}_2)\text{PPh}_3$. The compound shows an unusually intense band at 2000 cm^{-1} for the hydride. The high-field line of the more soluble methyldiphenylphosphine compound is centered 3.9 ppm upfield of Me_4Si, the deshielded position

$$\text{R}_2\text{P(H)O} \rightleftharpoons \text{R}_2\text{POH} \tag{6}$$

$$\text{Pt(PPh}_3)_3 + 2\text{Ph}_2\text{P(H)O} \rightarrow \text{PtH(OPPh}_2)(\text{HOPPh}_2)\text{PPh}_3 + 2\text{PPh}_3 \tag{7}$$

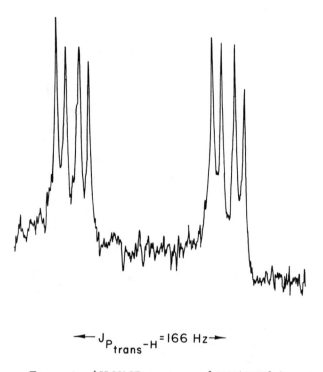

$$\longleftarrow J_{P_{trans}-H} = 166 \text{ Hz} \longrightarrow$$

Figure 4. ¹H NMR spectrum of PtH(OPPh₂)-
(HOPPh₂)PMePh₂

being a consequence of the high trans influence of the phosphinito ligand (22). This spectrum is shown in Figure 4 and the high-field line shows coupling to one trans (^2J(PH) = 166 Hz) and two cis (^2J(PH) = 24 Hz, 10 Hz) phosphines. The spectrum also shows a peak 13.4 ppm downfield of Me₄Si which is the phosphinous acid hydrogen. The compound contains a hydridic hydrogen bonded to Pt(II) that can be removed by addition of strong mineral acid and a protonic hydroxylic hydrogen that can be removed by titration with strong base (23). When the compound is treated with a solution of boron trifluoride in diethyl ether, the acidic hydrogen is replaced by a BF₂ group (Reaction 8). Figure 5 shows the ³¹P NMR spectra of PtH(OPPh₂)(HOPPh₂)PMePh₂ and PtH(OPPh₂)-(F₂BOPPh₂)PMePh₂. The chemical shift and coupling data obtained from these spectra are given in Table I. The broadening effect of the ¹¹B quadrupole on

$$
\begin{array}{c}
\text{H} \diagdown \quad\quad \overset{R_2}{\underset{}{P{=}O}} \diagdown \\
\quad \text{Pt} \quad\quad\quad\quad \text{H} \\
\text{L} \diagup \quad\quad \underset{R_2}{\overset{}{P{=}O}} \diagup
\end{array}
\ + \ BF_3 \ \longrightarrow \
\begin{array}{c}
\text{H} \diagdown \quad\quad \overset{R_2}{\underset{}{P{=}O}} \diagdown \\
\quad \text{Pt} \quad\quad\quad\quad BF_2 \\
\text{L} \diagup \quad\quad \underset{R_2}{\overset{}{P{=}O}} \diagup
\end{array}
\ + \ HF \quad\quad (8)
$$

(R = Ph; L = PPh₃ PMePh₂)

Table I. ^{31}P NMR Data for PtH(OPPh$_2$)(HOPPh$_2$)PMePh$_2$ ^{a}

P_1 $\delta = 93.2$ ^1J (P$_1$Pt) = 2302 Hz, ^2J(P$_1$P$_2$) = 28 Hz, ^2J(P$_1$P$_3$) = 17 Hz.

P_2 $\delta = 80.2$ ^1J(P$_2$Pt) = 2907 Hz, ^2J(P$_2$P$_3$) = 371 Hz.

P_3 $\delta = 5.8$ ^1J(P$_3$Pt) = 2224 Hz.

^{a} δ is given in ppm downfield from 85% H$_3$PO$_4$.   P$_1$ is trans to hydride and P$_2$ is trans to PMePh$_2$.

a ^{31}P nucleus two bonds away is apparent, but there is little broadening on the ^{31}P nucleus of PMe$_2$Ph that is four bonds away.

Although these complexes formally contain a diphenylphosphinito and a diphenylphosphinous acid ligand with respective penta-valent and tri-valent phosphorus centers, in reality it is more correct to consider the molecules as being conjugated, with the proton hydrogen bonded between the pair of oxygens.   Such a situation was suggested initially by Dixon (*24*), based on the position of ν(OH) in the ir spectrum.   We have verified this idea by observing that the ^{31}P NMR spectrum of both Pt(OPPh$_2$)$_2$(HOPPh$_2$)$_2$ and Pt[OP(OMe)$_2$]$_2$[HOP(OMe)$_2$]$_2$ show

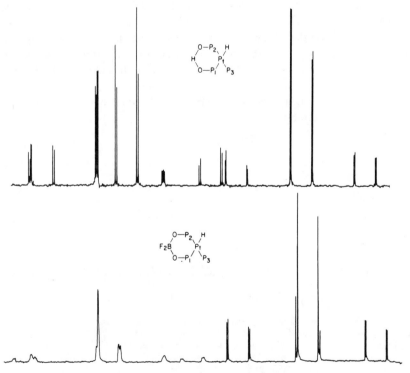

Figure 5. ^{31}P *NMR spectrum of PtH(OPPh$_2$)(HOPPh$_2$)PMePh$_2$ and PtH(OPPh$_2$)(F$_2$BOPPh$_2$)PMePh$_2$*

only a single chemical shift. The respective values for these lines are 72.5 and 90.7 ppm downfield of 85% H_3PO_4.

These hydrides form from diphenylphosphine oxide because of the strong coordination of the phosphorus-bonded diphenylphosphino ligands to the platinum metal group of elements. Substitution of the additional tertiary phosphine from either $Pt(PPh_3)_3$ or $Pt(PMePh_2)_3$ is related to the chelate effect noted earlier with the thiolate ligands, except that now the additional ligand is bonded to the first one by a hydrogen bond rather than by an alkyl chain. Since diphenylphosphine oxide exists as a hydrogen-bonded aggregate, it is possible that it is this adduct that reacts as a dimer with the phosphinous acid molecule already attached; however, until kinetic measurements are made on these systems, such questions will remain speculative.

Platinum hydrides are significant in that they show the range of product types that are expected from protonation reactions and also the correlation of reactivity with acid HX and the ligands around platinum. This knowledge then can be used to suggest guidelines for designing homogeneous catalytic systems that facilitate the addition of protonic acids to unsaturated hydrocarbons. Of particular importance is the activation of Si–H, N–H, P–H, O–H, and maybe C–H bonds to the addition of alkenes and alkynes. Currently, considerable work has been published on the hydrosilylation of alkenes and alkynes, but much less has been published on the synthesis and use of highly basic d^8 and d^{10} complexes for the catalyzed addition of the these other functional groups. The major applied work currently in progress is the formation of adiponitrile from butadiene and HCN by using complexes $Ni[P(OR)_3]_4$ as catalysts, but it is likely that the range and scope of such catalytic processes will continue to expand, especially for such transformations as the conversion of alkenes into amines and ethers from ammonia and alcohols, respectively.

Acknowledgment

This research was funded by the National Science Foundation and Cities Service Oil Company, Cranbury, New Jersey.

Literature Cited

1. Roundhill, D. M., "Advances in Organometallic Chemistry," Vol. 13, p. 273, Academic, New York, 1975.
2. Thomas, K., Dumler, J. T., Renoe, B. W., Nyman, C. J., Roundhill, D. M., *Inorg. Chem.* (1972) **11**, 1795.
3. Cariati, F., Ugo, R., Bonati, F., *Inorg. Chem.* (1966) **5**, 1128.
4. Tripathy, P. B., Roundhill, D. M., *J. Organomet. Chem.* (1970) **24**, 247.
5. Bird, P., Harrod, J. F., Than, K. A., *J. Am. Chem. Soc.* (1974) **96**, 1222.
6. Caputo, R. E., Mak, D. K., Willett, R. D., Roundhill, S. G. N., Roundhill, D. M., *Acta Crystallogr.* (1977) **B33**, 215.
7. Dingle, T. W., Dixon, K. R., *Inorg. Chem.* (1974) **13**, 846.
8. Roundhill, D. M., *Inorg. Chem.* (1970) **9**, 254.

9. Roundhill, D. M., Tripathy, P. B., Renoe, B. W., *Inorg. Chem.* (1971) **10**, 727.
10. Chatt, J., Shaw, B. L., *J. Chem. Soc.* (1962) 5075.
11. Dumler, J. T., Roundhill, D. M., *J. Organomet. Chem.* (1971) **30**, C35.
12. Roundhill, D. M., Jonassen, H. B., *J. Chem. Soc., Chem. Commun.* (1968) 1233.
13. Rasmussen, S. E., Mariezcurrena, R. A., *Acta Chem. Scand.* (1973) **27**, 2678.
14. Anderson, D. W. W., Ebsworth, E. A. V., Rankin, D. W. H., *J. Chem. Soc., Dalton Trans.* (1973) 854.
15. Ebsworth, E. A. V., Edward, J. M., Rankin, D. W. H., *J. Chem. Soc., Dalton Trans.* (1976) 1667.
16. Ebsworth, E. A. V., Edward, J. M., Rankin, D. W. H., *J. Chem. Soc., Dalton Trans.* (1976) 1673.
17. Rauchfuss, T. B., Roundhill, D. M., *J. Am. Chem. Soc.* (1975) **97**, 3386.
18. Morelli, D., Segre, A., Ugo, R., LaMonica, G., Cariati, S., Conti, F., Bonati, F., *J. Chem. Soc., Chem. Commun.* (1967) 524.
19. Murro, D., *Chem. Br.* (1977) **13**, 100.
20. Rauchfuss, T. B., Shu, J. S., Roundhill, D. M., *Inorg. Chem.* (1976) **15**, 2096.
21. Kosolapoff, G. M., Maier, L., "Organic Phosphorus Compounds," Vol. 4, p. 491, Wiley, New York, 1972.
22. Chatt, J., Heaton, B. T., *J. Chem. Soc. A.* (1968) 2745.
23. Sperline, R. P., Roundhill, D. M., *Inorg. Chem.* (1977) **16**, 2612.
24. Dixon, K. R., Rattray, A. D., *Can. J. Chem.* (1971) **49**, 3996.

RECEIVED July 19, 1977.

13

Osmium Hydridoalkyls and Their Elimination Mechanisms

JACK R. NORTON, WILLIE J. CARTER, JOHN W. KELLAND, and STANLEY J. OKRASINSKI

Department of Chemistry, Princeton University, Princeton, NJ 08540

Eliminations from $Os(CO)_4RR'$ occur by dinuclear mechanisms only if either R or R' is H. A hydride on one metal is necessary to interact with a vacant coordination site on the other in the dinuclear transition state. With $Os(CO)_4H_2$, the vacant site is created by dissociation of CO. With $Os(CO)_4$-$(H)CH_3$, the vacant site is created by a facile rate-determining isomerization which we suggest is to an acetyl hydride. The unique instability of hydridoalkyl carbonyls thus is explained. The synthesis and properties of $Os(CO)_4(H)C_2H_5$ and various polynuclear ethyl osmium derivatives show that β-hydrogens have no significant effect on these elimination mechanisms. Dinuclear hydridoalkyls are excellent starting points for the synthesis of more complex polynuclear alkyls.

Three years ago, while we were considering possible reasons for the general inability of transition metals to insert into and activate C–H bonds, our attention turned to the question of the instability of transition metal alkyl hydride complexes. We have listed the few alkyl hydride complexes of which we are aware (1) (one additional case (2) recently came to our attention) as well as some of the only slightly more numerous cases of substituted alkyl hydrides stabilized by chelation (3). In contrast, there are enormous numbers of polyalkyls (4, 5) and polyhydrides (6). While rarity does not logically imply instability, it does suggest it, so we considered possible mechanistic explanations for the assumed rapid decomposition of cis-$ML_n(R)(H)$ relative to cis-ML_nR_2 and cis-ML_nH_2. We have focused on octahedral complexes since they are both more important and more numerous.

Construction of an orbital correlation diagram (7) discloses that concerted cis elimination of R–R' from cis-ML_4RR' is not forbidden by symmetry considerations; thus, there appears to be no advantage for the asymmetric elimination

0-8412-0390-3/78/33-167-170/$05.00/0

R–H over the symmetric eliminations R–R and H–H. Bond strength considerations suggest that if rates were determined entirely by the strengths of the bonds made and broken, R–H elimination might be fastest. If one assumes a constant M–H bond strength X greater than a constant M–R bond strength Y and considers that the strength of the C–H bond being formed in R–H (about 100 kcal/mol) is more than halfway between the bond strengths of R–R (85 kcal/mol) and H–H (104 kcal/mol), $\Delta H^0 (X + Y - 100)$ for R–H elimination can, for certain values of X–Y, be lower and more favorable than ΔH^0 for R–R elimination $(2Y - 85)$ or H–H elimination $(2X - 104)$.

This difference would be reflected in $\Delta H^{\ddagger}$ and in reaction rates only if the elimination transition state strongly resembled the products. To the extent that the elimination transition state resembles the starting complex cis-ML_4RR', one would expect relative rates to be largely a function of the energies of the bonds being broken. There is then no reason to expect elimination of R–H from the mixed species to be faster than elimination of both R–R from ML_4R_2 and H–H from ML_4H_2.

We thus began to question the traditional belief that concerted cis elimination of R–R' (simple intramolecular reductive elimination) was in fact the mechanism:

$$ML_4RR' \rightarrow R-R' + ML_4$$

of organic ligand elimination from ML_4RR'. Our first working hypothesis was that attention should be paid to the energy of the fragment complex ML_4 that would remain after such an elimination. If this energy were sufficiently great, other elimination processes would be observed rather than simple intramolecular reductive elimination.

The literature already contained some results that supported this hypothesis. While organoplatinum(IV) complexes such as $PtL_2I(CH_3)_3$ readily undergo ethane elimination, leaving stable trans-$PtL_2(CH_3)I$ (8), the superficially similar $(\pi\text{-}C_5H_5)Pt(CH_3)_3$ does not, giving methyl radicals and eventually methane instead (9). The difference is explained reasonably as a consequence of the high energy of the hypothetical $C_5H_5PtCH_3$ compared with the stable, square-planar fragment trans-$PtL_2(CH_3)I$.

$$PtL_2I(CH_3)_3 \xrightarrow{\Delta} C_2H_6 + trans\text{-}PtL_2(CH_3)I$$

$$(\pi\text{-}C_5H_5)Pt(CH_3)_3 \xrightarrow{\Delta} \!\!\!\!/\!\!\!\!\to \begin{array}{l} C_2H_6 + C_5H_5PtCH_3 \\ \longrightarrow \cdot CH_3 \rightarrow CH_4 \end{array}$$

To investigate the validity of this hypothesis, we looked at the series of complexes, cis-$Os(CO)_4RR'$. Several of its members were reported, and even an apparent hydridomethyl complex was observed spectroscopically in small quantities (10, 11). Furthermore, there was reason to believe that $Os(CO)_4$ would be a very high energy fragment and hence that mechanisms other than simple intramolecular reductive elimination might be observed. $Os(CO)_4$ had been generated in matrix isolation experiments and gave behavior qualitatively similar to that of the more extensively studied $Fe(CO)_4$. The latter appears to be extremely reactive toward various inert matrix materials and has a C_{2v} structure in most of them (12, 13). Although the question of whether such $M(CO)_4$ fragments are singlets or triplets is under active discussion (14, 15), it is clear that they are not stable, square-planar d^8 complexes like $trans$-$PtL_2(CH_3)I$.

A preliminary investigation showed that $Os(CO)_4H_2$, $Os(CO)_4(C_2H_5)_2$, and $Os(CO)_4(CH_3)_2$ do not decompose appreciably below $100°C$ whereas $Os(CO)_4$-$(H)CH_3$ decomposes rapidly at $40°C$. In addition to providing a concrete example of the relative instability of alkyl hydrides, these results suggested strongly that more than one elimination mechanism was operating in this series of compounds and thus that something other than simple intramolecular reductive elimination must be occurring.

Elimination Mechanisms of $Os(CO)_4(CH_3)_2$ and $Os(CO)_4H_2$

Further work has shown that the most obvious alternative, metal–carbon bond homolysis to form free radicals, occurs with $Os(CO)_4(CH_3)_2$. The methyl radicals that were produced (slowly even at $162.5°C$) attack a variety of solvents. Further, quantitative verification of their intermediacy was obtained by measuring k_H/k_D in mixtures of n-$C_{12}H_{26}$ and n-$C_{12}D_{26}$. The ratio, 5.3, is characteristic of the known selectivity for methyl radicals at that temperature (16).

The first hydride examined was naturally $Os(CO)_4H_2$. The cis structure presumed for this compound (and for all the other $Os(CO)_4RR'$) from ir data was confirmed by Raman observation (methylcyclohexane solution) of ν_{Os-H} at 1971 (A_1) and 1942 (B_2) cm^{-1} and by an electron diffraction study on the compound in the gas phase (17). The structure departs only slightly from octahedral geometry, with the CO (axial)–Os–CO–(equatorial) angle being $96.3 \pm 0.7°$. The structure of the $Os(CO)_4$ unit is similar to those of the same units in $Os_3(CO)_{12}$ (18).

Deuterium labeling experiments proved that the initial reaction upon thermolysis at $125.8°C$:

$$2\ Os(CO)_4H_2 \rightarrow H_2Os_2(CO)_8 + H_2$$

does not proceed via simple intramolecular reductive elimination to form $Os(CO)_4$ (which might then insert in an Os–H bond in $Os(CO)_4H_2$), but rather is dinuclear, i.e., a mixture of $Os(CO)_4H_2$ and $Os(CO)_4D_2$ gives HD as well as H_2 and D_2 (19). Dinuclear elimination is defined as being the formation of R–R' from ML_4RR' with R and R' originating on different molecules of ML_4RR'. The process can be identified only by labeling studies of the type described.

With $Os(CO)_4H_2$ and other complexes to be discussed later, dinuclear elimination is kinetically first order. The detailed mechanism for $Os(CO)_4H_2$ appears (19) to be:

$$Os(CO)_4H_2 \rightarrow Os(CO)_3H_2 + CO \text{ (rate determining)}$$

$$Os(CO)_3H_2 + Os(CO)_4H_2 \xrightarrow{\text{fast}} H_2Os_2(CO)_7 + H_2$$

(the actual dinuclear elimination step)

$$H_2Os_2(CO)_7 + CO \xrightarrow{\text{fast}} H_2Os_2(CO)_8$$

When the reaction is run in the presence of ^{13}CO and stopped after one half-life, not only is the label incorporation into recovered starting material negligible, as predicted by the mechanism, but the label incorporation into $H_2Os_2(CO)_8$ exceeds that required by the mechanism.   This shows that it is undergoing carbonyl exchange under the reaction conditions and suggests that such dinuclear species are much more labile than their mononuclear counterparts.

In view of the structural similarity between the $Os(CO)_4$ units in $Os(CO)_4H_2$ and $Os_3(CO)_{12}$ noted above, it is interesting to compare the $Os(CO)_4H_2$ thermolysis rate, for which we said carbonyl dissociation is rate determining, with the known rate of dissociative carbonyl exchange for $Os_3(CO)_{12}$ (*20*).   The exchange rate per $Os(CO)_4$ unit extrapolated to 125.8°C is 25×10^{-5} sec^{-1}; the rate of $Os(CO)_4H_2$ thermolysis at that temperature is 6×10^{-5} sec^{-1}. The similarity of these numbers is final evidence that carbonyl dissociation from $Os(CO)_4H_2$ does occur in the rate-determining step.   A vacant coordination site apparently is needed before the actual dinuclear elimination step can occur.

Synthetic Aspects of Os(CO)₄(H)(R) Chemistry

Next our interest focused on *cis*-$Os(CO)_4(H)CH_3$.   First it was necessary to devise a synthesis capable of producing this rather unstable complex pure and in high yield.   Initial attempts to methylate $[HOs(CO)_4]^-$ were thwarted by side reactions caused by proton transfer from the product $Os(CO)_4(H)CH_3$ onto anion not yet methylated.   The use of CH_3OSO_2F to increase the methylation rate solved this problem (*1*) and permitted the synthesis of 99% pure *cis*-$Os(CO)_4$-$(H)CH_3$ in yields up to 90%

We recently extended this approach to the synthesis of the ethyl analog *cis*-$Os(CO)_4(H)C_2H_5$.

$$Na_2Os(CO)_4 \xrightarrow[\text{tetraglyme}]{CF_3CO_2H} Na^+ [HOs(CO)_4]^-$$

$$Na^+[HOs(CO)_4]^- \xrightarrow[\text{tetraglyme}]{C_2H_5OSO_2F} \textit{cis}\text{-}Os(CO)_4(H)(C_2H_5)$$

Two equivalents of ethyl fluorosulfonate must be used (one ethylates the trifluoroacetate ion) to avoid complications arising from ethylation of ether oxygens in the tetraglyme solvent.   The product must be distilled out of the reaction mixture on a vacuum line within 5 min after addition is complete.   The hydri-

Table I. ^{1}H NMR Spectra of Ethyl Osmium Complexes[a]

Complex	Coupling Constants (Hz)	Chemical Shifts (τ)	
cis-Os(CO)$_4$(H)(C$_2$H$_5$) (in CD$_2$Cl$_2$)	$J_{CH_2-CH_3} = 7$ $J_{CH_2-H} = 1.5$	Os–CH$_2$CH$_3$ Os–CH$_2$CH$_3$ Os–H	8.28 8.70 17.90
cis-Os(CO)$_4$(Cl)(C$_2$H$_5$) (in CDCl$_3$)	$J_{CH_2-CH_3} = 7$	Os–CH$_2$CH$_3$ Os–CH$_2$CH$_3$	8.24 8.16
CH$_3$Os(CO)$_4$Os(CO)$_4$C$_2$H$_5$ (in C$_6$D$_6$)	$J_{CH_2-CH_3} = 7$	Os–CH$_3$ Os–CH$_2$CH$_3$ Os–CH$_2$CH$_3$	9.67 8.60 8.24

[a] Parameters are taken from computer simulations of the observed complex multiplets.

doethyl complex, a colorless liquid that decomposes in a few hours at room temperature, is somewhat less stable than its methyl analog.   Its ^{1}H NMR spectrum is given in Table I and its ir spectrum in Table II.   To our knowledge it is the first reported hydridoethyl carbonyl complex of a transition metal.   Its existence suggests that β-hydrogens do not have an important effect on stability in these systems.   For further characterization, the hydridoethyl complex was treated with CCl$_4$ and converted to the stable, but extraordinarily volatile (it sublimes readily at room temperature), white crystalline solid cis-Os(CO)$_4$-(Cl)(C$_2$H$_5$).   Spectroscopic data for it are given also in Tables I and II.

Cis-Os(CO)$_4$(H)CH$_3$ decomposes rapidly at 40°C or over several days at room temperature, evolving about one-half equivalent of methane.   The initial product formed is HOs(CO)$_4$Os(CO)$_4$CH$_3$, but the system eventually produces substantial Os$_3$(CO)$_{12}$(CH$_3$)$_2$ as well.   The structures shown below were assigned originally on the basis of ir and NMR spectra of these and other di-osmium and triosmium compounds (1).   The structure assigned to Os$_3$(CO)$_{12}$(CH$_3$)$_2$ is supported now by x-ray crystallographic results (21) on Os$_3$(CO)$_{12}$I$_2$; their ir spectra suggest they are iso-structural.   Although the two iodine atoms in Os$_3$(CO)$_{12}$I$_2$ adopt a trans configuration in the crystal whereas the methyl groups in

Table II.   Carbonyl Region ir Spectra (in Pentane) of Ethyl Osmium Complexes

Complex	Position of Bands
cis-Os(CO)$_4$(H)-(C$_2$H$_5$)	2133(w), 2060(s), 2043(vs), 2030(s)
cis-Os(CO)$_4$(Cl)-(C$_2$H$_5$)	2155(w), 2078(vs), 2023(s)
CH$_3$Os(CO)$_4$Os-(CO)$_4$C$_2$H$_5$	2130(vw), 2088(vs), 2054(s), 2044(vs), 2038(vs), 2032(s), 2027(m), 2016(m), 2007(w)
ClOs(CO)$_4$Os-(CO)$_4$C$_2$H$_5$	2139(w), 2098(s), 2062(sh), 2057(vs), 2044(m), 2030(vs), 2014(m), 2008(m)
Os$_3$(CO)$_{12}$(C$_2$H$_5$)$_2$	2132(vw), 2093(s), 2065(w), 2051(m), 2034(vs), 2025(s), 2005(m), 1990(w)

$Os_3(CO)_{12}(CH_3)_2$ are drawn in a cis configuration, in both cases a variety of conformers are doubtless present in solution.

Preparative TLC on the mixture resulting from the decomposition of $Os(CO)_4(H)CH_3$ allows the separation of $HOs(CO)_4Os(CO)_4CH_3$ from $Os_3(CO)_{12}(CH_3)_2$; both compounds can be isolated normally in 25–30% yield. The compound, $HOs(CO)_4Os(CO)_4CH_3$, is a colorless, light-sensitive liquid that is converted into the stable white crystalline solid $ClOs(CO)_4Os(CO)_4CH_3$ when treated with CCl_4. On the basis of the similarity of their ir spectra, the latter compound is assigned the same structure as its precursor, with both substituents cis to the Os–Os bond (*1*).

The above synthetic results raise the question of whether or not $HOs(CO)_4$-$Os(CO)_4CH_3$ is intermediate in the formation of $Os_3(CO)_{12}(CH_3)_2$. We have now confirmed that it is, although the question was more complex than expected. Small quantities of $Os_3(CO)_{12}(CH_3)_2$ are formed during the thermolysis of pure $HOs(CO)_4Os(CO)_4CH_3$ at 74°C. However, much greater amounts of trinuclear product are formed when a 1:1 mixture of dinuclear and mononuclear hydridomethyl complexes is allowed to react at 49°C. Furthermore, when $HOs(CO)_4Os(CO)_4CD_3$ is made from $Os(CO)_4(H)CD_3$ (*see* next section) and reacts with one equivalent of $Os(CO)_4(H)CH_3$, the resulting $Os_3(CO)_{12}R_2$ shows considerable deuterium incorporation ($R = CD_3$ as well as $R = CH_3$). Thus, it appears that the intermediate $HOs(CO)_4Os(CO)_4CH_3$ forms $Os_3(CO)_{12}(CH_3)_2$ primarily through reaction with another mononuclear unit.

$HOs(CO)_4Os(CO)_4CH_3 + Os(CO)_4(H)CH_3 \rightarrow$

$$CH_3Os(CO)_4Os(CO)_4Os(CO)_4CH_3$$

The apparent reactivity of the hydride end of the dinuclear hydridomethyl species is in line with the considerable substitutional lability noted above for $H_2Os_2(CO)_8$. This reactivity contrasts with the comparative inertness (mentioned above) of the mononuclear hydride $Os(CO)_4H_2$ and of $Os_3(CO)_{12}$. Its origin is unclear, but it makes the dinuclear hydrides and hydridoalkyls excellent starting points for the synthesis of more complex polynuclear alkyls. For example,

$HOs(CO)_4Os(CO)_4CH_3$ reacts with ethylene (1 atm, 75°C, in heptane) to form a novel dinuclear mixed dialkyl. (The NMR and ir spectra of $C_2H_5Os(CO)_4$-$Os(CO)_4CH_3$ are given in Tables I and II, respectively.)

$$HOs(CO)_4Os(CO)_4CH_3 + C_2H_4 \rightarrow C_2H_5Os(CO)_4Os(CO)_4CH_3$$

In comparison, $Os(CO)_4H_2$ does not react with ethylene while $Os(CO)_4(H)CH_3$, rather than forming an ethyl osmium complex, reacts with ethylene to form $Os(CO)_4(C_2H_4)$ and methane in a reaction to be discussed below.

The most important question about the reactivity of $HOs(CO)_4Os(CO)_4CH_3$ is whether or not it can form methane by a 1,2 elimination. It does evolve methane; after several days at 74°C, analysis of the gaseous products from its decomposition shows approximately one equivalent of CH_4 and one equivalent of CO. However, the reaction is extremely complicated. Via $Os(CO)_4(D)CH_3$ and $Os(CO)_4(H)CD_3$ (described below), it is possible to prepare $DOs(CO)_4$-$Os(CO)_4CH_3$ and $HOs(CO)_4Os(CO)_4CD_3$ separately; when they are mixed and heated to 74°C, CD_4 is found in the resulting methane. Therefore, the reaction is intermolecular, and there is no evidence for a 1,2-elimination. It can be argued that it is unreasonable to expect a concerted 1,2-elimination ever to occur in a system of this sort where the methyl group and hydrogen are attached to osmium atoms almost 3 Å apart, and where migration or bridging by CH_3 or H seems likely to be required prior to elimination.

Preliminary investigation suggests once again that the presence of β-hydrogens makes little difference and that cis-$Os(CO)_4(H)C_2H_5$ has a chemistry like that of its methyl counterpart—it evolves ethane in a few hours at room temperature. Its decomposition products appear to be $Os_3(CO)_{12}(C_2H_5)_2$ and $HOs(CO)_4Os(CO)_4(C_2H_5)$. The latter has not been isolated, but its presence is inferred from the fact that treatment of the reaction mixture with CCl_4 permits the isolation of $ClOs(CO)_4Os(CO)_4(C_2H_5)$ (characterized by mass spectrometry). The ir spectra of $ClOs(CO)_4Os(CO)_4C_2H_5$ and $Os_3(CO)_{12}(C_2H_5)_2$ appear in Table II.

Mechanism of Dinuclear Elimination from cis-$Os(CO)_4(H)CH_3$

The diagnostic reaction, thermolysis of a $Os(CO)_4(D)CH_3$ and $Os(CO)_4$-$(H)CD_3$ mixture, yields CD_4 and CH_4 as well as CH_3D and CD_3H. Therefore, this is classified as a dinuclear elimination reaction. Appropriately labeled starting materials are easily obtained by using CF_3CO_2D or CD_3OSO_2F at the appropriate place in the synthesis. The rate of the elimination reaction:

$$2\, Os(CO)_4(H)CH_3 \rightarrow CH_4 + HOs(CO)_4Os(CO)_4CH_3$$

is, as with the analogous reaction with $Os(CO)_4H_2$, first order:

$$\frac{d[Os(CO)_4(H)CH_3]}{dt} = -k_1[Os(CO)_4(H)CH_3]$$

with $k_1 = 1.38 \times 10^{-4}$ sec^{-1} at 49°C. However, unlike Os(CO)$_4$H$_2$, carbonyl dissociation is not the rate-determining step. The value of $\Delta S^{\ddagger}$ is −8 eu, and when the reaction is carried out under ^{13}CO, no label is incorporated into the product or the recovered starting material (3).

Our presumption that the rate-determining step somehow produced a vacant coordination site led to our investigating the reaction in the presence of triethylphosphine. Methane is still eliminated, Os(CO)$_4$(Et$_3$P) is formed, and the disappearance rate of Os(CO)$_4$(H)CH$_3$ is still first order in that material, but the rate constant (6.4 × 10^{-5} sec^{-1}) is one-half that observed in the absence of Et$_3$P. The rate is completely independent of the concentration of Et$_3$P. Furthermore, reaction of a Os(CO)$_4$(H)CD$_3$ and Os(CO)$_4$(D)CH$_3$ mixture with Et$_3$P gives only CD$_3$H and CH$_3$D. These observations require the following mechanism (3):

$$\text{Os(CO)}_4\text{(H)CH}_3 \xrightarrow{\text{rate determining}} [\text{Os(CO)}_4\text{(H)CH}_3]^*$$

$$[\text{Os(CO)}_4\text{(H)CH}_3]^* \begin{array}{l} \xrightarrow{\quad L \quad} \text{Os(CO)}_4\text{L} + \text{CH}_4 \\[1ex] \xrightarrow{\;\text{Os(CO)}_4\text{(H)CH}_3\;} \text{HOs(CO)}_4\text{Os(CO)}_4\text{CH}_3 + \text{CH}_4 \end{array}$$

where [Os(CO)$_4$(H)CH$_3$]* is an isomerized, reactive form of the starting material. The course of the reaction reflects the outcome of the competition for this intermediate between external nucleophiles L and the original hydridomethyl compound.

Several considerations (such as the activation parameters and small solvent dependence of the reaction in which it is formed) suggest that [Os(CO)$_4$(H)CH$_3$]*

$$\text{O}$$
$$\|$$

is the unsolvated, five-coordinate acyl hydride Os(CO)$_3$(H)CCH$_3$. Attempts to trap this intermediate with the acyl ligand intact were not successful—either dinuclear elimination or rapid methane evolution occurs. One observes the reaction:

$$\textit{cis-}\text{Os(CO)}_4\text{(H)CH}_3 + \text{L} \rightarrow \text{Os(CO)}_4\text{L} + \text{CH}_4$$

for L = Et$_3$P, (CH$_3$O)$_3$P, pyridine, and C$_2$H$_4$. The compound, Os(CO)$_4$(C$_2$H$_4$), has ir bands in pentane at 2111(w), 2023(s), and 1993(s) and a ^{1}H NMR signal at 8.56 τ (C$_6$D$_6$).

If the intermediate is a five-coordinate acyl hydride, the above observations require that:

$$\text{O}$$
$$\|$$
$$\text{Os(CO)}_3\text{(CCH}_3\text{)(H)L} \rightarrow \text{Os(CO)}_4\text{L} + \text{CH}_4$$

be faster than the reaction of L with the intermediate. While the thermodynamic driving force for methane elimination is obvious [$Os(CO)_4L$ can be generally isolated as stable complexes, whereas acetaldehyde elimination would leave the high-energy fragment $Os(CO)_3L$], a mechanism for methane elimination is not readily apparent. It cannot involve CO dissociation—carrying out the reaction of $Os(CO)_4(H)CH_3$ with L = Et_3P under ^{13}CO leaves neither the product ($Os(CO)_4L$) nor the recovered starting material with any label incorporation.

Conceivable mechanisms for methane elimination from $Os(CO)_3(\overset{\displaystyle O}{\overset{\displaystyle \|}{C}}CH_3)(H)L$ involve either a seven-coordinate transition state or direct migration of CH_3 onto H.

The relative ability of various nucleophiles to compete for the reactive intermediate requires comment. The reactivity of Os–H in $Os(CO)_4(H)CH_3$ (or in $Os(CO)_4H_2$—*see* below) is so high that only strong nucleophiles such as Et_3P can compete successfully with it. ($CH_3O)_3P$ is somewhat less reactive and pyridine is much less reactive than Et_3P, as measured by their ability to divert the reaction from $HOs(CO)_4Os(CO)_4CH_3$ production into $Os(CO)_4L$ production. The low reactivity of CO that might be inferred from the results of the labeling experiments (lack of ^{13}CO incorporation) is incorrect since the partial pressure of ^{13}CO in such experiments is much less than 1 atm, and the concentration of ^{13}CO in solution is $<10^{-3}M$. The Os–H bonds in starting material are present at concentrations of $>10^{-2}M$; therefore, their concentration relative to that of ^{13}CO is high enough that no conclusions can be drawn regarding the relative reactivity of Os–H and CO toward the intermediate. The proposed alkyl migration in the rate-determining step accounts nicely for the instability of $Os(CO)_4(H)C_2H_5$ compared with $Os(CO)_4(H)CH_3$, as ethyl groups generally migrate onto carbonyls more rapidly than methyl groups (*22*).

Conclusion

Proposed General Explanation for Alkyl Hydride Instability. The advantage of suggesting a five-coordinate acyl hydride as the reactive intermediate in $Os(CO)_4(H)CH_3$ thermolysis is that it leads to a general explanation of elimination processes of this type. That general explanation contains the following elements:

(1) Dinuclear elimination processes are possible only when at least one of the ligands to be eliminated is a hydride. This is supported by our observations (dinuclear elimination from $Os(CO)_4(H)CH_3$ and $Os(CO)_4H_2$ but not from $Os(CO)_4(CH_3)_2$) and is explained reasonably by the unique ability of a hydride to bridge a pair of transition metal atoms. The interaction of Os–H with a vacant coordination site on another Os to form a dinuclear species appears to be an essential part of the dinuclear elimination process.

(2) Alkyl carbonyl complexes can create vacant coordination sites by alkyl migration to give acyls far more readily than hydride carbonyl complexes can

by forming formyl complexes. The chemical literature does not contain a single example of the direct generation of a formyl complex from a hydridocarbonyl.

(3) Combination of both of the above elements in a single molecule such as $Os(CO)_4(H)CH_3$ gives rise to facile dinuclear elimination. The dihydride is capable of dinuclear elimination but must rely on the comparatively high-energy process of carbonyl dissociation to provide the necessary vacant coordination site. The dimethyl compound has the necessary vacant site easily available but no hydride to interact with it. The hydridomethyl compound has both elements and is uniquely unstable.

We conclude that our working hypothesis is valid, and that when ML_4 is a sufficiently unstable fragment, both dinuclear elimination and metal–carbon bond homolysis can occur instead of simple reductive elimination of R–R′ from $ML_4RR′$. We conclude further that the involvement of a hydride ligand is necessary for dinuclear elimination from such $ML_4RR′$. On that basis, we propose the above general mechanism to explain the instability of hydridoalkyls of this type.

The above conclusions and suggestions, if valid, require than an alkyl carbonyl (because it can easily generate a vacant coordination site) and a hydride should be able to carry out a facile dinuclear alkane elimination. Thus we predict that dinuclear eliminations between $Os(CO)_4H_2$ and $Os(CO)_4(CH_3)_2$ should occur more rapidly than the decomposition of either compound separately, and we have confirmed this prediction.

$$Os(CO)_4H_2 + Os(CO)_4(CH_3)_2 \rightarrow HOs(CO)_4Os(CO)_4CH_3 + CH_4$$

Some acetaldehyde is formed with prolonged heating of the reaction mixture in a sealed tube.

Similarly, we predict that the Os–H bonds in $Os(CO)_4H_2$ should be able to compete with those in $Os(CO)_4(H)CH_3$ for the reactive acyl hydride intermediate formed in the decomposition of the latter. We have also confirmed this prediction.

$$Os(CO)_4H_2 + Os(CO)_4(H)CH_3 \rightarrow H_2Os_2(CO)_8 + CH_4$$

We are now examining other hydride and methyl complexes to see whether or not the reactions predicted on the basis of our conclusions and proposals actually occur.

Literature Cited

1. Evans, J., Okrasinski, S. J., Pribula, A. J., Norton, J. R., *J. Am. Chem. Soc.* (1976) **98**, 4000.
2. Strope, D., Shriver, D. F., *J. Am. Chem. Soc.* (1973) **95**, 8197.
3. Okrasinski, S. J., Norton, J. R., *J. Am. Chem. Soc.* (1977) **99**, 295.
4. Schrock, R. R., Parshall, G. W., *Chem. Rev.* (1976) **76**, 243.
5. Davidson, P. J., Lappert, M. F., Pearce, R., *Chem. Rev.* (1976) **76**, 219.
6. "Transition Metal Hydrides," Muetterties, E. L., Ed., Marcel Dekker, Inc., New York 1971.

7. Braterman, P. S., Cross, R. J., *Chem. Soc. Rev.* (1973) **2**, 271.
8. Ruddick, J. D., Shaw, B. L., *J. Chem. Soc. A* (1969) 2969.
9. Egger, K. W., *J. Organomet. Chem.* (1970) **24**, 501.
10. L'Eplattenier, F., Pelichet, C., *Helv. Chim. Acta* (1970) **53**, 1091.
11. George, R. D., Knox, S. A. R., Stone, F. G. A., *J. Chem. Soc., Dalton* (1973) 972.
12. Poliakoff, M., Turner, J. J., *J. Chem. Soc., Dalton* (1973) 1351.
13. Poliakoff, M., Turner, J. J., *J. Chem. Soc., Dalton* (1974) 2276.
14. Burdett, J. K., *J. Chem. Soc., Faraday 2* (1974) **70**, 1599.
15. Elian, M., Hoffman, R., *Inorg. Chem.* (1975) **14**, 1058.
16. Evans, J., Okrasinski, S. J., Pribula, A. J., Norton, J. R., *J. Am. Chem. Soc.* (1977) **99**, 5835.
17. Robiette, A. G., Hedberg, K., unpublished data.
18. Churchill, M. R., DeBoer, B. G., *Inorg. Chem.* (1977) **16**, 878.
19. Evans, J., Norton, J. R., *J. Am. Chem. Soc.* (1974) **96**, 7577.
20. Cetini, G., Gambino, O., Sappa, E., Vaglio, G. A., *Atti Accad. Sci. Torino* (1967) **101**, 855.
21. Cook, N., Smart, L., Woodward, P., *J. Chem. Soc., Dalton Trans.* (1977) 1744.
22. Wojcicki, A., *Adv. Organomet. Chem.* (1973) **11**, 88.

RECEIVED July 18, 1977. We thank the National Science Foundation for supporting this work.

Photochemistry of Transition Metal Hydride Complexes

GREGORY L. GEOFFROY, MARK G. BRADLEY, and RONALD PIERANTOZZI

Department of Chemistry, The Pennsylvania State University, University Park, PA 16802

The photochemical studies of transition metal hydride complexes that have appeared in the chemical literature are reviewed, with primary emphasis on studies of iridium and ruthenium that were conducted by our research group. The photochemistry of the molybdenum hydride complexes $[Mo(\eta^5\text{-}C_5H_5)_2M_2]$ and $[MoH_4(dppe)_2]$ $(dppe = Ph_2PCH_2CH_2PPh_2)$, which eliminate H_2 upon photolysis, is discussed in detail. The photoinduced elimination of molecular hydrogen from di- and polyhydride complexes of the transition elements is proposed to be a general reaction pathway.

Transition metal hydride complexes have become an important class of compounds in inorganic and organometallic chemistry, and the field has expanded tremendously since the 1955 report (1) of the first thermally stable hydride complex, $[Re(\eta^5\text{-}C_5H_5)_2M]$. Many reviews describing the properties of metal hydrides were published (2–7), and in a recent literature survey (8) of the three catalytically important metals, ruthenium, rhodium, and iridium, over 2000 known hydride complexes were found. Transition metal hydrides are essential in many homogeneous catalytic reactions, are useful synthetic intermediates, are promising as hydrogen and energy storage systems, and have been proposed as important intermediates for obtaining molecular hydrogen from water (9).

Although there are many known hydride complexes that are important in homogeneous catalysis, relatively few photochemical studies were conducted on these compounds before we began our investigations. Because transition metal hydride complexes are used in homogeneous catalysis and might be used in the photoassisted production of H_2 from water, we initiated a study of their photochemical properties.

The studies that were conducted by other workers are reviewed, and our previously published work on ruthenium and iridium hydrides is presented. This

0-8412-0390-3/78/33-167-181/$05.00/0
© American Chemical Society

is followed by a discussion of the photochemical properties of the molybdenum hydrides $[Mo(\eta^5\text{-}C_5H_5)_2H_2]$ and $[MoH_4(dppe)_2]$ ($dppe = Ph_2PCH_2CH_2PPh_2$), that show interesting photochemistry and lead to the photogeneration of very reactive complexes.

One of the first reports concerning the photochemical properties of metal hydrides was a statement by Sacco and Aresta (10) that in sunlight $[FeH_2(N_2)\text{-}(PEtPh_2)_3]$ undergoes reversible loss of hydrogen (see Reaction 1). It was proposed that irradiation induces formation of coordinatively unsaturated $[Fe(N_2)\text{-}(PEtPh_2)_3]$, and in a subsequent step, iron inserts into an ortho C–H bond of PPh$_3$ (orthometallation). In a separate report, Koerner von Gustorf and co-workers presented evidence that N_2, not H_2, is lost upon photolysis, but they gave few details (11). Darensbourg (12) reported that irradiation of $[FeH_2(N_2)(PEtPh_2)_3]$ in the presence of excess carbon monoxide yields $[Fe(CO)_4(PEtPh_2)]$ and trans-$[Fe(CO)_3(PEtPh_2)_2]$ (see Reaction 2). Although the photochemistry of this complex is not resolved, the complex probably loses H_2 and N_2 upon irradiation, and the nature of the final product probably depends upon reaction conditions.

Green and co-workers showed that uv irradiation of $[W(\eta^5\text{-}C_5H_5)_2H_2]$ solutions results in $[W(\eta^5\text{-}C_5H_5)_2H(R)]$ or $[W(\eta^5\text{-}C_5H_5)_2R_2]$ formation, where R is derived from the solvent ($13, 14, 15$). For example, irradiation of the dihydride complex in benzene produces $[W(\eta^5\text{-}C_5H_5)_2H(C_6H_5)]$ (14), and in methanol, $[W(\eta^5\text{-}C_5H_5)_2H(OMe)]$ and $[W(\eta^5\text{-}C_5H_5)_2Me(OMe)]$ are formed (15). These reactions presumably occur through photoinduced H_2 elimination and yield reactive tungstenocene, which inserts into a C–H or O–H bond of a solvent molecule.

In 1971 Kruck and co-workers showed that uv irradiation of the monohydride $[IrH(PF_3)_4]$ produced H_2 and $[Ir_2(PF_3)_8]$ (16). Although this reaction apparently occurs with a very low quantum yield, $[Ir_2(PF_3)_8]$ reacts with water to regenerate $[IrH(PF_3)_4]$ (16), which completes a cycle for the photochemical generation of H_2 from water. These workers showed that irradiation of the cobalt analog $[CoM(PF_3)_4]$ produces the hydride- and phosphide-bridged complex shown in Reaction 3 (17). Elimination of H_2 from $[CoH_2(Chel)(PR_3)_2]^+$ (Chel = 2,2'-bypyridine,1,10-phenanthroline) is light accelerated (18), and photolysis of $[Fe(\eta^5\text{-}C_5H_5)_2H(CO)_2]$ leads to $[Fe(\eta^5\text{-}C_5H_5)_2(CO)_2]_2$ formation (19). Ellis and co-workers recently demonstrated that although replacement of H_2 by CO in $[VH_3(CO)_3(diars)]$ (diars = ortho-$(Me_2As)_2C_6H_4$) does not occur thermally, irradiation readily yields $[VH(CO)_4(diars)]$ (see Reaction 4).

$$[FeH_2(N_2)(PEtPh_2)_3] \xrightarrow{h\nu} H_2 + [\overline{FeH(C_6H_4PEtPh)}(N_2)(PEtPh_2)_2] \quad (1)$$

$$[FeH_2(N_2)(PEtPh_2)_3] + CO \xrightarrow{h\nu} [Fe(CO)_4(PEtPh_2)]$$

$$+ trans\text{-}[Fe(CO)_3(PEtPh_2)_2] \quad (2)$$

$$2\,[CoH(PF_3)_4] \xrightarrow{h\nu} (PF_3)_3\,Co \overset{\displaystyle H}{\underset{\displaystyle P}{\diagup\diagdown}} Co(PF_3)_3 \quad (3)$$

$$\overset{}{\underset{F\quad F}{}}$$

$$[vh_3(CO)_3(\text{diars})] + CO \xrightarrow{h\nu} H_2 + [VH(CO)_4(\text{diars})] \quad (4)$$

In connection with a study of the photochemical properties of several dioxygen complexes of iridium, Geoffroy, Gray, and Hammond (21) observed that irradiation of argon-purged solutions of $[IrH_2(Ph_2PCH_2CH_2PPh_2)_2]^+$, $[IrH_2(Ph_2PCHCHPPh_2)_2]^+$, and $[IrClH_2(CO)(PPh_3)_2]$ eliminated H_2 and formed the stable Ir(I) complexes $[Ir(Ph_2PCH_2CH_2PPh_2)_2]^+$, $[Ir(Ph_2PCHCHPPh_2)]^+$, and $[IrCl(CO)(PPh_3)_2]$. Of the dihydrides studied, only $[IrClH_2(CO)(PPh_3)_2]$ loses hydrogen thermally, and photolysis is the only known method for effecting H_2 elimination from the diphosphine complexes. The nature of the active excited state in these complexes was not identified, and the mechanism of H_2 elimination was not unambiguously determined.

Results and Discussion

$[IrClH_2(PPh_3)_3]$ and *mer-* and *fac-*$[IrH_3(PPh_3)_3]$. Our first objective was to test the generality of photoinduced elimination of molecular hydrogen from stable di- and polyhydride transition element complexes, and we first examined the well-characterized triphenylphosphine complexes, $[IrClH_2(PPh_3)_3]$, *mer-*$[IrH_3(PPh_3)_3]$, and *fac-*$[IrH_3(PPh_3)_3]$. $[IrClH_2(PPh_3)_3]$ was first prepared by Vaska, who reported it to be an air stable, light-sensitive white solid (22) whose configuration was determined by NMR and ir analysis (*see* Structure 1) (23). The

PPh_3

H ⋅⋅⋅⋅⋅⋅⋅→ PPh_3

Ir

H ⋅⋅⋅⋅⋅⋅⋅ Cl

PPh_3

1

complex also was prepared by Bennett and Milner (24) by the irreversible addition of H_2 to $[IrCl(PPh_3)_3]$. Compounds *mer-*$[IrH_3(PPh_3)_3]$ and *fac-*$[IrH_3(PPh_3)_3]$ are easily prepared (25) by heating $Na_2[IrCl_6]$, PPh_3, and $NaBH_4$ in ethanol, and have Structures 2 and 3 (26). All three complexes are resistant to thermal loss of hydrogen. For example, we showed that there is no H_2 loss when solutions are purged with an inert gas or when solid samples are heated to 150°C for 24 hr under vacuum.

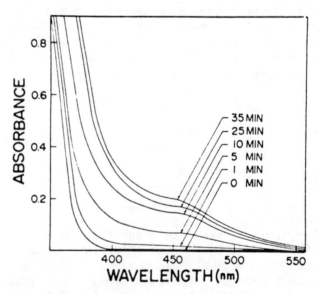

mer-[IrH$_3$(PPh$_3$)$_3$] fac-[IrH$_3$(PPh$_3$)$_3$]

We observed (27) that [IrClH$_2$(PPh$_3$)$_3$] is quite photosensitive, and when solid samples or solutions of the complex are irradiated with sunlight or fluorescent light, a rapid color change from white to orange occurs.   Mass spectral analysis of the gases above the irradiated solid samples shows a large amount of hydrogen. The electronic absorption spectral changes that occur during 366-nm photolysis of a degassed 2.3 × 10^{-2}M CH$_2$Cl$_2$ solution are shown in Figure 1.   The spectrum of [IrClH$_2$(PPh$_3$)$_3$] is featureless below 300 nm, and as irradiation proceeds, a new band appears at 449 nm.   This band is identical in position and shape to that displayed by a [IrCl(PPh$_3$)$_3$] sample prepared by the reaction of [IrCl(N$_2$)(PPh$_3$)$_2$] with PPh$_3$ (28).   When photolysis is followed in the ir spectral region, solutions show a steady decrease in intensity of the metal hydride vibrations at 2215 and 2110 cm^{-1}, and no new bands appear between 1800 and 2300 cm^{-1}.

The mass spectral analysis and the electronic and ir spectral changes demonstrate that hydrogen is eliminated from [IrClH$_2$(PPh$_3$)$_3$] upon photolysis and

Figure 1.   Electronic absorption spectral changes accompanying 366-nm photolysis of a 2.3 × 10^{-2}M degassed CH$_2$Cl$_2$ solution of [IrClH$_2$(PPh$_3$)$_3$] (27)

that [IrCl(PPh$_3$)$_3$] is the primary photoproduct.   If the photolysis is not prolonged (<30 min), [IrCl(PPh$_3$)$_3$] can be isolated as an orange solid from the irradiated solutions by solvent evaporation.   However, the 16-valence electron complex is quite reactive, and if hydrogen is removed from the solution or if photolysis is prolonged (forcing a buildup in its concentration), the complex undergoes the ortho-metallation reaction that was described by Bennett and Milner (*24*).

The entire photolysis sequence can be reversed readily by H$_2$, as summarized in Scheme 1 (*27*).   A sealed, degassed benzene solution of [IrClH$_2$(PPh$_3$)$_3$], from which the photoreleased H$_2$ is not allowed to escape, can be cycled repeatedly (>50 cycles) through the photoinduced H$_2$ elimination–thermal H$_2$ addition reactions without any observable loss of complex.   Elimination of H$_2$ can be induced by irradiation with λ < 400 nm.   The quantum yield of elimination, measured at 254 nm by monitoring the growth of the 449-nm band of [IrCl(PPh$_3$)$_3$], is 0.56 ± 0.03.

Scheme 1

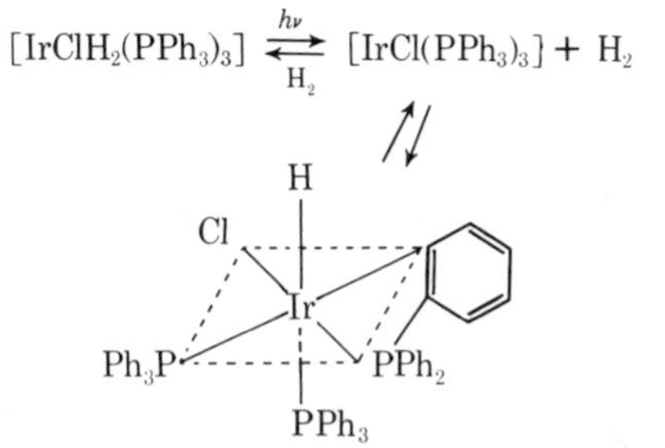

$$[IrClH_2(PPh_3)_3] \underset{H_2}{\overset{h\nu}{\rightleftarrows}} [IrCl(PPh_3)_3] + H_2$$

Thermal elimination of H$_2$ from [IrH$_2$(CO)$_2$(PMePh$_2$)$_2$]$^+$ proceeds in a concerted fashion, and thermal elimination of H$_2$ from other polyhydride complexes of iridium probably occurs by a similar mechanism.   We found that photolysis (λ = 366 nm; 15 min irradiation) of a degassed CH$_2$Cl$_2$ solution containing equimolar amounts of [IrClH$_2$(PPh$_3$)$_3$] and [IrClD$_2$(PPh$_3$)$_3$] gave H$_2$ and D$_2$, and no HD was detected by mass spectrometry (*27*).   The absence of HD in the gases above the irradiated solution of the mixture indicates that the photoinduced elimination of H$_2$ is also concerted since elimination of H$^-$(D$^-$), H·(D·), or H$^+$(D$^+$) would lead to detectable amounts of HD.   Further evidence for the concerted pathway of photoinduced elimination comes from previous studies (*21*) where irradiation of [IrClH$_2$(CO)(PPh$_3$)$_2$] in toluene did not yield bi-benzyl, an expected product if hydrogen atoms were formed in the photoprocess.

Preparation of [IrH$_3$(PPh$_3$)$_3$] according to literature procedures (*25*) gives a mixture of facial and meridional isomers that are separated by recrystallization from benzene–methanol.   Irradiation of a degassed benzene solution of the synthetic mixture yields a rapid decrease in intensity of the 1740 cm^{-1} $\nu_{Ir\text{-}H}$ of the meridional isomer, a slow decrease of the 2080 cm^{-1} $\nu_{Ir\text{-}H}$ of the facial isomer, and no new $\nu_{\text{Ir-H}}$ vibrations. Formation of H_2 was verified by mass spectral analysis of the gases above the irradiated solutions. Irradiation of a degassed benzene solution of pure mer-[IrH$_3$(PPh$_3$)$_3$] shows electronic absorption spectral changes similar to those observed for photolysis of [IrClH$_2$(PPh$_3$)$_3$], with new absorption shoulders at 375 and 430 nm. The expected primary photoproduct from both isomers, [IrH(PPh$_3$)$_3$], should be more reactive toward ortho-metallation than [IrCl(PPh$_3$)$_3$]. Evaporation of solvent from irradiated solutions of mer-[IrH$_3$(PPh$_3$)$_3$] yields an orange solid (see Structure 4) that we characterized as an ortho-metallated derivative [Ir(C$_6$H$_4$PPh$_2$)(PPh$_3$)$_2$], formed through photoelimination of a second H_2 molecule (see Scheme 2) (27). Further support for the formulation of Structure 4 is that the photoproduct can be converted quantitatively into [IrH$_3$(PPh$_3$)$_3$] by stirring a benzene solution of the complex under a H_2 atmosphere (27). Evidence for the initial photoproduction of [IrH(PPh$_3$)$_3$] is that irradiation of mer- and fac-[IrH$_3$(PPh$_3$)$_3$] under a CO atmosphere leads to formation of [IrH(CO)(PPh$_3$)$_3$] (27). This reaction does not occur thermally; therefore, [IrH(CO)(PPh$_3$)$_3$] probably is formed by [IrH(PPh$_3$)$_3$] scavenging CO.

Scheme 2

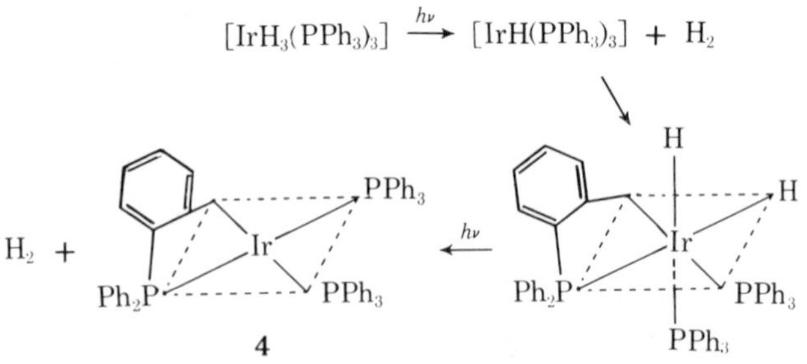

$$[\text{IrH}_3(\text{PPh}_3)_3] \xrightarrow{h\nu} [\text{IrH}(\text{PPh}_3)_3] + \text{H}_2$$

If photolysis of mer- or fac-[IrH$_3$(PPh$_3$)$_3$] is conducted under a H_2 atmosphere, a different product is obtained (27). Prolonged photolysis of [IrH$_3$(PPh$_3$)$_3$] in benzene under a H_2 purge yields a white precipitate with a single $\nu_{\text{Ir-H}}$ at 1948 cm^{-1} (KBr). This white solid is insoluble in water, acetone, benzene, chloroform, and dichloromethane, and further characterization proved impossible. A complex of similar color and solubility, with an identical ir spectrum, was prepared by Chatt and co-workers (26). Originally, it was formulated as [IrH$_3$(PPh$_3$)$_2$] but later was proposed (30) to be [IrH$_5$(PPh$_3$)$_2$]. Irradiation under H_2 suppresses the principal photoreaction pathway (H_2 loss) and allows a second pathway (PPh$_3$ loses) to be observed.

$$[\text{IrH}_3(\text{PPh}_3)_3] \xrightarrow[\text{H}_2]{h\nu} [\text{IrH}_5(\text{PPh}_3)_2] + \text{PPh}_3 \qquad (5)$$

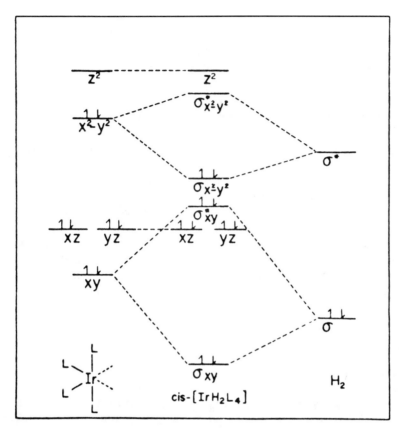

Figure 2. Molecular orbital energy level diagram for six-coordinate, cis-dihydride complexes of iridium. The metal d orbitals for an IrL_4 complex in octahedral geometry with vacant cis sites are shown on the left, and the H_2 molecular orbitals are indicated on the right (27).

To determine the nature of the photoactive excited state in these hydride complexes, we examined the electronic absorption spectra of a series of iridium hydride complexes. The spectra showed only a few shoulders on a rising absorption into the uv, and no definitive excited state assignments could be made.

However, a molecular orbital diagram for an iridium dihydride complex was drawn (*see* Figure 2). The orbitals on the left are d orbitals in proper ordering for an IrL_4 complex that is distorted to form an octahedron with vacant cis sites. The σ and σ^* orbitals of molecular hydrogen are shown on the right. In a *cis*-$[IrH_2L_4]$ complex, hydrogen bonding occurs through interactions of the H_2-σ orbital with d_{xy} and the H_2-σ^* orbital with $d_{x^2-y^2}$ (bonding and antibonding combinations are formed in each case). The exact position of the H_2-σ and H_2-σ^* orbitals, relative to the metal d orbitals, influences the final molecular orbital order although the positioning in Figure 2 appears reasonable in view of the low electronegativity of H_2. Similarly, if hydrogen is added as two distinct H atoms rather than a H_2 molecule, an electronic transition that depopulates $\sigma_{x^2-y^2}$ or populates $\sigma^*_{x^2-y^2}$ should weaken the metal-hydrogen bonding. The transition $\sigma_{x^2-y^2} \rightarrow \sigma^*_{x^2-y^2}$ leaves the complex with zero net bonding between H_2 and iridium, and elimination of H_2 is predicted. Although the exact nature of the photoactive excited state was not elucidated from spectral measurements, we view the complex in terms of the MO diagram shown in Figure 2 and propose that $\sigma^*_{x^2-y^2}$ is populated in the photoactive state. A similar molecular orbital diagram based on semi-empirical calculations was drawn for analogous O_2, S_2, and Se_2 adducts of iridium (31). Electrochemical studies of these complexes suggest that during the reduction of the adducts, the electron goes into an orbital of the $\sigma^*_{x^2-y^2}$ type (32). Note that irradiation of these di-oxygen (21) and di-sulfur (33) adducts leads to efficient elimination of O_2 and S_2.

The $[IrClH_2(PPh_3)_3]-[IrCl(PPh_3)_3]$ system serves as a model storage system for hydrogen and energy (27). $[IrCl(PPh_3)_3]$ readily takes up H_2 to store it as $[IrClH_2(PPh_3)_3]$, and then releases it on demand by irradiation with uv light or with sunlight. When H_2 adds to $[IrCl(PPh_3)_3]$, approximately 15–20 kcal/mol of energy is released (34): $[IrCl(PPh_3)_3] + H_2 \rightarrow [IrClH_2(PPh_3)_3] + 15$–$20$ kcal/mol. Irradiation of $[IrClH_2(PPh_3)_3]$ with 366 nm (78 kcal/mol) gives H_2 and $[IrCl(PPh_3)_3]$ that can be stored separately. When the components react to reform $[IrClH_2(PPh_3)_3]$, approximately 15–20 kcal/mol of energy is released and this amount of energy was stored. Because of iridium's high cost, a system like this can serve only as a model.

$[RuClH(CO)(PPh_3)_3]$, $[RuH_2(CO)(PPh_3)_3]$, and $[RuClH(CO)_2(PPh_3)_2]$. We examined (35) the well-characterized ruthenium compounds $[RuClH(CO)-(PPh_3)_3]$ (see Structure 5), $[RuH_2(CO)(PPh_3)_3]$ (see Structure 6), and [Ru-ClH(CO)_2(PPh_3)_2] (see Structure 7). Structures 5 and 6 are ideal for photo-

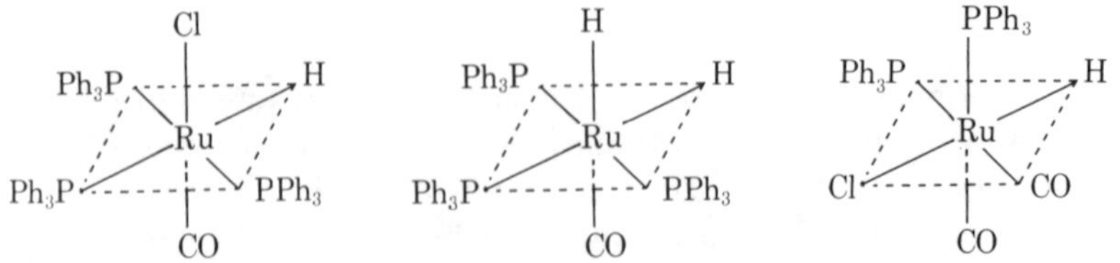

| 5 | 6 | 7 |

chemistry comparison since **6** is derived from **5** by hydride substitution of chloride. The three complexes are easily prepared and relatively air stable in the solid state. Solutions, however, slowly decompose when exposed to air. The dihydride, $[RuH_2(CO)(PPh_3)_3]$, is resistant to H_2 loss under thermal conditions, and no reaction was observed when the complex was heated for prolonged periods under vacuum.

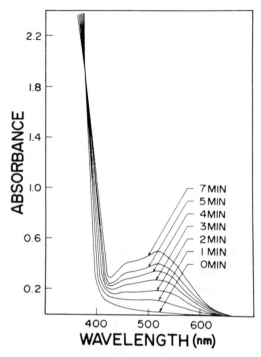

Figure 3.   Electronic absorption spectral changes resulting from 366-nm irradiation of a 10^{-3}M CH_2Cl_2 solution of [RuClH(CO)-(PPh$_3$)$_3$] (35)

The complex [RuClH(CO)(PPh$_3$)$_3$] is light sensitive, and irradiation of a degassed $10^{-3}M$ benzene solution of [RuClH(CO)(PPh$_3$)$_3$] with 366 nm results in the electronic absorption spectral changes shown in Figure 3.   As irradiation proceeds, the solution changes from yellow to purple, and new absorption bands appear and increase in intensity at 520 and 470 nm.   The final spectrum (*see* Figure 3) is identical to that of an authentic [RuClH(PPh$_3$)$_3$] sample prepared by the reaction of [RuCl$_2$(PPh$_3$)$_4$] with Et$_3$N and H$_2$ (36), which suggests that the photochemical reaction expressed in Reaction 6 occurs.   Similar spectral changes were obtained in degassed toluene and dichloromethane solutions, but irradiation in the presence of oxygen leads to rapid decomposition because of the air sensitivity of [RuClH(PPh$_3$)$_3$].

Confirmation of [RuClH(PPh$_3$)$_3$] photogeneration comes from a catalysis experiment.   [RuClH(PPh$_3$)$_3$] was reported (37) to be one of the most efficient homogeneous catalysts for the hydrogenation of terminal olefins.   A deoxygenated benzene solution, $10^{-3}M$ in [RuClH(CO)(PPh$_3$)$_3$] and $3 \times 10^{-2}M$ in 1-hexene, under a H$_2$ atmosphere showed no H$_2$ uptake before irradiation, but rapid hydrogen uptake occurred upon photolysis with 366 nm.

The ir changes that occur during photolysis of [RuClH(CO)(PPh₃)₃] indicate that the overall reaction is not as simple as that expressed in Reaction 5. Although a smooth decrease in the $\nu_{C\equiv O}$ of [RuClH(CO)(PPh₃)₃] at 1925 cm⁻¹ occurs, two new weak bands appear at 1965 and 2040 cm⁻¹. These two bands are characteristic (38) of [RuClH(CO)₂(PPh₃)₂], which probably is formed by the reaction of [RuClH(CO)(PPh₃)₃] with photoreleased CO (see Reaction 7). The dicarbonyl complex is photosensitive but does not lose carbon monoxide. Instead it appears to undergo photoisomerization (35) as do other [RuX₂(CO)₂L₂] complexes (39). Consequently, it was not possible to achieve quantitative conversion of [RuClH(CO)(PPh₃)₃] into the catalyst [RuClH(PPh₃)₃], and the best yield we obtained was about 85% (35).

The quantum yield measured at 313 nm for CO elimination from [RuClH(CO)(PPh₃)₃] is 0.06 ± 0.02. Because of the air sensitivity of [RuClH(PPh₃)₃], the yield was determined by irradiating a CH₂Cl₂ solution of [RuClH(CO)-(PPh₃)₃] in a degassed and sealed uv cell. Since the reaction vessel was sealed, reverse reaction with CO was not prevented, and the measured quantum yield should be considered a lower limit.

Irradiation of a degassed benzene solution of [RuH₂(CO)(PPh₃)₃] results in the electronic absorption spectral changes shown in Figure 4. As irradiation proceeds, the solutions change from colorless to yellow, and a shoulder appears

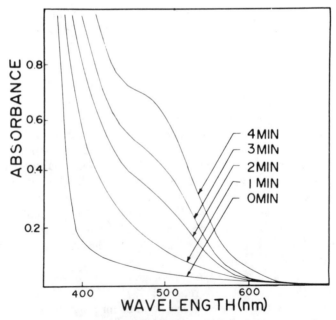

Figure 4. Electronic absorption spectral changes resulting from 366-nm irradiation of a 10⁻³M benzene solution of [RuH₂CO(PPh₃)₃] (35)

$$[RuClH(CO)(PPh_3)_3] \xrightarrow{h\nu} CO + [RuClH(PPh_3)_3] \tag{6}$$

$$[RuClH(CO)(PPh_3)_3] + CO \rightarrow [RuClH(CO)_2(PPh_3)_2] + PPh_3 \tag{7}$$

$$[RuH_2(CO)(PPh_3)_3] \xrightarrow{h\nu} [Ru(CO)(PPh_3)_3] \tag{8}$$

$$[RuH_2(CO)(PPh_3)_3] \xrightarrow[CO]{h\nu} H_2 + [Ru(CO)_3(PPh_3)_2] + PPh_3 \tag{9}$$

and increases in intensity at 400 nm. Mass spectral analysis of the gases above an irradiated, degassed benzene solution showed substantial amounts of molecular hydrogen and no evidence of CO, which suggests that irradiation of $[RuH_2(CO)(PPh_3)_3]$ leads to photoinduced H_2 elimination (*see* Reaction 8). The primary photoproduct expected from H_2 loss, $[Ru(CO)(PPh_3)_3]$, should be extremely reactive. The only material isolated from irradiated, degassed benzene solutions was an amorphous yellow solid that showed ir bands suggestive of an ortho-metallated product (35).

Although we could not isolate and characterize a pure product from photolysis of $[RuH_2(CO)(PPh_3)_3]$ in benzene, we could trap the proposed intermediate by irradiation under carbon monoxide. When irradiation was conducted under CO, the ir and ^{31}P NMR spectral changes showed that $[Ru(CO)_3(PPh_3)_2]$ was produced quantitatively (*see* Reaction 9) (35). The product could be isolated pure by solvent concentration.

Three ruthenium hydride complexes were examined, and each showed a different photochemistry (35). Irradiation of $[RuClH(CO)(PPh_3)_3]$ leads to CO loss, $[RuH_2(CO)(PPh_3)_3]$ undergoes photoinduced reductive elimination of molecular hydrogen, and $[RuClH(CO)_2(PPh_3)_2]$ undergoes photoisomerization. Unfortunately, the electronic absorption spectra of the three complexes are not clearly resolved and therefore reveal little about the nature of the excited states that give the three different reactivities. It is significant, however, that the replacement of chloride by hydride when going from $[RuClH(CO)(PPh_3)_3]$ to $[RuH_2(CO)(PPh_3)_3]$ completely changed the photochemistry. We believe that photoinduced elimination of H_2 is such a favored reaction that when a complex has two hydrogens in cis positions it will undergo H_2 loss, regardless of other possible reaction pathways.

$[Mo(\eta^5\text{-}C_5H_5)_2H_2]$ and $[MoH_4(dppe)_2]$. Our studies of the di- and trihydride complexes of ruthenium and iridium, described above and published previously (27, 35), and those of other workers (discussed at the beginning of this chapter), indicate that photoinduced elimination of molecular hydrogen is a common reaction pathway for di- and polyhydride complexes. To demonstrate the photoreaction's generality and its utility for generating otherwise unattainable, extremely reactive metal complexes, we have begun to study the photochemistry of polyhydride complexes of the early transition metals. We focused initially on molybdenum and examined $[Mo(\eta^5\text{-}C_5H_5)_2H_2]$ and $[MoH_4(dppe)_2]$, and reexamined the previously studied $[W(\eta^5\text{-}C_5H_5)_2H_2]$ (13, 14, 15).

The $[Mo(\eta^5\text{-}C_5H_5)_2H_2]$ and $[W(\eta^5\text{-}C_5H_5)_2H_2]$ complexes are prepared by the reaction of $MoCl_5$ and WCl_6, respectively, with $NaBH_4$ and NaC_5H_5, and are isolated as air-sensitive yellow solids (40). Both complexes are resistant to thermal hydrogen loss and can be recovered unchanged from sublimation at 100°C under vacuum. $[Mo(\eta^5\text{-}C_5H_5)_2H_2]$ is photosensitive, and irradiation of a degassed iso-octane solution of the complex with 366-nm light gives a rapid color change from yellow to brown. A red-brown precipitate is obtained after prolonged photolysis. Accompanying the irradiation is a smooth decrease in intensity of the characteristic ν_{M-H} of $[Mo(\eta^5\text{-}C_5H_5)_2H_2]$ at 1847 cm^{-1}, and no new vibrations appear in the ν_{M-H} region. The ^1H NMR spectrum of the red-brown precipitate and its color suggest its identity as the polymeric $[Mo(\eta^5\text{-}C_5H_5)_2]_x$ species previously described by Thomas (41, 42). Mass spectral analysis of the gases above the irradiated solutions shows a considerable amount of H_2. The apparent reaction sequence is outlined in Reaction 10. In contrast to $[W(\eta^5\text{-}C_5H_5)_2H_2]$, photolysis of $[Mo(\eta^5\text{-}C_5H_5)_2H_2]$ in aromatic solvents does not produce any products arising from C–H insertion, and only the polymeric material is observed.

Further evidence for initial generation of molybdenocene (see Reaction 10) comes from trapping experiments. Irradiation under a carbon monoxide or an acetylene purge, for example, leads to near quantitative formation of the previously characterized $[Mo(\eta^5\text{-}C_5H_5)_2CO]$ (42) and $[Mo(\eta^5\text{-}C_5H_5)_2(C_2H_2)]$ adducts (43). These can be separated and purified by fractional sublimation and identified by their ir, NMR, and mass spectra.

$$[Mo(\eta^5\text{-}C_5H_5)_2H_2] \xrightarrow{h\nu} H_2 + [Mo(\eta^5\text{-}C_5H_5)_2] \rightarrow [Mo(\eta^5\text{-}C_5H_5)_2]_x \quad (10)$$

Irradiation of $[Mo(\eta^5\text{-}C_5H_5)_2H_2]$ with excess PPh_3 or PEt_3 leads to formation of the new tertiary phosphine adducts, $[Mo(\eta^5\text{-}C_5H_5)_2PR_3]$. Electronic absorption spectral changes are obtained when a $1.1 \times 10^{-4}M$ hexane solution of $[Mo(\eta^5\text{-}C_5H_5)_2H_2]$ is irradiated with 366 nm with excess PPh_3 (see Figure 5), and the isosbestic points at 285 and 270 nm (not shown) suggest a clean conversion. The phosphine adducts can be isolated from the photolysis mixture by fractional sublimation. $[Mo(\eta^5\text{-}C_5H_5)_2PPh_3]$ and $[Mo(\eta^5\text{-}C_5H_5)_2PEt_3]$ sublime at 90° and 80°C (10^{-3} mm Hg), respectively, whereas unreacted $[Mo(\eta^5\text{-}C_5H_5)_2H_2]$ sublimes at 50°–60°C. The PPh_3 adduct often contains small quantities of PPh_3 impurity, but the PEt_3 adduct can be isolated pure. Both adducts were characterized by their NMR, ir, and mass spectra. $[Mo(\eta^5\text{-}C_5H_5)_2PEt_3]$, for example, shows a doublet at 6.17 τ ($J_{P-H} = 5$ Hz) in its ^1H NMR spectrum assigned to the $\eta^5\text{-}C_5H_5$ protons, and a singlet at 34.9 ppm in its ^{31}P NMR spectrum assigned to coordinated PEt_3. $[Mo(\eta^5\text{-}C_5H_5)_2PPh_3]$ shows corresponding resonances at 6.18 τ ($J_{P-H} = 4$ Hz) and 18.9 ppm.

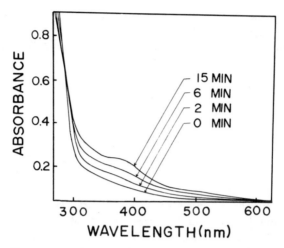

Figure 5. Electronic absorption spectral changes obtained during 366-nm irradiation of a 1.1 × 10^{-4}M hexane solution of $[Mo(\eta^5\text{-}C_5H_5)_2H_2]$ and excess PPh$_3$

The phosphine complexes provide a thermal route to other molybdenocene adducts since the molybdenum–phosphorus bond appears to be labile. When solutions of $[Mo(\eta^5\text{-}C_5H_5)_2PEt_3]$ react with CO or diphenylacetylene, formation of the corresponding adduct results (*see* Reaction 11).

$$[Mo(\eta^5\text{-}C_5H_5)_2PR_3] + L \xrightarrow{27°C} [Mo(\eta^5\text{-}C_5H_5)_2L] + PR_3 \qquad (11)$$

The quantum yield for H_2 elimination from $[Mo(\eta^5\text{-}C_5H_5)_2H_2]$, measured at 366 nm in hexane solution in a degassed and sealed spectrophotometer cell, is 0.10. This value should be treated as a lower limit since reverse reaction of H_2 with photogenerated molybdenocene was not prevented. This compares with a value of 0.01 that we obtained for H_2 elimination from $[W(\eta^5\text{-}C_5H_5)_2H_2]$ under similar photolysis conditions.

We conducted experiments to probe the mechanism of H_2 elimination from $[Mo(\eta^5\text{-}C_5H_5)_2H_2]$. Mass spectral analysis of gases above irradiated toluene-d_8 solutions showed predominant H_2 production with less than 10% HD. Since toluene is an efficient hydrogen atom scavenger, the absence of HD indicates that free hydrogen atoms are not produced to a large extent (if at all) during photolysis; therefore, an intramolecular elimination process is suggested. Irradiation of $[Mo(\eta^5\text{-}C_5H_5)_2D_2]$ in C_6D_6, C_6H_6, or C_6D_6 solutions containing excess PPh$_3$ gave D_2/HD mixtures in an approximate 3:2 ratio. Since the toluene solution experiments showed that free hydrogen atoms are apparently not produced, and since heterolytic cleavage of a Mo–H bond is unlikely, substantial D_2 production should arise from concerted D_2 elimination. The HD presumably comes from secondary thermal reactions of photogenerated $[Mo(\eta^5\text{-}C_5H_5)_2]$ with un-reacted $[Mo(\eta^5\text{-}C_5H_5)_2D_2]$. These reactions could give dimeric intermediates containing $\eta^1,\eta^5\text{-}C_5H_5$ rings (e.g., Structure 8), which could lose HD to give di-meric products similar to those described by Green and co-workers (45).

8

The electronic absorption spectra of $[Mo(\eta^5\text{-}C_5H_5)_2H_2]$ and $[W(\eta^5\text{-}C_5H_5)_2H_2]$ reveal little about the nature of the active excited state. They each show an in-tense peak at 270 nm ($\epsilon \approx 5000$) which may be attributed to a metal-to-C_5H_5 charge transfer, and a shoulder near 315–330 nm ($\epsilon \approx 800$–2000) that cannot be assigned readily. Neither complex shows the ligand field transitions at low energy characteristic of $[Mo(\eta^5\text{-}C_5H_5)_2Cl_2]$ (45), but this is expected because of the high ligand field strength of the hydride ligand. The photochemistry is, however, consistent with a molecular orbital diagram calculated for $[Mo(\eta^5\text{-}C_5H_5)_2H_2]$ by Dahl and co-workers (46) (see Figure 6). Significantly, the highest occupied molecular orbital, $8a_1$, is the principal bonding orbital between Mo and H_2. The $6b_2$ orbital, which could easily be populated by excitation, is strongly antibonding between Mo and H_2. Depopulation of $8a_1$ or population of $6b_2$ should greatly weaken the Mo-H_2 bonding, and H_2 loss would be expected from such excited states.

The complex $[MoH_4(dppe)_2]$ is synthesized easily by reaction of $[MoCl_4(dppe)]$ with $NaBH_4$ in the presence of excess dppe (47), and can be iso-lated as a pale yellow solid which is relatively stable to thermal loss of H_2. De-composition does occur, however, on prolonged heating under N_2 or vacuum, but clean conversion to a single product does not occur. Although the structure of $[MoH_4(dppe)_2]$ has not been determined by x-ray diffraction, the analogous $[MoH_4(PMe_2Ph)_4]$ complex has been shown to possess the MoH_4P_4 core below. It is best described as two inter-penetrating tetrahedra with the hydrogen atoms forming an elongated tetrahedron and the phosphine ligands forming a flattened tetrahedron. Spectral evidence indicates that the structure of $[MoH_4(dppe)_2]$ is similar (48). The ^1H and ^{31}P NMR spectral data suggest that $[MoH_4(dppe)_2]$ is fluxional at or near room temperature (48).

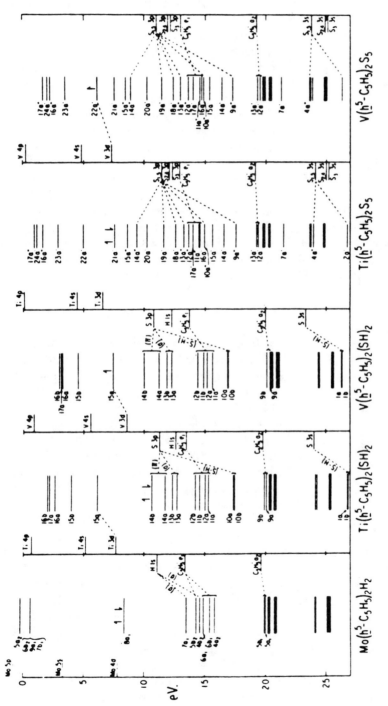

Figure 6. Molecular orbital energy level diagram for [Mo(η⁵-C₅H₅)₂H₂] and related complexes. The half-arrows designate the number of electrons in the highest occupied molecular orbital (46).

Irradiation of degassed benzene solutions of the complex with 366-nm light induces a rapid color change from bright yellow to orange. This color change is accompanied by a steady decrease in intensity of the molybdenum–hydride stretch at 1745 cm^{-1} and the hydride resonance at $\tau = 13.6$ ppm. The initial singlet at -83.5 ppm in the ^{31}P NMR spectrum of the complex in benzene solution decreases in intensity as the photolysis proceeds. Mass spectra, gas chromatographic, and Toepler-pump analysis of the gases above the irradiated solutions show H_2 formation with two moles released per mole of complex irradiated. Solvent evaporation from the irradiated solutions gave a very air-sensitive orange solid that showed no metal–hydride vibrations in the ir, and no hydride resonances in its ^{1}H NMR spectrum.

The absence of evidence for a ligating hydrogen suggests the formulation of the product as [Mo(dppe)$_2$] or its dimer. The relatively low solubility of the complex has thus far prevented molecular weight measurements. An ortho-metallated derivative, which might be expected, would be expected to show evidence for a molybdenum–hydride bond. The overall photochemical reaction that is consistent with this formulation and with the quantity of hydrogen released is given in Reaction 12.

$$[MoH_4(dppe)_2] \xrightarrow{h\nu} 2\,H_2 + [Mo(dppe)_2] \qquad (12)$$

Further evidence for the formulation given for the product comes from its reactivity. Irradiation of [MoH$_4$(dppe)$_2$] under a CO atmosphere (or addition of CO to a solution of the orange product) gave a mixture of cis- and trans-[Mo(CO)$_2$(dppe)$_2$], as shown by their characteristic (49, 50) ir spectra with bands at 1853 and 1770 cm^{-1} (cis) and 1812 cm^{-1} (trans). Photolysis in the presence of ethylene gave an orange solid that displayed a ^{1}H NMR spectrum indicative of an ethylene complex but not identical to that reported (51) for [Mo(C$_2$H$_4$)$_2$-(dppe)$_2$]. Further characterization of this complex is under way.

Irradiation of [MoH$_4$(dppe)$_2$] under an N$_2$ atmosphere results in virtually quantitative formation of [Mo(N$_2$)$_2$(dppe)$_2$]. In a typical experiment, 225 mg of [MoH$_4$(dppe)$_2$] in 100 mL of N$_2$-saturated benzene was irradiated with 366 nm for 12 hr. Concentration of the solution to 10 mL, followed by addition of 75 mL of methanol, gave 221 mg of [Mo(N$_2$)$_2$(dppe)$_2$] (93% yield). The bis-dinitrogen complex was characterized by its ^{31}P NMR spectrum ($\delta = -64.9$ ppm) and its ir spectrum (52) ($\nu_{N\equiv N} = 1972$ cm^{-1}) that showed no remaining [MoH$_4$(dppe)$_2$]. This synthetic procedure represents a considerable improvement in yield and convenience over previously reported methods (52, 53) that gave yields from 13–60%, depending on starting material and reaction conditions. Lengthy isolation and purification steps are not required, and the overall yield from [MoCl$_4$(dppe)] through [MoH$_4$(dppe)$_2$] is about 70%.

No experiments have been conducted to determine the elimination mechanism (although it is likely concerted), and poor spectral properties have precluded quantum yield measurements. The reaction does appear efficient, however, since 250 mg of $[MoH_4(dppe)_2]$ can be converted entirely to $[Mo(N_2)_2$-$(dppe)_2]$ with a few hours of irradiation.

Summary

Although only relatively few hydride complexes have been examined, several aspects concerning their photoreactivity are becoming clear.   First, it appears that irradiation of complexes containing only one hydride ligand will lead to typical photochemical reactions.   For example, photolysis of $[RuClH(CO)$-$(PPh_3)_3]$ leads to CO loss as do many metal carbonyls, and irradiation of $[Ru$-$ClH(CO)_2(PPh_3)_2]$ gives photoisomerization analogous to that observed (*39*) for a series of $[RuX_2(CO)_2L_2]$ (X = halide; L = tertiary phosphine) complexes.   In particular, we do not believe that photoelimination of H^+, H·, or H^- is important. Even Adamson's photochemical rules (*54*) predict that loss of H^+, H·, or H^- should not occur often since the weakest ligand-field axis is rarely the hydride-containing axis.

We provided strong evidence that a complex with two or more hydrogens will lose H_2 when irradiated regardless of how thermally resistant it is to H_2 loss (*27, 35*).   We believe that this is a general reaction for di- and polyhydrides of all the transition elements, and it has been observed for V, Mo, W, Fe, Ru, Co, and Ir.   We are investigating a series of polyhydride complexes to test this general concept and to explore the possibility of generating extremely reactive metal complexes that cannot be obtained thermally but which might be derived through photoinduced H_2 elimination.

Experimental

The complexes $[Mo(\eta^5\text{-}C_5H_5)_2H_2]$ (*40*), $[Mo(\eta^5\text{-}C_5H_5)_2D_2]$ (*40*), $[W(\eta^5$-$C_5H_5)_2H_2]$ (*40*), $[MoCl_4(dppe)]$ (*47*), and $[MoH_4(dppe)_2]$ (*47*) were prepared by published procedures.   Triphenylphosphine was obtained from Aldrich Chemical Co. and was recrystallized from benzene/ethyl alcohol before use.   All other chemicals were reagent grade and were used without further purification. All solvents were dried by standard methods and rigorously degassed before use. Manipulations and reactions with air-sensitive compounds were carried out under an argon atmosphere unless otherwise stated.

General Irradiation Procedures.   Irradiations were conducted at 366 nm using a 450-W Hanovia, medium-pressure Hg lamp equipped with Corning Glass 0–52 and 7–37 filters ($I \approx 10^{-7}$ einstein/min), a 100 W Blak–Ray B100A lamp equipped with a 366-nm narrow bandpass filter, or a 350-nm Rayonet photoreactor.   The complex to be studied was placed in an evacuable quartz uv cell or in a Schlenk tube.   After degassing, the appropriate solvent was distilled onto the sample.   Solutions for ir studies were transferred in an inert atmosphere glovebox to 0.5-nm path length, NaCl solution ir cells.   Solutions were irradiated with the appropriate lamp, and electronic and ir spectra were recorded periodically.   Samples for ^1H and ^{31}P NMR spectra were prepared similarly, and the NMR tubes sealed under vacuum.   Lamp intensities were measured by ferrioxalate actinometry.

Photolysis of [Mo(η^5-C$_5$H$_5$)$_2$H$_2$]. Irradiation of degassed iso-octane, benzene, and hexane solutions of [Mo(η^5-C$_5$H$_5$)$_2$H$_2$] with uv light led to change in color from yellow to brown. Prolonged photolysis led to a red-brown precipitate that showed ir bands of coordinated η^5-C$_5$H$_5$. The ^{1}H NMR spectrum of this precipitate showed a complex pattern in the cyclopentadienyl region but no evidence for a M–H resonance. The quantum yield of H$_2$ loss from [Mo(η^5-C$_5$H$_5$)$_2$H$_2$] was determined by 366-nm irradiation ($I = 1.42 \times 10^{-6}$ einstein/min) of thoroughly degassed n-hexane solutions placed in sealed uv cells. The photoreaction was monitored by measuring the decrease in absorbance of the [Mo(η^5-C$_5$H$_5$)$_2$H$_2$] band at 270 nm.

Trapping Experiments with CO and HC$\equiv$CH. Photolysis of deoxygenated iso-octane solutions of [Mo(η^5-C$_5$H$_5$)$_2$H$_2$] under a CO purge gave a color change from yellow to green. Solvent evaporation at $-10°$C yielded a green solid that was purified by sublimation at $25°$–$30°$C (10^{-3} mm Hg). The green sublimed product showed a single $\nu_{C\equiv O}$ at 1910 cm^{-1} in its ir spectrum, a singlet at 5.70 τ in its ^{1}H NMR spectrum, and a parent ion at 258 m/e in its mass spectrum. These data implied that the product was identical to the previously characterized [Mo(η^5-C$_5$H$_5$)$_2$CO] (42).

Photolysis of deoxygenated iso-octane solutions of [Mo(η^5-C$_5$H$_5$)$_2$H$_2$] in the presence of an HC$\equiv$CH purge gave a yellow-to-orange color change. Solvent evaporation yielded an orange-brown solid that was purified by sublimation at $25°$–$35°$C (10^{-3} mm Hg) to give orange crystals. The ^{1}H NMR spectrum of the product showed two singlets (at 2.32 and 5.70τ) identical to those of the previously characterized [Mo(η^5-C$_5$H$_5$)$_2$(C$_2$H$_2$)] (43).

Preparation of [Mo(η^5-C$_5$H$_5$)$_2$PEt$_3$] and [Mo(η^5-C$_5$H$_5$)$_2$PPh$_3$]. Irradiation of a solution of 100 mg of [Mo(η^5-C$_5$H$_5$)$_2$H$_2$] and a tenfold excess of PPh$_3$ in 75 mL of degassed iso-octane gave a color change from yellow to red-orange. Removal of the solvent yielded a red-orange solid that sublimed at $80°$–$90°$C (10^{-3} mm Hg), and that was shown to be [Mo(η^5-C$_5$H$_5$)$_2$PPh$_3$] by its spectral data. (The product is often contaminated with co-sublimed PPh$_3$ and [Mo(η^5-C$_5$H$_5$)$_2$-H$_2$].)

Photolysis of a solution containing 100 mg of [Mo(η^5-C$_5$H$_5$)$_2$H$_2$] and a tenfold excess of PEt$_3$ in 75 mL of iso-octane gave a color change similar to that of the PPh$_3$ reaction. Solvent removal followed by prolonged pumping to remove the excess PEt$_3$ yielded a red-orange material that sublimed at $70°$–$80°$C. This substance was characterized as [Mo(η^5-C$_5$H$_5$)$_2$PEt$_3$]. The product is collected more easily by solvent evaporation and removal of unreacted [Mo(η^5-C$_5$H$_5$)$_2$H$_2$] by sublimation at $50°$C (10^{-3} mm Hg). This method essentially gave quantitative conversion based on reacted dihydride.

Mechanistic Experiments. Sample solutions of [Mo(η^5-C$_5$H$_5$)$_2$H$_2$] and [Mo(η^5-C$_5$H$_5$)$_2$D$_2$] were prepared by distilling appropriate degassed solvent into Schlenk tubes that contained the complex. The sample solutions were subjected to short-term photolysis (<5 min) in a 350-nm Rayonet reactor. The solutions were frozen immediately after photolysis, and the mass spectrum of the gases above the frozen solutions was recorded.

Photolysis of [MoH$_4$(dppe)$_2$]. Photolysis with 366 nm of an argon-saturated benzene solution of [MoH$_4$(dppe)$_2$] gave a color change from yellow to bright orange. Benzene removal by vacuum evaporation gave an orange solid that was washed with pentane and dried in vacuo. If the photolysis was sufficiently prolonged, this orange solid was free of unreacted [MoH$_4$(dppe)$_2$] according to ir measurements. The spectral data for this material (discussed earlier) suggests its formulation as [Mo(dppe)$_2$]. Irradiation of a degassed benzene solution containing 0.05 g (0.055 mmol) of $[MoH_4(dppe)_2]$ released 0.111 mmol of H_2, as shown by Toepler pump analysis of the gases above the irradiated solution.

Preparation of $[Mo(N_2)_2(dppe)_2]$. A solution of 225 mg of $[MoH_4(dppe)_2]$ dissolved in 100 mL of benzene was irradiated with 366 nm for 12 hr under a continuous N_2 purge. The solution changed from yellow to orange, and with a decrease in concentration to 10 mL followed by the addition of 75 mL of dried, degassed MeOH, the solution gave a 220 mg precipitate of $[Mo(N_2)_2(dppe)_2]$, a yield of 93%.

Spectral Measurements. The ir spectra were recorded on a Perkin-Elmer 621 grating ir spectrophotometer using KBr disks prepared from ir spectroquality powder (MCB) or 0.5 mm path length, NaCl solution ir cells. Electronic absorption spectra were recorded with a Cary 17 spectrophotometer using 1-cm quartz spectrophotometer cells. Mass spectra were recorded with an AEI MS902 mass spectrometer.

Acknowledgment

We thank the donors of the Petroleum Research Fund administered by the American Chemical Society and The National Science Foundation for support of this research. The assistance of A. Freyer and R. Minard in obtaining NMR and mass spectra is greatly appreciated.

Literature Cited

1. Wilkinson, G., Birmingham, J. M., *J. Am. Chem. Soc.* (1955) **77**, 3421.
2. Chatt, J., *Science* (1968) **160**, 723.
3. Ginsberg, A. P., *Transition Met. Chem.* (1965) **1**, 111.
4. Green, M. L. H., *Endeavour*, (1967) **26**, 129.
5. Green, M. L. H., Jones, D. J., *Adv. Inorg. Chem. Radiochem.* (1965) **7**, 115.
6. Kaesz, H. D., Saillant, R. B., *Chem. Rev.* (1972) **72**, 231.
7. Roundhill, D. M., *Adv. Organomet. Chem.* (1975) **13**, 273.
8. Geoffroy, G. L., Lehman, J. R., *Adv. Inorg. Chem. Radiochem.* (1977) **20**, 189.
9. Balzani, V., Moggi, L., Manfrin, M. F., Bolletta, F., Gleria, M., *Science* (1975) **189**, 852.
10. Sacco, A., Aresta, M., *J. Chem. Soc., Chem. Commun.* (1968) 1223.
11. Koerner von Gustorf, E., Fischler, I., Leitish, J., Dreeskamp, H., *Angew. Chem. Int. Ed. Engl.* (1972) **11**, 1088.
12. Darensbourg, D. J., *Inorg. Nucl. Chem. Lett.* (1972) **8**, 529.
13. Elmit, K., Green, M. L. H., Forder, R. A., Jefferson, I., Prout, K., *J. Chem. Soc., Chem. Commun.* (1974) 747.
14. Giannotti, C., Green, M. L. H., *J. Chem. Soc., Chem. Commun.* (1972) 1114.
15. Fairugia, L., Green, M. L. H., *J. Chem. Soc., Chem. Commun.* (1975) 416.
16. Kruck, T., Sylvester, G., Kunau, I. P., *Angew. Chem. Int. Ed. Engl.* (1971) **10**, 725.
17. Kruck, T., Sylvester, G. S., Kunau, I. P., *Z. Naturforsch., Teil B* (1973) **28**, 38.
18. Camus, A., Cocevar, C., Mestroni, G., *J. Organomet. Chem.* (1972) **39**, 355.
19. Nesmeyanov, A. N., Chapovsky, Y. A., Ustynyuk, Y. A., *J. Organomet. Chem.* (1967) **9**, 345.
20. Ellis, J. E., Faltynek, R. A., Hentges, S. G., *J. Am. Chem. Soc.* (1977) **91**, 626.
21. Geoffroy, G. L., Gray, H. B., Hammond, G. S., *J. Am. Chem. Soc.* (1975) **97**, 3933.
22. Vaska, L., *J. Am. Chem. Soc.* (1961) **83**, 756.
23. Taylor, R. C., Young, J. F., Wilkinson, G., *Inorg. Chem.* (1966) **5**, 20.

24. Bennett, M. A., Milner, D. L., *J. Am. Chem. Soc.* (1969) **91**, 6983.
25. Ahmad, N., Robinson, S. D., Uttley, M. F., *J. Chem. Soc., Dalton Trans.* (1972) 843.
26. Chatt, J., Coffey, R. S., Shaw, B. L., *J. Chem. Soc.* (1965) 7391.
27. Geoffroy, G. L., Pierantozzi, R., *J. Am. Chem. Soc.* (1976) **98**, 8054.
28. Collman, J. P., Kubota, M., Vastine, F. D., Sun, J. Y., Kang, J. W., *J. Am. Chem. Soc.* (1968) **90**, 5430.
29. Mays, M. J., Simpson, R. N. F., Stefanini, F. P., *J. Chem. Soc. A* (1970) 3000.
30. Clerici, M. G., DiGioacchio, S., Maspero, F., Penotti, E., Zanobi, A., *J. Organomet. Chem.* (1975) **84**, 379.
31. Ginsberg, A. P., Teo, B. K., presented at the 169th American Chemical Society National Meeting, Philadelphia, Pa., April 1975.
32. Teo, B. K., Ginsberg, A. P., Calabrese, J. C., *J. Am. Chem. Soc.* (1976) **98**, 3027.
33. Ginsberg, A. P., private communication.
34. Vaska, L., Werneke, M. F., *Trans. N.Y. Acad. Sci.* (1971) **33**, 70.
35. Geoffroy, G. L., Bradley, M. G., *Inorg. Chem.* (1977) **16**, 744.
36. Schunn, R. A., Wonchoba, E. R., *Inorg. Synth.* (1972) **13**, 131.
37. Hallman, P. S., McGarvey, B. R., Wilkinson, G., *J. Chem. Soc. A* (1968) 3143.
38. James, B. R., Markham, L. D., Hui, B. C., Rempel, G. L., *J. Chem. Soc., Dalton Trans* (1973) 2247.
39. Barnard, C. F. J., Daniels, J. A., Jeffery, J., Mawby, R. J., *J. Chem. Soc., Dalton Trans.* (1976) 953.
40. Green, M. L. H., McCleverty, J. A., Pratt, L., Wilkinson, G., *J. Chem. Soc.* (1961) 4854.
41. Thomas, J. L., *J. Am. Chem. Soc.* (1973) **95**, 1838.
42. Thomas, J. L., Brintzinger, H. H., *J. Am. Chem. Soc.* (1972) **94**, 1386.
43. Tang Wong, K. L., Thomas, J. L., Brintzinger, H. H., *J. Am. Chem. Soc.* (1974) **96**, 3694.
44. Cooper, N. J., Green, M. L. H., Couldwell, C., Prout, F., *J. Chem. Soc., Chem. Commun.* (1977) 145.
45. Cooper, R. L., Green, M. L. H., *J. Chem. Soc. A* (1967) 1155.
46. Peterson, J. L., Lichtenberger, D. L., Fenske, R. F., Dahl, L. F., *J. Am. Chem. Soc.* (1975) **97**, 6433.
47. Pennella, F., *Inorg. Synth.* (1976) **15**, 42.
48. Meaken, P., Guggenberger, L. J., Peet, W. G., Muetterties, E. L., Jesson, J. P., *J. Am. Chem. Soc.* (1973) **95**, 1467.
49. Chatt, J., Watson, H. R., *J. Chem. Soc.* (1961) 4980.
50. George, T. A., Seibold, C. D., *Inorg. Chem.* (1973) **11**, 2548.
51. Byrne, J. W., Blaser, H. U., Osborn, J. A., *J. Am. Chem. Soc.*, (1975) **97**, 3871.
52. George, T. A., Seibold, C. D., *Inorg. Chem.* (1973) **11**, 2544.
53. Midai, M., Tominori, K., Uchida, Y., Mesono, A., *Inorg. Synth.* (1976) **15**, 25.
54. Adamson, A. W., *J. Phys. Chem.* (1967) **71**, 798.

RECEIVED July 19, 1977.

Molecular Orbital Analysis of Bonding in ReH_9^{2-} and $[ReH_8\{PH_3\}]^-$

A. P. GINSBERG

Bell Laboratories, Murray Hill, NJ 07974

Self-consistent field Xα-scattered wave calculations were carried out for the hydride complex ReH_9^{2-}, its monophosphine derivative, equatorial $[ReH_8\{PH_3\}]^-$, and the ligand array H_9^{2-}. The results lead to the following significant conclusions. (1) H–H bonding interactions contribute to the stabilization of the complexes. (2) Replacing an equatorial hydrogen of ReH_9^{2-} with PH_3 causes the net atomic charges to change from -0.19 to $+0.17$ for Re, from -0.24 to -0.11 for apical hydrogen, and from -0.12 to -0.09 for equatorial hydrogen. The PH_3 group takes on a net negative change (-0.31). (3) Re 5d $\rightarrow$ P 3d π backbonding is a minor effect compared with P 3p $\rightarrow$ Re 5d σ donation. The net transfer of negative charge in PH_3 largely results from Re–H $\rightarrow$ P σ donor interactions.

Compounds containing hydrogen and tertiary phosphine ligands bound to a rhenium atom comprise a large class of hydride complexes (*1, 2*). These may be regarded as derivatives of the enneahydriderhenate ion, ReH_9^{2-}, many of which may be synthesized by reaction of ReH_9^{2-} with the appropriate tertiary phosphine (*3, 4*). Electronic structure and bonding in these hydride complexes is not well understood; in particular, the following interesting questions have not been answered.

(1) What is the contribution of H–H interaction to the bonding in the polyhydrido complexes?

(2) What is the effect on the charge distribution of replacing a hydride ligand by a tertiary phosphine group?

(3) Does the phosphine group function simply as a σ donor, or is there a significant amount of Re 5d $\rightarrow$ P 3d π backbonding?

To investigate these questions, I carried out self-consistent field Xα-scattered wave (SCF-Xα-SW) calculations (*5, 6*) on ReH_9^{2-}, its monophosphine derivative, equatorial $[ReH_8\{PH_3\}]^-$, and the ligand array H_9^{2-}. ($[ReH_8\{PH_3\}]^-$ serves as a model for the known anions $[ReH_8L]^-$ L = $(C_6H_5)_3P$, $(C_2H_5)_3P$, and (*n*-

0-8412-0390-3/78/33-167-201/$05.00/0

$C_4H_9)_3P$. In terms of the shift in CO stretching frequencies in phosphine metal carbonyl complexes, PH_3 is less basic (or more π-acid) than trialkyl or triaryl phosphines (7, 8, 9).) The results, reported in this chapter, lead to the following significant conclusions.

(1) H–H bonding interactions contribute to the stabilization of the complexes.

(2) Replacing an equatorial hydrogen of $ReH_9{}^{2-}$ with PH_3 causes the net atomic charges to change from -0.19 to $+0.17$ for Re, from -0.24 to -0.11 for apical hydrogen, and from -0.12 to -0.09 for equatorial hydrogen. The PH_3 group takes on a net negative charge (-0.31).

(3) Re $5d \rightarrow$ P $3d$ π backbonding is a minor effect compared with P $3p \rightarrow$ Re $5d$ σ donation. The net transfer of negative charge to PH_3 largely results from Re–H $\rightarrow$ P σ donor interactions.

Procedure for Calculations

SCF-Xα-SW calculations were executed in double precision on a Honeywell 6000 computer using current versions of the programs written originally by K. H. Johnson and F. C. Smith.

Figure 1 shows the coordinate axes and the geometry of the systems studied. $ReH_9{}^{2-}$ was assumed to have D_{3h} symmetry, as found by neutron diffraction on K_2ReH_9 (10). The average measured Re–H distance (1.68 Å) and the H-Re-H angle shown in Figure 1 were used to determine hydrogen coordinates in atomic units (1 bohr = 0.52917 Å). The $H_9{}^{2-}$ calculation utilized the same symmetry and coordinates as those for $ReH_9{}^{2-}$. The calculation for $[ReH_8\{PH_3\}]^-$ was carried out for the equatorial isomer with C_s symmetry. (The nine-coordinated

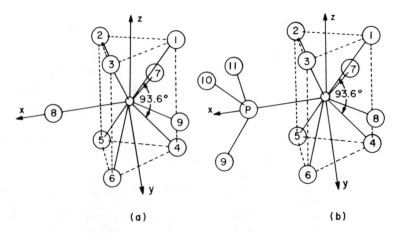

(a)   (b)

Figure 1.   Coordinate system, geometry, and hydrogen labeling scheme. (a) $ReH_9{}^{2-}$ and $H_9{}^{2-}$ (D_{3h}); (b) $[ReH_8\{PH_3\}]^-$ (C_s; σ_n in the xz plane).

Table I.   Exchange Parameters and Sphere Radii (Atomic Units) for ReH_9^{2-}, H_9^{2-}, and $[ReH_8\{PH_3\}]^-$ [a]

System	Region	α	R
ReH_9^{2-}	Re	0.69325	2.481
	H_{Re}	0.77725	1.488
	OUT	0.77725	4.365
	INT	0.72879	—
H_9^{2-}	H	0.77725	2.352
	OUT	0.77725	5.057
	INT	0.77725	—
$[ReH_8\{PH_3\}]^-$	Re	0.69325	2.481
	H_{Re}	0.77725	1.488
	P	0.72620	2.220
	H_P	0.77725	1.407
	OUT	0.77725	6.272
	INT	0.72990	—

[a] OUT refers to the outer sphere surrounding the entire cluster, and INT refers to the regions between the atomic spheres and inside the outer sphere. H_{Re} represents hydrogen bound to rhenium, and H_P represents hydrogen bound to phosphorus.

$[ReH_8\{PR_3\}]^-$ anions presumably retain the ReH_9^{2-} structure with the phosphine substituted in an equatorial or an apical position.   Low temperature ir studies (4) on $[ReH_8\{PPh_3\}]^-$ suggest that it may be a mixture of both isomers.   X-ray diffraction studies on $ReH_7\{PMe_2Ph\}_2$ (11) show the phosphorus atoms in positions consistent with equatorial substitution in the ReH_9^{2-} structure.)   Rhenium and rhenium-bonded hydrogen coordinates were the same as those for ReH_9^{2-}; the phosphorus atom was located on the x-axis 2.34 Å from the rhenium.   (In $H_8Re_2(PEt_2Ph)_4$, with essentially the same terminal Re–H distance as ReH_9^{2-}, the Re–P distance is 2.335 Å (12).   A similar Re–P distance has been found in $ReH_3(diphos)_2$ (13).)   The phosphorus-bonded hydrogen coordinates were derived from the known structure of PH_3 (P–H = 1.415 α and angle H-P-H = 93.45°) (14, 15).

Table I summarizes the α-exchange parameters and sphere radii used in the calculations.   The α value for hydrogen is that recommended by Slater (16); the same value was used for the extramolecular region.   For phosphorus, α_{HF} was taken from Schwarz's table (17) while the α value for rhenium was determined by computing α_{VT} as described by Schwarz (17).   In the intersphere regions, a weighted average α was used where the weights were the number of valence electrons in the atoms.   Overlapping atomic sphere radii for ReH_9^{2-} were determined by Norman's Procedure (18).   The same radii were used for $[ReH_8\{PH_3\}]^-$ together with the P and H_P sphere radii used for the free PH_3 molecule (19).   Overlapping atomic sphere radii for the H_9^{2-} system were obtained by increasing the radii of touching atomic number spheres (18) by 25% (20) without attempting optimization with respect to the virial ratio. In each case, the outer sphere was taken to be tangent to the outermost touching atomic number spheres, which gave an overlapping outer sphere for the actual atomic sphere radii (18). The outer sphere was centered at the origin for ReH_9^{2-} and H_9^{2-}, for $[ReH_8\{PH_3\}]^-$ it was centered at the valence electron-weighted average of the atom positions. A Watson sphere (21), concentric with the outer sphere

Table II. Ground-State Valence Level Eigenvalues, Ionization Energies, and Charge Distributions for H_9^{2-} and ReH_9^{2-}

H_9^{2-}

Level	Eigenvalue (Ry)	Charge Distribution (%)[a]			
		H1–6	H7–9	OUT	INT
$2a'_1$	−0.172	23	53	14	9
$1e''$	−0.302	84	0	10	6
$2e'^c$	−0.302	34	45	5	16
$1e'$	−0.452	46	38	6	9
$1a''_2$	−0.574	91	0	7	2
$1a'_1$	−0.807	66	29	4	0

ReH_9^{2-}

Level	Eigenvalue (Ry)	I.E. (eV)	Charge Distribution (%)[a]					Major Re Spher. Harmonic[b]
			Re	H1–6	H7–9	OUT	INT	
$11a'_1$	−0.005		2	0	2	95	1	
$9e'^c$	−0.556	10.6	28	37	5	11	19	$p_{x,y}, d_{x^2-y^2,xy}$
$6a''_2$	−0.573	10.8	11	53	0	17	19	p_z
$10a'_1$	−0.619	11.6	52	12	23	5	8	d_{z^2}
$8e'$	−0.625	11.6	20	3	35	2	40	$d_{x^2-y^2,xy}$
$4e''$	−0.700	12.8	49	42	0	5	4	$d_{xz,yz}$
$9a'_1$	−0.839	14.5	18	34	19	11	17	s
$7e'$	−3.143	46.7	98	1	0	0	0	$p_{x,y}$
$5a''_2$	−3.149	46.8	98	2	0	0	0	p_z
$8a'_1$	−5.008	72.3	99	0	0	0	0	s

[a] Percentage of the total population of the given level located within the indicated regions. H1–6 refers to the combined prism corner (apical) hydrogen spheres (*see* Figure 1); H7–H9 refers to the combined equatorial hydrogen spheres (*see* Figure 1); OUT and INT refer, respectively, to the extramolecular and intersphere regions.
[b] The major spherical harmonic basis functions in the Re sphere.
[c] Highest occupied orbital.

and bearing a charge equal in magnitude but opposite in sign to the cluster charge, was used to simulate the electrostatic interaction of the cluster with its surrounding crystal lattice.

The highest-order spherical harmonics used to expand the valence level wave functions were $l = 3$ in the extramolecular region, $l = 2$ in the Re and P spheres, and $l = 0$ in the H spheres. All ground-state SCF calculations converged to better than ±0.0004 Ry for each variance level. Core levels were relaxed in the ReH_9^{2-}

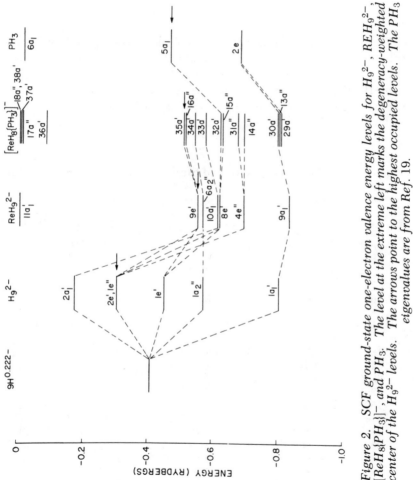

Figure 2. SCF ground-state one-electron valence energy levels for H_9^{2-}, ReH_9^{2-}, $[ReH_8\{PH_3\}]^-$, and PH_3. The level at the extreme left marks the degeneracy-weighted center of the H_9^{2-} levels. The arrows point to the highest occupied levels. The PH_3 eigenvalues are from Ref. 19.

calculation and were converged to ±0.0026 Ry or better. All core levels (Re 1–4s, 2–4p, 3d, 4d, and 4f; P 1s, 2s, and 2p) were frozen in the [ReH$_8${PH$_3$}]$^-$ calculation. The converged ground-state potentials were used to search for excited state levels up to a maximum energy of −0.002 Ry. For ReH$_9^{2-}$, the Slater transition state procedure (5, 6) was used to calculate the ionization energies of the valence levels.

Results

The calculated ground-state one-electron valence energy levels and charge distributions for H$_9^{2-}$, ReH$_9^{2-}$, and [ReH$_8${PH$_3$}]$^-$ are summarized in Tables II and III. Calculated valence-level ionization energies for ReH$_9^{2-}$ are also given in Table II. Experimental ionization energies for comparison are not available presently. Figure 2, a diagram of the eigenvalues in Tables II and III, shows the

Table III.   Ground State Valence Level Eigenvalues

Level	Eigen-value (Ry)	Charge Distribution (%)^a					
		Re	P	H1	H2 + H3	H4	H5 + H6
18a″	−0.0055	2	2	0	0	0	0
38a′	−0.007	2	1	1	0	1	0
37a′	−0.009	9	4	0	1	0	2
17a″	−0.014	14	4	0	2	0	0
36a′	−0.085	19	8	0	1	0	1
35a′^{c}	−0.512	19	6	14	6	15	6
16a″	−0.518	13	2	0	20	0	19
34a′	−0.547	12	1	10	15	11	16
33a′	−0.581	49	10	2	4	1	3
32a′	−0.627	39	26	1	0	1	0
15a″	−0.634	46	1	0	2	0	2
31a′	−0.681	44	4	12	7	12	7
14a″	−0.700	47	0	0	20	0	23
30a′	−0.802	19	6	5	14	6	13
13a″	−0.806	2	39	0	1	0	0
29a′	−0.812	4	37	1	1	0	3
28a′	−1.280	1	62	0	0	0	0
27a′	−3.098	98	0	0	0	0	0
12a″	−3.101	98	0	0	0	0	0
26a′	−3.103	98	0	0	1	0	1
25a′	−4.952	99	0	0	0	0	0

^{a} Percentage of the total population of the given level located within the indicated regions. The hydrogen sphere numbers refer to Figure 1. OUT and INT refer, respectively, to the extramolecular and intersphere regions.

correlation between levels in the different systems. Wave function contour maps of selected orbitals are shown in Figures 3, 4, 5, and 6. Each map was generated from numerical values of the wave function at 6561 grid points within a 10 × 10 (H$_9^{2-}$ and ReH$_9^{2-}$) or 16 × 16 ([ReH$_8${PH$_3$}]$^-$) bohr^2 area.

The ground-state total energies and total charge distributions are given in Table IV. If the total charge in the intersphere and extramolecular regions is partitioned among the atomic spheres in proportion to the charge density at the sphere surface, the charge distributions in Table IV lead to the net atomic charges listed in Table V. In the case of TcH$_9^{2-}$, net atomic charges derived by this method from the SCF-Xα-SW total charge distribution (4) can be compared with the net charges given by an SCF Gaussian orbital calculation (22). As shown in Table V, the agreement is excellent. (Partitioning the intersphere and extramolecular charge among the atoms proportionally to the number of valence electrons within each atomic sphere (23) leads to results similar to those in Table

and Charge Distributions for [ReH$_8${PH$_3$}]$^-$

	Charge Distribution (%)[a]				Major Spher. Harmonic[b]	
H7 + H8	H9	H10 + H11	OUT	INT	Re	P
1	0	0	82	13		
0	0	0	80	14		
0	1	0	72	11		
0	0	1	42	35	d_{yz}	
1	0	1	40	29	$d_{x^2-y^2}$, d_{z^2}	
3	1	1	3	26	p_x, $d_{x^2-y^2}$	p_x
11	0	1	2	30	p_y	d_{xy}
0	1	1	2	30	p_z	d_{xz}
16	1	1	1	12	d_{z^2}	p_x
14	2	4	2	12	$d_{x^2-y^2}$	p_x
35	0	2	1	10	d_{xy}	d_{xy}
0	5	2	1	5	d_{xz}	d_{xz}, p_z
0	0	0	0	8	d_{yz}	
11	2	1	1	23	s	p_x
0	0	48	3	7	d_{xy}	p_y, d_{xy}
0	28	15	3	8	d_{xz}	p_z, d_{xz}
0	10	20	2	3		s
0	0	0	0	0	p_x	
1	0	0	0	0	p_y	
0	0	0	0	0	p_z	
0	0	0	0	0	s	

he major Re and P spherical harmonic basis functions.
ighest occupied orbital.

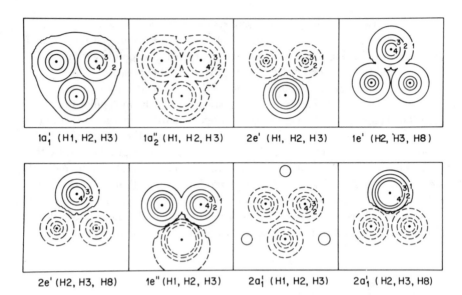

$1a_1'$ (H1, H2, H3)	$1a_2''$ (H1, H2, H3)	$2e'$ (H1, H2, H3)	$1e'$ (H2, H3, H8)
$2e'$ (H2, H3, H8)	$1e''$ (H1, H2, H3)	$2a_1'$ (H1, H2, H3)	$2a_1'$ (H2, H3, H8)

Figure 3. Wavefunction contour maps of the H_9^{2-} molecular orbitals. Solid and broken lines denote contours of opposite sign having magnitudes indicated by the numerical labels: 1, 2, 3, 4 = 0.02, 0.04, 0.06, 0.08, respectively. Maps in the H1, H2, and H3 plane show apical–apical interactions; maps in the H2, H3, and H8 plane show equatorial–apical interactions. A · marks the position of H atom centers.

Table IV.   Ground-State Total Energies (Rydbergs) and Total Charge Distributions for H_9^{2-}, ReH_9^{2-}, and $[ReH_8\{PH_3\}]^-$

	H_9^{2-}	ReH_9^{2-}	$[ReH_8\{PH_3\}]^-$
Total energy	−8.4739	−31579.3048	−32274.4659
Kinetic energy, T	9.3033	31579.6832	32258.3258
Potential energy, V	−17.7772	−63158.9880	−64532.7917
$-2T/V$	1.047	1.000006	0.999750
Intersphere potential energy	−0.0948	−0.6072	−0.4064
Total charge (electrons) in various regions			
Re	—	73.39	73.79
apical H	6.01	5.34	5.59[a]
equatorial H	3.47	2.45	1.81
P	—	—	13.89
phosphine H	—	—	2.95[b]
extramolecular	0.63	1.38	0.45
intersphere	0.89	3.43	3.51

[a] The nonequivalent apical hydrogen spheres have similar total charges: H1, 0.92; H2 + H3, 1.85; H4, 0.92; H5 + H6, 1.86.
[b] Each of the phosphine hydrogen spheres has the same total charge.

Table V.   Net Atomic Charges in H$_9^{2-}$, ReH$_9^{2-}$, [ReH$_8${PH$_3$}]$^-$, and TcH$_9^{2-}$

Atom	H$_9^{2-}$	ReH$_9^{2-}$	[ReH$_8${PH$_3$}]$^-$	SCF-GO[a] TcH$_9^{2-}$	SCF-Xα[b] TcH$_9^{2-}$
Re or Tc	—	−0.19	+0.17	−0.37	−0.33
Apical H	−0.16	−0.24	−0.11	−0.20	−0.22
Equatorial H	−0.34	−0.12	−0.09	−0.14	−0.12
P	—	—	+0.32[c]	—	—
Phosphine H	—	—	−0.21[c]	—	—

[a] From a Hartree–Fock–Roothan SCF calculation in a contracted Gaussian orbital basis (*22*).

[b] From an SCF-Xα-SW calculation; intersphere and extramolecular charge partitioned by the same method described for ReH$_9^{2-}$ (*4*).

[c] The SCF-Xα-SW total charge distribution for the free PH$_3$ molecule (*19*) leads to the following net atomic charges if the intersphere and extramolecular charge are partitioned as described in the text: P, +0.51 and H, −0.17.

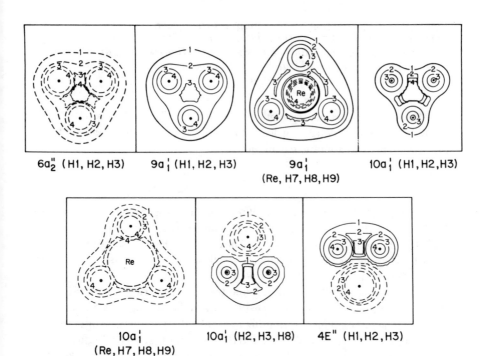

6a$_2''$ (H1, H2, H3)   9a$_1'$ (H1, H2, H3)   9a$_1'$ 10a$_1'$ (H1, H2, H3)
                                              (Re, H7, H8, H9)

10a$_1'$ 10a$_1'$ (H2, H3, H8)   4E$''$ (H1, H2, H3)
(Re, H7, H8, H9)

Figure 4.   Wavefunction contour maps for ReH$_9^{2-}$ molecular orbitals. Contour values and sign convention as in Figure 1. Interior contours close to the rhenium atom center have been omitted. H atom centers are marked by ●.

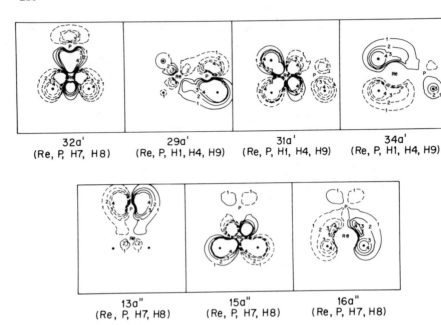

$$\begin{array}{cccc}
32a' & 29a' & 31a' & 34a' \\
(\text{Re, P, H7, H8}) & (\text{Re, P, H1, H4, H9}) & (\text{Re, P, H1, H4, H9}) & (\text{Re, P, H1, H4, H9})
\end{array}$$

$$\begin{array}{ccc}
13a'' & 15a'' & 16a'' \\
(\text{Re, P, H7, H8}) & (\text{Re, P, H7, H8}) & (\text{Re, P, H7, H8})
\end{array}$$

Figure 5. Wavefunction contour maps for the major Re–P σ bonding orbital (32a′) of [ReH₈{PH₃}]⁻ and the orbitals in which Re–P π bonding is allowed by symmetry. Contour values and sign convention as in Figure 1. Interior contours close to the Re and P atom centers have been omitted. H atom centers are marked by ·.

V. However, the partitioning method used to derive Table V is preferred since, for the C_3O_2 molecule (24), it leads to results in closer agreement with XPS measurements and ab initio calculations.)

Discussion

To separate H–H from Re–H interactions, it is appropriate to examine the interactions in the H_9^{2-} ligand array and to compare the molecular orbitals (MOs) for this system with those of ReH_9^{2-}. The choice of the -2 charge on the ligand array to be studied is consistent with the small value for the calculated net charge on Re (Table V).

From the large MO splittings found for H_9^{2-} (see Table II and Figure 2), it is evident that strong interactions exist between the hydrogen 1s orbitals in this system. The interactions contributing to each of the MOs may be inferred from the charge distributions in Table II and the contour diagrams in Figure 3. The major interactions take place between apical atoms within the triangular prism faces (apical–apical interaction; $d(\text{H–H}) = 1.99$ Å) and between an equatorial atom and the apical atoms in the nearby square prism face (equatorial–apical interaction; $d(\text{H–H}) = 1.93$ Å). In the lowest lying MO, $1a'_1$, only bonding interactions, both apical–apical and equatorial–apical occur.   In the $1a''_2$ orbital, the charge on the equatorial atoms is zero and only bonding apical–apical interactions take place.   Both bonding and antibonding interactions contribute to levels $1e'$ and $2e'$.   In both of these orbitals and apical–apical interactions are predominantly antibonding, but the equatorial–apical interactions are predominantly bonding in $1e'$ and antibonding in $2e'$.   The $1e''$ orbital, with zero charge on the equatorial atoms, has only apical–apical interactions and these are predominantly antibonding.   The highest lying orbital of the H$_9^{2-}$ system, $2a'_1$, is dominated by antibonding equatorial–apical interactions.

The correlation between H$_9^{2-}$ and ReH$_9^{2-}$ MOs (Figure 2) and comparison of the charge distribution and contour maps for corresponding orbitals (Table II and Figure 4) lead to the conclusion that H$_9^{2-}$ H–H bonding orbitals pre-

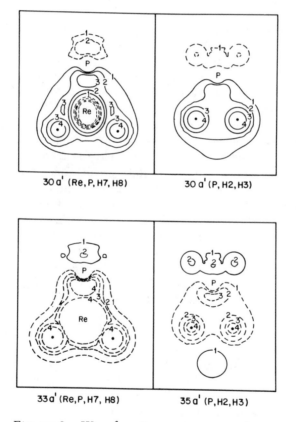

Figures 6.   *Wavefunction contour maps for orbitals in which Re–H—-P bonding interactions occur.   Contour values and sign convention is in Figure 1.   Interior contours close to the Re and P atom centers have been omitted.   H atom centers are marked by ·*

dominate over H–H antibonding orbitals in contributing to the ReH_9^{2-} MOs. It follows from this that a net H–H bonding interaction exists in the ReH_9^{2-} anion. In support of this conclusion, the two strongly H–H bonding orbitals of H_9^{2-} ($1a'_1$ and $1a''_2$) correlate with predominantly hydrogen-based orbitals in ReH_9^{2-}. It is evident from Figures 2, 3, and 4 that the $1a''_2$ (H_9^{2-}) orbital is perturbed only slightly by formation of the $6a''_2$ (ReH_9^{2-}) orbital; the contour maps suggest a slight enhancement of the H–H interaction. A more substantial perturbation occurs when the $1a'_1$ (H_9^{2-}) orbital interacts with a Re $6s$ orbital to form the $9a'_1$ (ReH_9^{2-}) MO; again the contour maps indicate enhanced H–H bonding in the complex (Figure 4). By contrast, the two most strongly H–H antibonding levels of H_9^{2-} ($1e''$ and $2a'_1$) correlate with predominantly metal-based levels in ReH_9^{2-} and are perturbed strongly by the interaction with rhenium. Comparison of the appropriate parts of Figures 3 and 4 shows how the $2a'_1$ (H_9^{2-}) orbital is altered by interaction with a Re $5d_{z^2}$ orbital to form the $10a'_1$ (ReH_9^{2-}) MO. The latter retains the equatorial–apical H–H antibonding interaction of its parent but has enhanced apical–apical and equatorial–equatorial H–H bonding interactions. Orbital $4e''$ (ReH_9^{2-}) retains, like its parent $1e''$ (H_9^{2-}), apical–apical H–H antibonding. The apical–apical antibonding interaction found in orbitals $1e'$ and $2e'$ of H_9^{2-} is present in $9e'$ (ReH_9^{2-}) but not $8e'$ (ReH_9^{2-}). Neither of the latter two orbitals show the equatorial–apical interactions of $1e'$ and $2e'$ (H_9^{2-}).

The effect on the charge distribution of replacing an equatorial hydrogen ligand with PH_3 is elucidated by Table V. One unit of net negative charge is lost in this replacement: $0.12e^-$ is associated with the hydrogen that is replaced and $0.88e^-$ comes from the remaining ReH_8 group. Since PH_3 in the complex has a net charge of -0.31 ($-0.21 \times 3 + 0.32$), there must also be a net transfer of $0.31e^-$ from the ReH_8 part of the molecule to PH_3. Replacing an equatorial hydrogen of ReH_9^{2-} with PH_3 therefore results in a loss of $1.19e^-$ from the remaining ReH_8 group. About 66% of the charge removed from the ReH_8 group comes from the apical hydrogen atoms, 30% is from the Re atom, and 4% is from the two equatorial hydrogen atoms. The hydrogen atoms remain negatively charged, with the apical and equatorial charge nearly the same, while the Re atom charge changes from small negative (-0.19) to small positive ($+0.17$).

The decreased magnitude of the calculated net hydrogen-atom charges in $[ReH_8\{PH_3\}]^-$ compared with ReH_9^{2-} is consistent with the smaller value of τ(Re–H) observed for phosphine octahydride complexes with the less basic phosphines. For example, τ(Re–H) for ReH_9^{2-} is 18.5 compared with 17.3 for $[ReH_8\{PPh_3\}]^-$ (3). In both ReH_9^{2-} and $[ReH_8\{PH_3\}]^-$, the magnitude of the calculated negative charge on the metal-bonded hydrogen is small. This agrees with an analysis of experimental measurements on a variety of hydride complexes which concluded that "the formally anionic hydride ligand is very strongly electron donating, being only slightly negative in its complexes" (25). The calculated phosphorus atom charge in $[ReH_8\{PH_3\}]^-$ agrees with the experimental indication that phosphorus in many tertiary phosphine complexes has a charge of $+0.3$ (25). However, the conclusion that PH_3 in $[ReH_8\{PH_3\}]^-$ functions as an electron-withdrawing group diverges from what is found for a tertiary alkyl or phenyl phosphine in complexes such as $MCl_n(PR_3)_n$, where the phosphine group serves as a good electron donor (25). Also, recent SCF-Xα calculations on $Pt(PH_3)_2L$ (23) lead to estimated net charges on P of +0.36 (L = O_2) and +0.33 (L = C_2H_4) and corresponding net charges for PH_3 of +0.09 and +0.06. The net electron-withdrawing behavior of PH_3 in [ReH$_8${PH$_3$}]$^-$ results mainly from charge transfer to the phosphine via P---H–Re interactions (*see* below).

The Re–P bond in [ReH$_8${PH$_3$}]$^-$ can be appraised by examining the charge distributions (*see* Table III) and contour maps (Figure 5) for the orbitals that contribute to this bond. By far the major contribution comes from the σ $Re(5d_{x^2-y^2})$–$P(3p_x)$ interaction in orbital $32a'$. Comparison of the charge distribution in this orbital with the charge distribution in the parent lone-pair $5a_1$ orbital of the free PH_3 molecule shows about a 30% loss of charge from PH_3 resulting from σ-donation. (The charge distribution in the $5a_1$ orbital of PH_3, calculated with the same sphere radii used for [ReH$_8 \rightarrow$ {PH$_3$}]$^-$, is: %P, 50; %3H, 12; %OUT, 16; and %INT, 22.) The charge lost by σ-donation is compensated for by transfer of charge on to the PH_3 molecule via interaction of the $P(3p_x)$ orbital with the apical and equatorial hydrogen atoms in MOs $30a'$, $33a'$, and $35a'$. Contour maps showing the P---H–Re interactions are exhibited in Figure 6. Charge transfer to PH_3 via π back donation from rhenium to phosphorus $3d$ orbitals is small. This is clear from the contour maps in Figure 5 that show all of the orbitals in which a Re–P π interaction is allowed by symmetry. Two of the orbitals in Figure 5, $29a'$ and $13a''$, show what can be described as a P–H $\rightarrow$ Re donor π interaction. Orbitals $31a'$, $34a'$, $15a''$, and $16a''$ encompass the Re $\rightarrow$ P π back donation and evidently do not make an important contribution to the Re–P bond. The negative charge on PH_3 in [ReH$_8${PH$_3$}]$^-$ is a consequence of Re–H $\rightarrow$ P σ donor interactions, not of Re $\rightarrow$ P π back donation.

Literature Cited

1. Kaesz, H. D., Saillant, R. B., *Chem. Rev.* (1972) **72**, 231.
2. Giusto, D., *Inorg. Chim. Acta, Rev.* (1972) **6**, 91.
3. Ginsberg, A. P., *Chem. Commun.* (1968) 857.
4. Ginsberg, A. P., unpublished data.
5. Johnson, K. H., *Annu. Rev. Phys. Chem.* (1975) **26**, 39.
6. Slater, J. C., "The Self-Consistent Field of Molecules and Solids: Quantum Theory of Molecules and Solids, Volume 4," McGraw–Hill, New York, 1974.
7. Fischer, E. O., Louis, E., Schneider, R. J. J., *Angew. Chem. Int. Ed. Engl.* (1968) **7**, 136.
8. Klanberg, F., Muetterties, E. L., *J. Am. Chem. Soc.* (1968) **90**, 3296.
9. Tolman, C. A., *J. Am. Chem. Soc.* (1970) **92**, 2953.
10. Abrahams, S. C., Ginsberg, A. P., Knox, K., *Inorg. Chem.* (1964) **3**, 558.
11. Bau, R., Carroll, W. E., Hart, D. W., Teller, R. G., Koetzle, T. F., ADV. CHEM. SER. (1978) **167**, 73.
12. Bau, R., Carroll, W. E., Teller, R. G., Koetzle, T. F., *J. Am. Chem. Soc.* (1977) **99**, 3872.
13. Albano, V. G., Bellon, P. L., *J. Organomet. Chem.* (1972) **37**, 151.
14. Kuchitsu, K., *J. Mol. Spectrosc.* (1961) **7**, 399.

15. Sirvetz, M. H., Weston, R. E., *J. Chem. Phys.* (1953) **21**, 898.
16. Slater, J. C., *Int. J. Quantum. Chem.* (1973) **7s**, 533.
17. Schwarz, K., *Phys. Rev. B:5* (1972) 2466.
18. Norman, J. G., Jr., *Mol. Phys.* (1976) **31**, 1191.
19. Norman, J. G., Jr., *J. Chem. Phys.* (1974) **61**, 4630.
20. Salahub, D. R., Messmer, R. P., Johnson, K. H., *Mol. Phys.* (1976) **31**, 529.
21. Watson, R. E., *Phys. Rev.* (1958) **111**, 1108.
22. Basch, H., Ginsberg, A. P., *J. Phys. Chem.* (1969) **73**, 854.
23. Norman, J. G., Jr., *Inorg. Chem.* (1977) **16**, 1328.
24. Ginsberg, A. P., Brundle, C. R., unpublished data.
25. Chatt, J., Elson, C. M., Hooper, N. E., Leigh, G. J., *J. Chem. Soc.* (1975) 2392.

RECEIVED August 16, 1977.

Hydrogen–Deuterium Exchange in Hydrido Transition Metal Cluster Complexes

Mass and Raman Spectrometric Characterization

MARK A. ANDREWS, STEPHEN W. KIRTLEY, and HERBERT D. KAESZ

Department of Chemistry, University of California, Los Angeles, CA 90024

The exchange of D_2O with a variety of hydrido-metal complexes is discussed. While the exchange of $HRe(CO)_5$ with D_2O is rapid, that of a variety of hydrido-metal cluster complexes is not observed. The exchange of D_2O with $H_2Re_2(CO)_8$ or $H_2Os_3(CO)_{10}$ on the other hand is found to be catalyzed by Florisil chromatographic absorbent. Mass and Raman spectrometric means of characterization are discussed.

Hydrido-metal carbonyl complexes are known to be thermodynamically acidic (*1, 2, 3, 4*). A broad range of acidity is observed with $HCo(CO)_4$, $HV(CO)_6$, and $H_2Fe_3(CO)_{11}$, reported to be strong acids ($pK_a \sim 1$); with $H_2Fe(CO)_4$ and $HMn(CO)_5$, reported to be weak acids ($pK_{a_1} \sim 4.4$ and $pK_a \sim 7.1$, respectively); and with $HRe(CO)_5$, reported as only a "very weak acid" (*1, 2, 3, 4*). The acidity of $H_3Re_3(CO)_{12}$ has been estimated by the slow equilibration of the corresponding anions, $HRe_3(CO)_{12}^{2-}$ and $H_2Re_3(CO)_{12}^{-}$, with a variety of acids; for $H_3Re_3(CO)_{12}$, pK_{a_1} is between 3 and 4 and pK_{a_2} is between 10 and 18 (*3*). Quite a contrast, however, is observed in the rates of proton exchange for mononuclear complexes compared with most polynuclear (cluster) complexes. The former with terminally bonded hydrogen atoms exchange protons fairly rapidly; for the manipulation of $DRe(CO)_5$ for instance, Beck, Hieber, and Braun (*4*) advised the use of an absolutely dry apparatus. In our experience (*5*) we found it impossible to bake out all components of a conventional chemical high vacuum line and therefore found it necessary to condition its inner surfaces with D_2O prior to evacuation and exposure to $DRe(CO)_5$. Failure to do so led to significant loss of deuterium content during vacuum transfer of the deuterio complex.

By contrast, the deuterio-metal cluster complexes $D_3Re_3(CO)_{12}$ (*6*) and $D_4Ru_4(CO)_{12}$ (*7*) proved to be stable toward proton exchange under chromato-

0-8412-0390-3/78/33-167-215/$05.00/0

graphic purification and other normal handling procedures. In these metal cluster complexes, the hydrogen atoms are in a bridging position rather than in terminal positions (8, 9, 10), and the coordinated carbonyl groups on adjacent metal atoms are situated at close to van der Waals contacts. The structure of $H_3Re_3(CO)_{12}$ has thus far eluded all attempts by x-ray crystallography since no suitable single crystals have been found. The structure of the manganese analog, $H_3Mn_3(CO)_{12}$, has recently been completed (8). A bridging position for hydrogen was found using Fourier difference methods. Other spectroscopic and structural evidence for bridging hydrogen in a variety of hydrido-metal cluster complexes is given in a recent review (9). Access of external reagents to the bridging hydrogen atoms is thus impeded (for further discussion, vide infra). It is understandable, therefore, that the rate of proton exchange in these complexes would be rather slow despite their thermodynamic acidity (3). This is, however, not necessarily the case in all cluster complexes, and our awareness of this developed over the series of observations described below.

Observations Leading to the Discovery of H/D Exchange in Hydrido-Metal Cluster Complexes

The deuterio-metal cluster complexes mentioned above were synthesized following the routes indicated in Reactions 1 and 2 (THF = tetrahydrofuran). The isotopic purity of the products was qualitatively checked by spectrophotometric means. We have found that Raman more frequently than ir spectra display characteristic features sufficiently discernible to distinguish between the protonated and deuterated derivatives (9). The Raman absorptions are usually broad but are observed to shift and to become more narrow upon deuteration (9), similar to what has been observed for other compounds containing hydrogen or deuterium in a bridging situation (11, 12). Examples of this can be seen in the spectra for $H_3Re_3(CO)_{12}$ and $D_3Re_3(CO)_{12}$ (Figure 1) or for $H_4Ru_4(CO)_{12}$ and $D_4Ru_4(CO)_{12}$ (Figure 2). The absorptions assigned to hydrogen or deuterium are shaded in the corresponding spectra. In $H_3Re_3(CO)_{12}$, the broad shaded absorption is centered at $1100\ cm^{-1}$. In the corresponding modes of the deuterium complex, the center is at $785\ cm^{-1}$, giving a ratio of $\nu_{MH}/\nu_{MD} = 1.40$, in good agreement with that expected for the isotope shift (1.414). Further discussions of the Raman spectral features are given in the latter portion of this chapter.

$$Re_2(CO)_{10} + NaBD_4/THF \xrightarrow{\text{reflux}} \text{intermediate anions}$$

$$\xrightarrow[\text{cyclohexane extract}]{D_3PO_4} D_3Re_3(CO)_{12} \quad (1)$$

$$Ru_3(CO)_{12} + D_2\ (1\ atm) \xrightarrow[\text{octane sol'n}]{90°C\ 1\ hr} D_4Ru_4(CO)_{12} \quad (2)$$

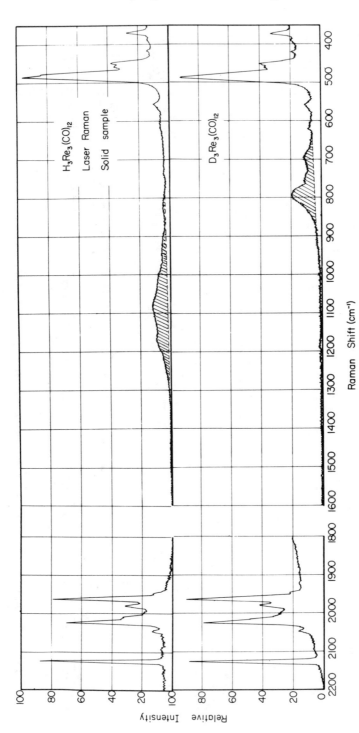

Figure 1. Raman spectra, Cary 81, He/Ne laser exciting line 15 803 cm^{-1}; microcrystalline samples (25)

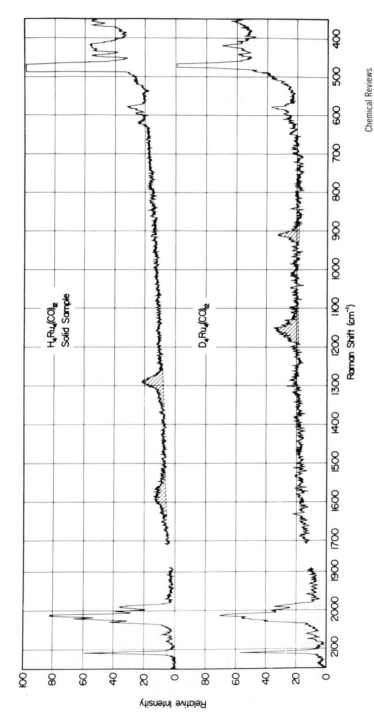

Figure 2.   Raman spectra, Cary 81, He/Ne laser exciting line 15 803 cm⁻¹; microcrystalline samples (9)

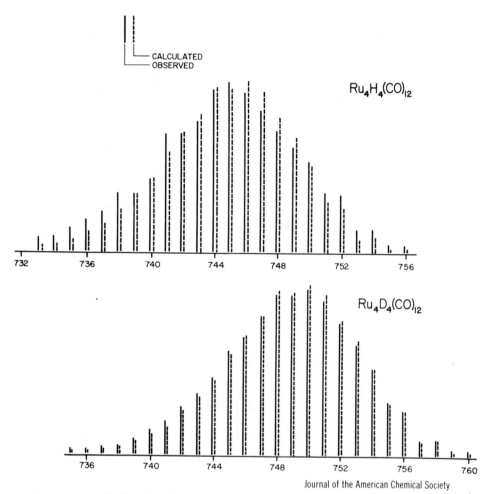

Journal of the American Chemical Society

Figure 3.   Calculated and average observed peaks in the parent ion multiplets; AE1 MS9 Spectrometer (7)

We also have used mass spectrometry to characterize these derivatives. Analysis by this method, however, is complicated because of partial hydrogen loss on ionization, which leads to overlapping isotopic multiplets for the parent ions.   An example of this can be seen in the mismatch of calculated and observed peaks in the parent ion multiplet shown for $H_4Ru_4(CO)_{12}$ (top part of Figure 3). To deal with this, a computer program, MASPAN, was written to deconvolute the spectra in terms of the various hydrogen-loss species, i.e. $H_n Ru_4(CO)_{12}$ (n = 4, 3, 2, 1, and 0), whose lines contribute to the parent ion multiplet.   For details of the program and its use *see* Refs. 7, *13*, and *14*.   The effect of this deconvolution is an improved fit between calculated and observed spectra expressed by an R factor:

$$R = \Sigma \frac{|I_o - I_c|}{I_c}$$

Thus an R factor of 17.1% is obtained for the match between observed and calculated multiplets prior to correction for hydrogen loss, as shown at the top of Figure 3. Deconvolution of the observed spectrum, however, improved the fit to $R = 6.5\%$ for the composition $H_n Ru_4(CO)_{12}$; $n = 4$, 65%; $n = 3$, 19%; $n = 2$ and $n = 1$, 0%; $n = 0$, 16% (7, 14). $D_4Ru_4(CO)_{12}$, on the other hand, undergoes much less fragmentation on ionization, as indicated by the match between observed and calculated multiplets shown in the lower scan of Figure 3. For this, an R factor of 5.3% is obtained without deconvolution, reflecting undoubtedly the lower zero-point energy in the bonding of deuterium as compared with that of the hydrogen with the metal atoms of the cluster. Similarly, only slight fragmentation of hydrogen occurs in the osmium hydrides $H_2Os_3(CO)_{10}$ and $H_4Os_4(CO)_{12}$, paralleling greater stability in the bonding of hydrogen to third-row transition metal atoms (9). A better fit of observed-to-calculated parent ion multiplets is obtained for these two hydrides, as shown in Figure 4; the R factor for these two spectra are 4.6% and 5.2%, respectively (7, 14), without deconvolution.

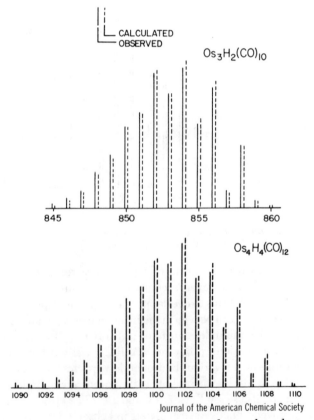

Journal of the American Chemical Society

Figure 4. Calculated and average observed peaks in the parent ion multiplets; AE1 MS9 Spectrometer (7)

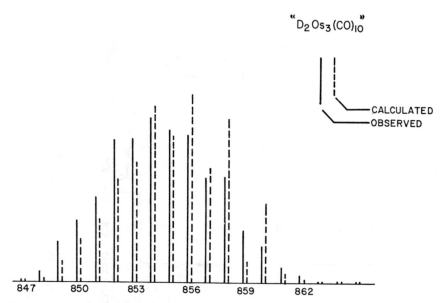

*Figure 5.   Calculated and average observed peaks in the parent ion multiplet;
AE1 MS9 Spectrometer (7)*

We were thus taken by surprise when the mass spectrum of a sample believed to be $D_2Os_3(CO)_{10}$ (Figure 5) showed wide variation between the observed and calculated intensities (7). The sample had been prepared from D_2 and $Os_3(CO)_{12}$ and purified by chromatography; deconvolution analysis ($R = 2.9\%$) indicated the sample consisted of a mixture of 9% $D_2Os_3(CO)_{12}$, 42% $HDOs_3(CO)_{10}$, and 49% $H_2Os_3(CO)_{10}$. We subsequently learned that Keister and Shapley (15) had been able to isolate pure $D_2Os_3(CO)_{10}$ from the same reaction of D_2 and $Os_3(CO)_{12}$; however, their material had been purified by crystallization. H/D exchange in our sample was thus indicated to occur during the chromatographic work-up rather than during the synthesis, as we had earlier suggested (7).

Two further instances of H/D exchange, these for $D_2Re_2(CO)_8$ came to mind; both were traceable to a chromatographic purification step. One of these occurred in the photochemical reaction of D_2 with $Re_2(CO)_{10}$ (13). Three products were isolated by us in that reaction; namely, $DRe(CO)_5$ (distilled out of the reaction mixture with the solvent) and $DRe_3(CO)_{14}$ and $H_2Re_2(CO)_8$ (isolated by column chromatography). A parallel and independent study of the photochemical reaction of H_2 with $Re_2(CO)_{10}$, giving the same products together with some $H_3Re_3(CO)_{12}$, was reported by Byers and Brown (16, 17). The isotopic content of the products isolated in our work was determined by ir and mass spectra (13). The loss of deuterium in the product $H_2Re_2(CO)_8$, which was at first puzzling, was soon connected with a prior instance of H/D exchange in this derivative that was observed during a collaboration with W. A. G. Graham and co-workers on the Raman spectrophotometric characterization of the derivatives

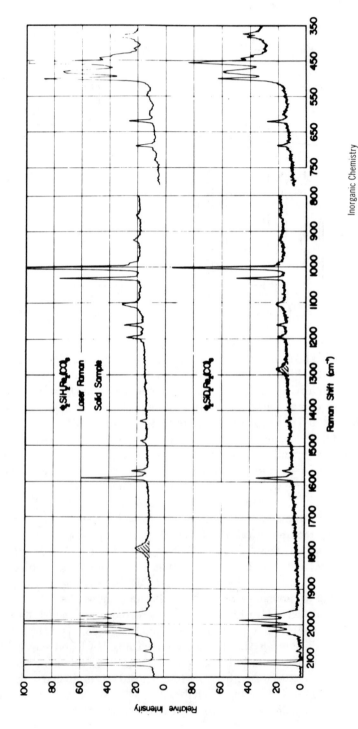

Inorganic Chemistry

Figure 6. Raman spectra, Cary 81, He/Ne laser exciting line 15 803 cm^{-1} (13)

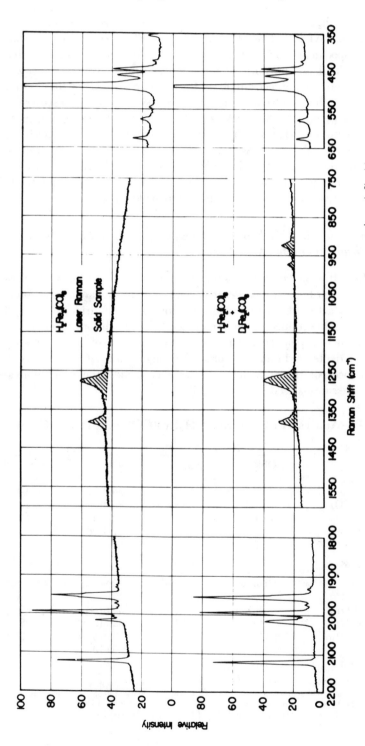

Figure 7. Raman spectra, Cary 81, He/Ne laser exciting line 15 803 cm^{-1} (13)

Inorganic Chemistry

$Re_2(CO)_8H_2SiPh_2$ and $H_2Re_2(CO)_8$ (*18*). The latter is formed during chromatography of the silyl derivative by loss of a diphenylsilylene group on the column. The Raman spectra of $Re_2(CO)_8H_2SiPh_2$ and its di-deuterio derivative are shown in Figure 6; a high isotopic purity in the latter is indicated by appearance of the band at 1280 cm^{-1} in its spectrum (lower trace Figure 6) replacing the absorption at 1790 cm^{-1} in the dihydrido species (upper scan). The two bands are related by the ratio $\nu_{MH}/\nu_{MD} = 1.398$, close to the value of 1.414 expected for such an isotopic substitution. The spectra of the dirheniumoctacarbonyl derivatives obtained from the silyl complexes are shown in Figure 7. The $D_2Re_2(CO)_8$ obtained from $Re_2(CO)_8D_2SiPh_2$ (lower scan Figure 7) shows the presence of an appreciable amount of $H_2Re_2(CO)_8$. Thus, experiments indicated in the title of the next section were undertaken further to explore these exchanges.

Hydrogen–Deuterium Exchange of $H_2Os_3(CO)_{10}$ and $H_2Re_2(CO)_8$ on Florisil Chromatographic Absorbent

We first attempted exchange of the above mentioned hydrido-metal clusters by contact of their hydrocarbon solutions with D_2O; no exchange was observed. Similarly, contact of the hydrocarbon solutions of the hydrides with D_3PO_4 produced the same results. Thus, we proceeded to conditions approximating those of chromatographic separation by contacting solutions of the hydrido-metal clusters with deuterated Florisil (Florisil exchanged with D_2O prior to its use). Significant incorporation of deuterium was observed under these conditions and the results are summarized in Table I (*13*).

Table I.   Analysis of Deuterated $H_2Re_2(CO)_8$ and $H_2Os_3(CO)_{10}$[a]

(A) $H_2Re_2(CO)_8$	Run 1		Run 2	
Species	Obs. %	Stat. %[b]	Obs. %	Stat. %[b]
$D_2Re_2(CO)_8$	5.5(10)	6	50.4(72)	46
$HDRe_2(CO)_8$	37.0(0)	38	35.4(47)	43
$H_2Re_2(CO)_8$	55.5(17)	56	12.2(44)	11
$HRe_2(CO)_8$	1.3(3)	—	1.4(26)	—
$Re_2(CO)_8$	0.8(4)	—	0.6(6)	—
(B) $H_2Os_3(CO)_{10}$	Run 1		Run 2	
Species	Obs. %	Stat. %[b]	Obs. %	Stat. %[b]
$D_2Os_3(CO)_{10}$	7.3(52)	1	65.3(58)	69
$HDOs(CO)_{10}$	0.0(0)	13	32.5(29)	28
$H_2Os_3(CO)_{10}$	92.6(67)	86	2.3(38)	3
$HOs_3(CO)_{10}$	0.0(0)	—	0.0(0)	—
$Os_3(CO)_{10}$	0.0(0)	—	0.0(0)	—

[a] For details *see* Ref. *13*; standard deviation of least significant digits given in parentheses.
[b] The total fractional deuterium content (TDC) is equal to the fraction of D_2 species plus one half of the fraction HD species. Random statistical distribution corresponds to (TDC)^2 for the D_2 species, 2(TDC)(1 − TDC) for the HD species and (1 − TDC)^2 for the H_2 species.

The complex $H_2Re_2(CO)_8$ exchanges about three times faster than $H_2Os_3(CO)_{10}$, in agreement with the visual observation of greater adsorption of the former on the Florisil.   Both exchanges follow random statistics within the experimental error of the analyses, as indicated in Table I.   This implies that adsorption and desorption on the support is competitive with H/D exchange.

One important characteristic of the two cluster hydrides which are observed to undergo H/D exchange on chromatographic support is their electronic unsaturation (*9, 10*).   Because of this they are both expected to be more susceptible than the usual cluster complexes to nucleophilic attack; indeed the facile addition of L to $H_2Os_3(CO)_{10}$, giving complexes of the type $H_2Os_3(CO)_{10}L$, recently have been reported (*21, 22*).   This latter complex is believed to have one bridging and one terminal hydrogen, and the presence of a terminal hydrogen in such an adduct could confer upon it a higher rate of H/D exchange rate, as observed for $HRe(CO)_5$ (discussed above).   Thus, the enhanced exchange rates for these hydrides may be attributed to reactions of the type (N = nucleophilic site on the support) in which the species with the terminal M–H groups were those in which H/D exchange with D_2O took place.

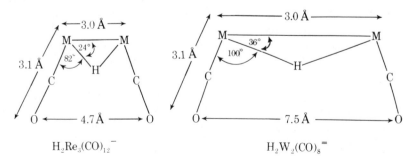

Alternatively, the higher rate of exchange of bridging hydrogen in the unsaturated clusters could result from their greater accessibility to external bases.   This could be caused by the greater interligand separation, as can be seen by comparison of the structures for $H_2Re_3(CO)_{12}^-$ (*23*) and $H_2W_2(CO)_8^{2-}$ (*24*), (*see* also structure of $H_2Re_2(CO)_8$) (*20*).   Since oxygen–oxygen van der Waal contacts are about 3 Å and even a small base such as water has a van der Waal diameter of 2.8–3.0 Å, it is clear that deprotonation in a complex of type $H_2Re_3(CO)_{12}^-$ requires a tunneling process (cf. the several hour half-life for deprotonation of $H_3Re_3(CO)_{12}$ by amines) (*3*).   In the type of complexes represented by $H_2W_2(CO)_8^{2-}$, direct contact between a base and the cluster-bonded hydrogen is possible, which would greatly facilitate exchange.

$H_2Re_3(CO)_{12}^-$ $H_2W_2(CO)_8^=$

Regardless of the details of the exchange mechanism, it is clear that careful control experiments are necessary when working with deuterated metal carbonyls, be they mononuclear or polynuclear. Also, D_2O exchange catalyzed by a chromatographic support may be a convenient method for synthesizing certain metal deuterio complexes.

Vibrational Analysis of the Bridging Hydrogen Absorptions

The vibrational data for hydrido-metal cluster complexes, although useful in characterization, present some difficulties in analysis. The absorptions for hydrogen bridging between metals usually is observed in the region 1600–800 cm^{-1} (9, 25) shifted and in some cases considerably broadened compared with the terminal metal–hydrogen stretching modes which usually are observed in the region 2200–1600 cm^{-1} (9). In some polynuclear hydride complexes, absorptions are seen above the bridging region, and it is possible that these contain a pronounced asymmetry in the position of the bridging hydrogen. This is discussed at greater length below.

The absorptions for bridging hydrogen also can display multiplicity, as in the pair of bands seen for $H_4Ru_4(CO)_{12}$ or $H_2Re_2(CO)_8$ (Figures 2 and 7, above). A discussion of these also is presented below. For the bridging absorptions: very broad bands can display a more complex pattern as evidenced by a series of minima in the spectrum of $H_3Re_3(CO)_{12}$ (Figure 1, above), or a pronounced complexity as evidenced in the spectra of $HRe_3(CO)_{14}$ (Figure 3, Ref. 9) or $HRe_2(CO)_8Cl$ (Figure 24, Ref. 25). Such patterns are too complex and spread over too broad a spectrum of energy to be accountable as fundamental vibrations of bridging hydrogen alone. Claydon and Sheppard (26) have discussed the effect of coincidences or near coincidences of overtones of lower-lying vibrations on the shape of the broad bands observed for hydrogen-bridging between non-metal atoms; possibly the same methodology can be applied to account at least in part for maxima (or minima) observed in the broader of the bridging hydrogen absorptions.

For $H_2Re_2(CO)_8$ (D_{2h} symmetry) (20), three M–H–M vibrational bands are expected in both the ir and Raman spectra, with no coincidences between the two because of the center of symmetry (shown on page 227). The $2A_g$ Re–Re stretching and B_{2u} Re–H–Re bending modes would be expected to occur at low energy (<300 cm^{-1}). The Raman spectrum (Figure 7) shows two bands at 1382 and 1275 cm^{-1}, which shift as expected to 973 and 922 cm^{-1} upon deuteration. These are thus assigned to the $2A_g$ and B_{2g} modes. The ir spectrum shows a somewhat broad band at 1249 cm^{-1}, which is probably a composite of both the B_{1u} and B_{3u} modes.

For $H_4Ru_4(CO)_{12}$, a larger number of fundamental absorptions are expected: five Raman-active hydrogen stretching modes (A_1, B_1, B_2, and $2E$) and three ir-active modes (B_2 and $2E$) coincident with the Raman bands. As mentioned above, the observed spectra consist of only two bands in both the Raman and ir, which are noncoincident [1585 and 1290 cm^{-1} in Raman (Figure 2), above, and 1605 and 1272 cm^{-1} in ir] (6). The spectra of $D_4Ru_4(CO)_{12}$ are analogous, showing Raman bands at 1153 and 895 cm^{-1} and ir bands at 1153 and 909 cm^{-1}. Since the Raman spectrum of $H_2D_2Ru_4(CO)_{12}$ is nearly an exact superposition of that of $H_4Ru_4(CO)_{12}$ and $D_4Ru_4(CO)_{12}$ (bands at 1587, 1291, 1156, and 909 cm^{-1}) (7) it is tempting to suggest that the various Ru–H–Ru oscillators are completely uncoupled. In this case, two hydrogen modes are predicted, $2A_1$ and B_1 (the $2A_1$ mode would be expected below 300 cm^{-1}). In this event, the two hydrogen absorptions should be coincident in the Raman and in the ir and they are not. Clearly this system merits further attention.

A further dimension to the problem of the bridging hydrogen modes has been introduced by the observation, now by two groups (27, 28), of a moderately strong high-energy absorption appearing only in low-temperature spectra, i.e. at 77° and 1680 cm^{-1} for [Et$_4$N][HW$_2$(CO)$_{10}$] (27) or at 10°K and 1274 cm^{-1} for [Et$_4$N][DCr$_2$(CO)$_{10}$] (28). These absorptions broaden and disappear at higher temperature, and discussions as to their interpretation are still in progress. The hydrido-bridged dinuclear anions exhibit variable geometry influenced by the nature of the counter ion (29, 30). Relatively shallow potential minima can thus be inferred for structural tautomers of the anions, and the change observed in the vibrational spectra could well be attributed to a further structural deformation occurring at low temperature. Because of the relatively high-energy position of the low-temperature absorption, we infer a pronounced asymmetry in the

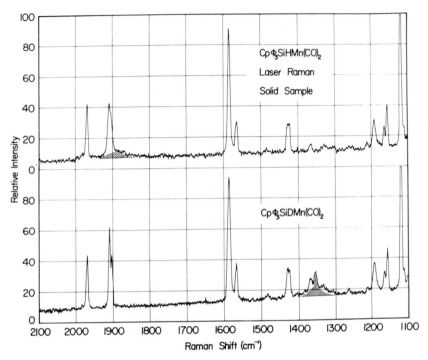

Figure 8.   Raman spectra, Cary 81, He/Ne laser exciting line 15 803 cm^{-1}
(25)

position of hydrogen (essentially equivalent to the proposal of a double-minimum potential forwarded by Harris and Gray (27)). In this, considerable terminal metal–hydrogen character would be expected in the stretching mode. Because of the special nature of these dinuclear anions, their low-temperature behavior can prove to be a singularity; it would thus be useful to obtain spectra at low temperatures for other hydrido-metal cluster complexes to investigate this matter further. We understand such work is in progress (28).

We conclude this presentation by calling attention to some unusual features in two hydrido-silyl complexes: namely, $Re_2(CO)_8H_2SiPh_2$ (18, 19, 20) and $(\eta^5-C_5H_5)Mn(H)(SiPh_3)(CO)_2$ (structure determinations: Refs. 32, 33). The hydrogen atoms might be expected in a bridging position in the former and the single hydrogen in a terminal position in the latter. The dirhenium complex shows a relatively high-lying absorption at 1790 cm^{-1} (the shaded feature in the top scan of Figure 6). An even higher-lying absorption is seen for the manganese complex (the shaded feature centered at 1890 cm^{-1} in the top scan of Figure 8); the latter is extremely broad. Both bands can be attributed to hydrogen through their shift in the corresponding deuterio derivatives: the band at 1280 cm^{-1} for $Re_2(CO)_8D_2SiPh_2$ (Figure 6) and the band at 1355 cm^{-1} for $(\eta^5-C_5H_5)Mn(D)-(SiPh_3)(CO)_2$ (Figure 8). Based on x-ray study of a series of silyl-rhenium derivatives, a terminal position for hydrogen on rhenium has been suggested for $Re_2(CO)_8H_2SiPh_2$ (*34*). Because the band attributed to hydrogen shows some broadening, we believe that some interaction with a neighboring atom (silicon) must be taking place. We come back to this point later.

For the manganese derivative a pronounced broadening is observed in the Mn–H stretching mode, and we are led to conclude that the hydrogen atom must occupy some type of bridging position. This conclusion also was reached by Smith and Bennett (*34, 35*) who attributed the bridging to steric crowding on manganese as compared with that of an analogous rhenium derivative. Although structures containing bridging halogen in group IV–halide derivatives of the transition metals are known (*37, 38, 39*), a distinctly different valency problem is encountered for a situation involving bridging hydrogen. Hart–Davis and Graham (*31*) have discussed this problem in terms of resonance hybrids between representations (a) and (c) below. We believe the situation is better expressed in terms of a structural tautomer such as (b), which corresponds to an equilibrium position of the atoms arrested at some point along the reaction coordinate (a) to (c) representing the oxidative addition of an Si–H bond at a transition-metal center. Such an adduct would occur where steric repulsions on the metal atom render unfavorable the formation of two new σ-bonds on the metal involving a higher coordination number.

(a)　　　　　　　(b)　　　　　　　(c)

We return to the derivative $Re_2(CO)_8H_2SiPh_2$ to note the lack of multiplicity in its bridging hydrogen absorption compared with the well-separated pair of absorptions seen in $H_2Re_2(CO)_8$ (*see* Figures 6 and 7, respectively). There is, of course, a pronounced difference in the environments and positions of the two hydrogen atoms in these derivatives, which would be expected to effect the coupling of their vibrations. Still, keeping in mind the suggestion of an "arrested" oxidative addition mentioned above for $(\eta^5\text{-}C_5H_5)Mn(H)(SiPh_3)(CO)_2$, it is intriguing that the lack of splitting in the hydrogen mode for $Re_2(CO)_8H_2SiPh_2$

is reminiscent of that observed in the parent silanes, R_2SiH_2; in these the symmetric and anti-symmetric modes are so close together as to be indistinguishable in most cases (40). Thus, the vibrational features suggest that the structure of the silyl-di-rhenium complex (19) also should be regarded as an "arrested" oxidative adduct of Ph_2SiH_2 with $Re_2(CO)_8$.

Acknowledgment

We are grateful to W. A. G. Graham for providing various samples used in our Raman studies as cited in the text and to J. R. Johnson for assistance in investigations into the ir spectra of $H_2Re_2(CO)_8$ and its partially deuterated derivative. Thanks are also expressed to the National Science Foundation for support of this work. (Contribution No. 3942.)

Literature Cited

1. Hieber, W., *Adv. Organomet. Chem.* (1970) **8**, 8.
2. Ginsberg, A. P., *Transition Met. Chem.* (1965) **1**, 148.
3. Fontal, B., Dissertation, UCLA, 1969, pp. 88 and 125.
4. Beck, W., Hieber, W., Braun, G., *Z. Anorg. Allg. Chem.* (1961) **308**, 23.
5. Braterman, P. S., Harrill, R. W., Kaesz, H. D., *J. Am. Chem. Soc.*, (1967) **89**, 2851.
6. Huggins, D. K., Fellmann, W., Smith, J. M., Kaesz, H. D. *J. Am. Chem. Soc.* (1964) **86**, 4841.
7. Knox, S. A. R., Koepke, J. W., Andrews, M. A., Kaesz, H. D. *J. Am. Chem. Soc.* (1975) **97**, 3942.
8. Kirtley, S. W., Olsen, J. P., Bau, R., *J. Am. Chem. Soc.* (1973) **95**, 4532.
9. Kaesz, H. D., Saillant, R. B., *Chem. Rev.* (1972) **72**, 231.
10. Kaesz, H. D., *Chem. Br.* (1973) **9**, 344.
11. Pimentel, G. C., McClelland, A. L., "The Hydrogen Bond," W. H. Freeman & Co., San Francisco, 1960.
12. Pimental, G. C., McClelland, A. L., *Annu. Rev. Phys. Chem.* (1971) **22**, 347.
13. Andrews, M. A., Kirtley, S. W., Kaesz, H. D., *Inorg. Chem.* (1977) **16**, 1556.
14. Andrews, M. A., Dissertation, UCLA (1977).
15. Keister, J. B., Shapely, J. R., *J. Am. Chem. Soc.* (1976) **98**, 1056.
16. Byers, B. H., Brown, T. L., *J. Am. Chem. Soc.* (1975) **97**, 947 (preliminary communication).
17. *Ibid.* (1977) **99**, 2527.
18. Hoyano, J. K., Elder, M., Graham, W. A. G., *J. Am. Chem. Soc.* (1969) **91**, 4568.
19. Elder, M., *Inorg. Chem.* (1970) **9**, 762.
20. Bennett, M. J., Graham, W. A. G., Hoyano, J. K., Hutcheon, W. L., *J. Am. Chem. Soc.* (1972) **94**, 6232.
21. Deeming, A. J., Hasso, S., *J. Organomet. Chem.* (1975) **88**, C21.
22. Shapley, J. R., Keister, J. B., Churchill, M. R., DeBoer, B. G., *J. Am. Chem. Soc.* (1975) **97**, 4145.
23. Churchill, M. R., Bird, P. H., Kaesz, H. D., Bau, R., Fontal, B., *J. Am. Chem. Soc.* (1968) **90**, 7135.
24. Churchill, M. R., Chang, S. W. Y., *Inorg. Chem.* (1974) **13**, 2413.
25. Kirtley, S. W., Dissertation, UCLA (1972).
26. Claydon, M. F., Sheppard, N., *J. Chem. Soc., Chem. Commun.* (1969) 1431.
27. Harris, D. C., Gray, H. B., *J. Am. Chem. Soc.* (1975) **97**, 3073.
28. Cooper, C. B., Shriver, D. F., Onaka, S., ADV. CHEM. SER. (1978) **167**, 232.
29. Roziere, J., Williams, J. M., Stewart, R. P., Petersen, J. L., Dahl, L. F., *J. Am. Chem. Soc.* (1977) **99**, 4497.

30. Wilson, R. D., Graham, S. A., Bau, R., *J. Organomet. Chem.* (1975) **91,** C49.
31. Hart-Davis, A. J., Graham, W. A. G., *J. Am. Chem. Soc.* (1971) **93,** 4388.
32. Hutcheon, W. L., Dissertation, University of Alberta (1971).
33. *Chem. Eng. News* (1970) **48,** 75.
34. Cowie, M., Bennett, M. J., *Inorg. Chem.* (1977) **16,** 2321.
35. *Ibid.,* p. 2325.
36. Smith, R. A., Bennett, M. J., *Acta Crystallogr.* (1977) **B33,** 1113.
37. Elder, M., Graham, W. A. G., Hall, D., Kummer, D., *J. Am. Chem. Soc.* (1968) **90,** 2189.
38. Elder, M., Hall, D., *Inorg. Chem.* (1969) **8,** 1269 and 1273.
39. Cradwick, E. M., Hall, D., *J. Organomet. Chem.* (1970) **25,** 91.
40. Bellamy, L. J., "Infrared Spectra of Complex Molecules," 3rd ed., p. 380, Methuen, London (Wiley, New York) 1975.

RECEIVED December 19, 1977.

17

Vibrational Spectroscopy of Hydride-Bridged Transition Metal Compounds

C. B. COOPER III, D. F. SHRIVER, and S. ONAKA

Department of Chemistry, Northwestern University, Evanston, IL 60201

The fundamental vibrations have been assigned for the $M-H-M$ backbone of $HM_2(CO)_{10}^-$, $M = Cr$, Mo, and W. When it is observable, the asymmetric $M-H-M$ stretch occurs around 1700 cm^{-1} in low temperature ir spectra. One or possibly two deformation modes occur around 850 cm^{-1} in conjunction with overtones that are enhanced in intensity by Fermi resonance. The symmetric stretch, which involves predominantly metal motion, is expected below 150 cm^{-1}. For the molybdenum and tungsten compounds, this band is obscured by other low frequency features. Vibrational spectroscopic evidence is presented for a bent $Cr-H-Cr$ array in $[PPN][(OC)_5Cr-H-Cr(CO)_5]$. This structural inference is a good example of the way in which vibrational data can supplement diffraction data in the structural analysis of disordered systems. Implications of the bent $Cr-H-Cr$ array are discussed in terms of a simple bonding model which involves a balance between nuclear repulsion, $M-M$ overlap, and $M-H$ overlap. The literature on $M-H-M$ frequencies is summarized.

Ir spectroscopy has played a major role in the characterization of hydrogen-bonded compounds and hydride-bridged boron compounds, but similar progress has not been made with the vibrational spectra of hydride-bridged metal compounds. One of the main problems is that the M–H modes of hydride-bridged metal compounds are frequently weak and broad in both the ir and Raman. Indeed it is often not possible to detect the hydride stretch of a M–H–M system in the ir spectrum obtained at room temperature. Low temperature ir and Raman spectroscopy afford significant improvements in the ease of detecting bridging M–H modes. In the present work, these techniques provide sufficient information to permit assignment of the major fundamental vibrational modes of M–H–M systems and to supplement the structural information available from neutron diffraction.

0-8412-0390-3/78/33-167-232/$05.00/0

© American Chemical Society

Boron Hydride Bridged Systems

The three-centered, two-electron hydride bridge, which is prevalent in boron hydride chemistry, has been very well characterized in the ir and Raman spectroscopy of diborane and certain metal borohydrides.   A brief review of these data will be given at this point because they afford insight into hydride-bridged systems.

Diborane.   As shown in Table I, the B–H terminal stretching modes of diborane occur at an average frequency of 2563 cm^{-1}, whereas the average of the bridging B–H stretches (ν_2, ν_6, ν_{13}, and ν_{17}) occurs 28% lower, at 1847 cm^{-1}.   The existing normal coordinate vibrational analyses display some abnormalities, such as a near zero H_b–B–H_b bending force constant, but the general features of the force field for diborane are reasonably well understood.   The vibrational analyses indicate that the B–H_b force constant is about half of that for B–H_t (*1, 2, 3*).   This

Table I.   Some Fundamentals for Gas Phase $^{11}B_2H_6$ (*8*)

A_g	ν_1	B–H_t	stretch	2524 cm^{-1}
	ν_2	B–H_b	stretch	2104
	ν_4	B–B	stretch (ring breathing)	794
B_{1g}	ν_6	B–H_b	stretch (ring stretch)	1768[a]
B_{1u}	ν_8	B–H_t	stretch	2612
B_{2g}	ν_{11}	B–H_t	stretch	2591
B_{2u}	ν_{13}	B–H_b	stretch (ring stretch)	1915
B_{3u}	ν_{16}	B–H_t	stretch	2525
	ν_{17}	B–H_b	stretch (ring stretch)	1602

[a] Corrected for Fermi resonance.

result fits with the simple bonding models that distribute two bonding electrons between B–H in the terminal case and an average of one bonding electron between each B–H in a hydride bridge.   Also, a large B–B force constant is calculated (ca. 2.7 mdyn/Å), which the authors interpret as evidence for significant B–B bonding (*2*).   This interpretation is based on the reasonable assumption that the B–H–B bending force constant is very low and can be neglected.   However, it is important to point out that if a significant resistance to B–H–B bending is introduced, the B–B force constant would be calculated lower.   Unfortunately, in cyclic systems such as these, the redundancy between the various stretching and bending coordinates makes it impossible to prove the extent of direct B–B bonding.

Metal Borohydrides.   Several covalent borohydrides, Be(BH$_4$)$_2$ (*4, 5*), Al(BH$_4$)$_3$ (*6*), Zr(BH$_4$)$_4$ (*7*), and Hf(BH$_4$)$_4$, and U(BH$_4$)$_4$ (*7*), have been the topic of detailed ir and Raman spectroscopic investigation.   For one of these, $Al(BH_4)_3$ (4), the combined vibrational and electron diffraction data indicate an overall D_{3h} symmetry, with each BH_4 bonded by a double hydride bridge (Structure 2) to the central aluminum atom.   Cook and Nibler make a case that the Al–H–B bridge stretching vibrations are best described as individual B–H_b and Al–H_b stretches rather than ring stretches and expansions similar to those in diborane. Their argument is based on the presence of four equivalent B–H bridge bonds in diborane as opposed to the large inequivalence between B–H and Al–H in $Al(BH_4)_3$, yielding two bands that are primarily B–H_b stretches above 2000 cm^{-1} and two predominantly Al–B_b stretches below 1600 cm^{-1}.   These frequencies are significantly lower than typical B–H_t and Al–H_t stretching modes.   The AlB_3 symmetric stretch in this compound is assigned to a polarized Raman band at 495 cm^{-1}, and the doubly degenerate asymmetric stretch is found at 596 cm^{-1}.

As opposed to the double hydride bridges (Structure 2) discussed for $Al(BH_4)_3$, single hydride bridges (Structure 1) appear to exist in $B_2H_7^-$, $H(B(CH_3)_3)_2^-$, and $H(B(C_2H_5)_3)_2^-$, for which rather broad ir bands of moderate intensity (2050, 2100, and 1915 cm^{-1} respectively) are assigned to the bridging hydride stretching modes (6).   The compounds $Be(BH_4)_2$ (gas phase) (7), $Zr(BH_4)_4$ (9), $Hf(BH_4)_4$ (10), and $U(BH_4)_4$ (9) all contain the triple hydride bridge (Structure 3).   For $Hf(BH_4)_4$, the symmetric Hf–B stretch is assigned to a Raman feature at 552 cm^{-1}, and the symmetric B–H_t and B–H_b stretches are at 2572 and 2192 cm^{-1} respectively.   This pattern of higher terminal than bridge B–H stretching modes first was discussed in detail by Price (11, 12) and is most pronounced when the M–H_b bond is strong; thus, weakening the B–H_b bond.   Owing to this frequency decrease, it is generally easy to detect M–H–B linkages in metal borohydrides.   Vibrational spectroscopy also offers a convenient method of distinguishing double (see Structure 2) and triple (see Structure 3) hydride-bridged borohydrides (13).   One of the most characteristic features is the presence of two BH_{2t} stretches in the former and a single BH_t stretch for the latter.

Transition Metal Hydride-Bridged Systems

Terminal M–H stretching modes of transition metal hydrides are readily identified in the ir around 1900 ± 300 cm^{-1}, with intensities that usually are stronger than CH stretching modes (14).   With bridging hydrides, however, the bands often are not observed in the ir.   As demonstrated in the laboratories of both Jones (15) and Kaesz (14), M–H modes of bridging hydrides are more readily observable in the Raman spectrum, but even with this technique the bands are broad at room temperature. As will be described later, low temperatures lead to greatly decreased bandwidths. Unfortunately, most synthetic chemists have not used the Raman technique with newly prepared transition metal hydrides so there is, at present, a paucity of data in this area. Also, the assignments are not entirely clear for the series of Raman bands observed between 800–900 cm^{-1}. More discussion of this problem will be given later.

As with diborane and the metal borohydrides, the M–H–M stretching modes are lower in frequency than their terminal counterparts. Other chapters in this volume describe the various environments for bridging hydrides which range from nearly linear (Structure 4) and strongly bent (Structure 5) single hydride bridges (*16, 17, 18, 19*) to double (*20*), Structure 6, triple (*21*), Structure 7, and quadruple (*22*), Structure 8, bridges between two metals. In addition, hydride bridges in metal clusters can either bridge a triangular face (*14*) or be present at the center of an octahedral site (*23*). In this chapter, we discuss three of these bridging hydride configurations, Structures 4, 5, and 6.

Data for bridging hydride modes are collected in Table II. One of the striking aspects of the Raman data is the large number of bands in the 800–1100 cm^{-1} region for some of the hydrides. In several instances, the number of bands greatly exceeds the number of expected fundamentals (*14*).

Table II. Vibrational Data on Metal Hydride Bridged Complexes

Complex	*Vibrational Frequencies Associated with Hydride (Deuteride) Motion*		Ref.
	Ir (cm^{-1})	Ramam (cm^{-1})	
μ_3 Hydrides			
HFeCo$_3$(CO)$_{12}$	1114		24
DFeCo$_3$(CO)$_{12}$	813		24
[HCo(η^5C$_5$H$_5$)]$_4$	1052		25
	950		
	890		
μ_2 Hydrides			
HZn(C$_6$H$_5$)$_2^-$	1250–1650		26
DZn(C$_6$H$_5$)$_2^-$	900–1200		26
HZn(C$_6$F$_5$)$_2^-$	1300–1700		27
DZn(C$_6$F$_5$)$_2^-$	1000–1200		27
[H$_3$Fe$_2$((CH$_3$)C(CH$_2$PO$_2$)$_3$)$_2$]$^+$	1048		21, 28
[D$_3$Fe$_2$((CH$_3$)C(CH$_2$PO$_2$)$_3$)$_2$]$^+$	790		21, 28
[HTi(η^5C$_5$H$_5$)(C$_5$H$_4$)]$_2$	1230		29
[HTi(η^5C$_5$H$_5$)$_2$]	1450		30
[DTi(η^5C$_5$H$_5$)$_2$]	1260		30
	1060		
[H$_2$Re$_2$(CO)$_8$]		1382	31
		1275	
D$_2$Re$_2$(CO)$_8$		973	31
		922	

Table II. Continued

Vibrational Frequencies
Associated
with Hydride (Deuteride)
Motion

Complex	Ir (cm^{-1})	Raman (cm^{-1})	Ref.
$HRe_2(CO)_8Cl$		1449	31
		1311	
		1236	
		1213	
		1175	
$DRe_2(CO)_8Cl$		1051	31
		958	
		925	
		907	
		881	
$HMo_2X_8^{-3}$ [a]	1250		32
$DMo_2X_8^{-3}$ [a]	910		32
$H_3Re_3(CO)_{12}$		1100	15, 31, 33
		1076	
		1000	
$D_3Re_3(CO)_{12}$		792	15, 31, 33
		752	
		692	
$H_2Os_3(CO)_{12}$	1930		34
(terminal + μ_2)	1525		
$D_2Os_3(CO)_{12}$	1410		34
(terminal + μ_2)	1110		
$HRe_3(CO)_{14}$		1258	31
(possibly μ_2)		1184	
		1097	
		1041	
		952	
		904	
		850	
$DRe_3(CO)_{14}$		1122	31
		825	
		742	
$H_4Ru_4(CO)_{12}$	1605		35, 36
		1585	
		1290	
	1272		
$D_4Ru_4(CO)_{12}$		1153	35, 36
	1095		
		909	
	895		
$H_2D_2Ru_4(CO)_{12}$		1587	36
		1291	
		1156	
		909	
$H_2Re_3(CO)_{12}^{-}$		1102	31, 37
		1052	

Table II.   Continued

Vibrational Frequencies
Associated
with Hydride (Deuteride)
Motion

Complex	Ir (cm^{-1})	Raman (cm^{-1})	Ref.
$D_2Re_3(CO)_{12}{}^-$		803	*31, 37*
		740	
		632	
$H_6Re_4(CO)_{12}{}^{-2}$		1165	*31*
		1125	
$D_6Re_4(CO)_{12}{}^{-2}$		832	*31*
$H_3Ru_4(CO)_{12}{}^-$		1442	*38*
(possibly μ_2 and/or μ_3)			
Group VIB μ_2–Hydrides			
[Et$_4$N][HCr$_2$(CO)$_{10}$]	~1750		
[Et$_4$N][DCr$_2$(CO)$_{10}$]^b	~1274		*39*
[PPN][HCr$_2$(CO)$_{10}$]^b	~1755		*39*
		835	
		785	
[PPN][DCr$_2$(CO)$_{10}$]	1274		*39*
		550	
[Et$_4$N][HMo$_2$(CO)$_{10}$]		940	*39*
		855	
		795	
		760	
[Et$_4$N][DMo$_2$(CO)$_{10}$]		785	*39*
		643	
		540	
[PPN][HMo$_2$(CO)$_{10}$]		985	*39*
		915	
[PPN][DMo$_2$(CO)$_{10}$]		692	*39*
		645	
		576	
[Et$_4$N][HW$_2$(CO)$_{10}$]	1683		*39, 40*
		960	
		905	
		870	
		830	
		783	
		709	
[Et$_4$N][DW$_2$(CO)$_{10}$]		613	*39*
		603	
[PPN][HW$_2$(CO)$_{10}$]	1702		*39*
		960	
		731	
[PPN][DW$_2$(CO)$_{10}$]		720	*39*
		628	

^{a} X = Cl$^-$, Br$^-$.
^{b} Et$_4$N$^+$ = (CH$_3$CH$_2$)$_4$N$^+$, PPN$^+$ = [(C$_6$H$_5$)$_3$P]$_2$N$^+$.

M—H—M $\quad$ M $\nearrow$ H $\searrow$ M $\quad$ M $\diagdown\diagup$ H $\diagdown\diagup$ M $\quad$ M—H—M $\quad$ M $\nearrow$ H $\searrow$ M

4 $\qquad$ 5 $\qquad\qquad$ 6 $\qquad\qquad$ 7 $\qquad\qquad$ 8

(OC)$_5$M–H–M(CO)$_5$⁻ SALTS. The single hydride bridged anions of Cr, Mo, and W, having the formula indicated above, were chosen by us as a starting point in the detailed investigation of hydride-bridged metal systems. The structures of these complexes in the solid state are strongly cation dependent. Room temperature neutron diffraction data for the PPN⁺, bis(triphenylphosphine) iminium, salt of HCr$_2$(CO)$_{10}$⁻ display a rather large thermal motion of the hydride ligand in a plane perpendicular to the Cr–Cr axis, and consequently, it was not possible to determine if the Cr–H–Cr backbone is truly linear or is a composite of slightly bent structures (*16*). Room temperature neutron data for the Et$_4$N⁺ salt, however, clearly shows a bent backbone with a M–H–M angle of 158.6° (*16*). Similarly, low temperature (14°K) neutron data indicates that the M–H–M backbone of (Et$_4$N)(HW$_2$(CO)$_{10}$) is strongly bent with an angle of 137.2° (*46*). Analogy between room temperature x-ray data for (PPN)(HW$_2$(CO)$_{10}$) and neutron data for HW$_2$(CO)$_9$(NO) indicates that the PPN⁺ salt has an angle on the order of 150° (*18, 19*). The wide variation in M–H–M angles for these species is indicative of a highly compliant M–H–M backbone.

As a first approximation to the influence of structure on the vibrational frequencies, we shall concentrate on the M–H–M tri-atomic array. For a linear M–H–M unit, the normal modes are very simple. There is a Raman-active symmetric M–H–M stretch (Structure 9) that only involves motion of the massive metal and therefore occurs at much lower frequencies than the B–B and M–B stretching modes discussed in a previous section. In an intermediate frequency region, a doubly degenerate M–H deformation mode occurs that involves hydrogen motion (Structure 10), and at high frequencies, an asymmetric ir-active stretching vibration (Structure 11) should be observed.

If the M–H–M array is bent, the selection rules are relaxed from the linear case; all three bands are allowed in the ir and Raman. Also, the form of the normal modes changes, in that the two symmetric modes (those associated with ν_1 and ν_2) are mixed.

A qualitative view of the influence of the M–H–M angle on ν_1, ν_2, and ν_3 can be drawn from a simple valence force field calculation in which the force

ν_1 $\quad$ M ← H → M $\qquad\qquad\qquad\qquad$ **9**

ν_2 $\quad$ M $\downarrow$ $\uparrow$ M / H $\downarrow$ $\qquad\qquad\qquad$ **10**

ν_3 $\quad$ M → H → M $\qquad\qquad\qquad\qquad$ **11**

constants are fixed and only the M–H–M angle, θ, is varied from linear (180°) to strongly bent (90°) (Figure 1). The frequencies should not be taken literally because some variation in force constants with angle is to be expected, and because the force field does not include interactions between bond stretching or deformation coordinates (i.e., no off-diagonal force constants are included). One feature to note in Figure 1 is the small change in all frequencies between the linear (180°) and slightly bent cases (160°); thus, molecules possessing these geometries should display similar vibrational frequencies. As the M–H–M angle, θ, decreases (i.e., the backbone becomes more strongly bent), a rapid change is noted for the

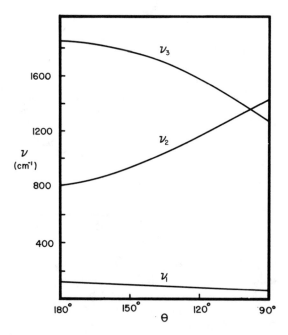

Figure 1. Results of a simple valence force field calculation for the Mo–H–Mo system showing the relationship among the frequencies of the three hydride modes and the Mo–H–Mo angle, θ

asymmetric stretch ν_3 and the deformation ν_2. As θ approaches 90°, it will be noted that the frequencies of ν_2 and ν_3 cross; thus, for molecules with a highly bent backbone, unambiguous assignment of ν_2 and ν_3 will be difficult. It is clear, however, that strongly bent M–H–M systems should display significantly lower ν_3 and higher ν_2 frequencies than their linear counterparts.

Next we shall discuss specific assignments for the $HM_2(CO)_{10}^-$ ions. The low frequency ν_1 mode is considered first, then the high frequency ν_3 M–H stretch, and finally the complex intermediate region which we believe to include ν_2.

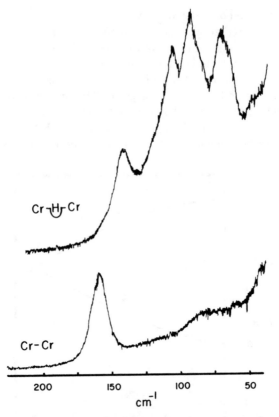

Figure 2. Low-frequency Raman spectra for solid (Et₄N)(HCr₂(CO)₁₀) (upper) and (Et₄N)₂-(Cr₂(CO)₁₀) (lower)

The ν_1 Region. A good set of reference points for establishing the ν_1 frequency range is provided by the simple metal–metal bonded analogs of the hydrides which we are considering, namely $Cr_2(CO)_{10}{}^{2-}$, $Mo_2(CO)_{10}{}^{2-}$, and $W_2(CO)_{10}{}^{2-}$. The M–M frequencies in these compounds are found at 160, 140, and 115 cm⁻¹ for Raman spectra of the respective tetraethylammonium salts in the solid state at room temperature (*19*). These frequencies should be upper bounds on the metal–metal stretches for the (OC)₅MHM(CO)₅⁻ complexes of Cr, Mo, and W. Careful inspection of the low frequency region of the Raman spectra for the hydrides reveals a medium strong peak at 144 cm⁻¹ for the Et₄N⁺ salt of the chromium compound, which is tentatively assigned as the ν_1 mode. In Figure 2, this feature is compared with the M–M stretch of its simple M–M bonded analog. It should be noted that the low frequency region of the hydrides (below about 115 cm⁻¹) is intense and complex. The presence of these strong bands can be responsible for our inability to locate the ν_1 frequency for the hydride-bridged molybdenum and tungsten compounds.   Even at 10°K where these features are sharpened considerably, it was not possible to pick out the ν_1 band. Another point of some importance is that ν_1 in [Et$_4$N][DCr$_2$(CO)$_{10}$] is identical to that in its hydrido counterpart.   Therefore, this hydride-bridged metal system does not appear to exhibit the very large anharmonicity of the type seen in some hydrogen-bonded systems, where the X--X stretching frequency can be quite different between X-H···X and X-D···X.

The ν_3 Region.   This mode will be ir-active for a linear or bent M–H–M array.   Accordingly, low temperature spectra for all 12 combinations of Et$_4$N$^+$ and PPN$^+$ with hydrido- and deuterio-bridged Cr, Mo, and W anions were examined.   For spectra recorded at 8°K, the H:D isotope shift allows positive identification of the ν_3 mode in the four chromium salts and two tungsten hydrides

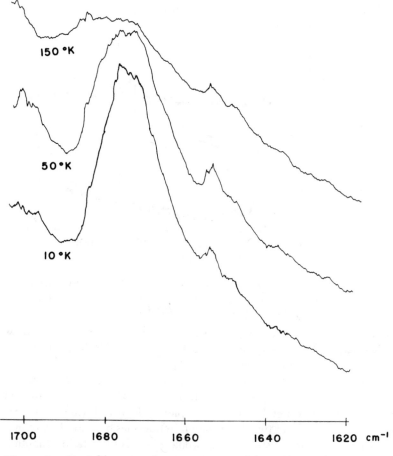

Figure 3.   Variable temperature ir spectra of (Et$_4$N)(HW$_2$(CO)$_{10}$).   The sample temperatures are probably higher than the measured temperatures shown here.

(*see* Table I).   These modes shift and broaden with increasing temperature, as illustrated in Figure 3.   Harris and Gray previously have observed this feature for $(Et_4N)(HCr_2(CO)_{10})$ (*40*).   Why these bands are so strongly temperature dependent and why they were not observed in all the salts is not known; however, failure to observe some of these modes, particularly in the deuterides, can be attributed to interferences from cation bands.   The Christiansen effect degraded the quality of some spectra and may be responsible for obscuring some of these modes.   The lack of a significant shift in ν_3 between Et_4N^+ and PPN^+ salts of $[HCr_2(CO)_{10}^-]$ implies that these compounds have similar Cr–H–Cr angles.

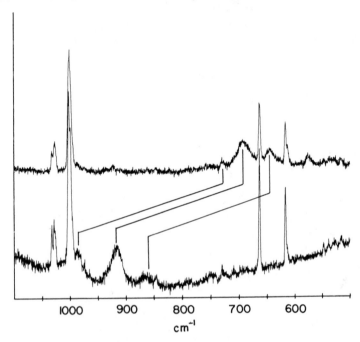

Figure 4.   Raman spectra of $(PPN)(HMo_2(CO)_{10})$ (upper) and $(PPN)(DMo_2(CO)_{10})$ (lower).   The connection lines suggest an approximate $\sqrt{2}$ shift in the frequencies.

Simple valence force field calculations predict a difference of 20 cm^{-1} between a truly linear Cr–H–Cr array and the slightly bent array (159.8°), which has been observed by neutron diffraction for the Et_4N^+ salt.

The ν_2 Region.   As discussed earlier, there should be one mode in this intermediate frequency range for a linear or bent M–H–M array.   However, in these more complex metal carbonyl complexes, the bent form of $(OC)_5M$–H–$M(CO)_5^-$ should exhibit two deformation frequencies because the presence of the carbonyl ligands transforms what would have been a rotational motion for an isolated M$\overset{H}{\diagup}$M into an internal vibrational mode.   Therefore, a maximum of two modes would be expected in the intermediate frequency region.   By contrast, a large number of features are seen between 700 and 960 cm^{-1} for the molybdenum and tungsten salts.   As illustrated in Figure 4, the substitution of D for H in (PPN)(HMo$_2$(CO)$_{10}$) shifts several bands roughly by the $\sqrt{2}$ factor. The existence of more than two bands can arise from Fermi resonance with overtones and combinations of the MCO and M–C modes.   Experiments utilizing ^{18}O-substituted compounds are being pursued to clarify this point.   The observation of features in the 700–900 cm^{-1} region (*see* Table I) for (PPN)-(HCr$_2$(CO)$_{10}$) would be anomalous if this anion were truly linear because ν_2 should be Raman inactive.   The observation of ν_2 in the Raman along with the similarity in ν_3 values for the Et$_4$N$^+$ and PPN$^+$ salts is, in our opinion, good evidence that in both salts, HCr$_2$(CO)$_{10}$$^-$ is bent.   As previously mentioned, the neutron diffraction data for the hydrogen in the PPN$^+$ salt show a large thermal amplitude perpendicular to the Cr–Cr axis.   However, observation of similar ν_3 values for the Et$_4$N$^+$ and PPN$^+$ salts, along with similar temperature dependence for the linewidth of ν_3, indicate that the Cr–H–Cr array in both salts is bent to approximately the same degree.   Apparently the large thermal ellipsoid for the PPN$^+$ salt is the result of a composite of bent M–H–M arrays in the crystal.

The bent structure for the M–H–M moiety in [Et$_4$N][HCr$_2$(CO)$_{10}$] and particularly that in [PPN][HCr$_2$(CO)$_{10}$] where there is little or no externally imposed distortion of the anion, indicate that there may be inherent stability in a bent array.   In previous discussions of related bent systems, it has been postulated that an energetic advantage is achieved by bending, which allows closer approach of the metal atoms and thus stronger M–M interaction (*16, 19, 22*).   It is our purpose here to point out the central role that nuclear repulsion must play in such a bonding argument.

Since hydrogen possesses no inner-core electrons, the central proton in a three-center, two-electron bond will exert only a simple electrostatic repulsion for the metal cores.   Close approach of the metal atoms, which would maximize M–M overlap, will be resisted by this electrostatic repulsion.   Thus, it can be energetically favorable for the proton to move off the M–M axis simultaneously with the closer approach of the metal atoms.   However, a very large off-axis excursion will reduce M–H overlap.   Thus, the nuclear repulsion, the M–M overlap, and the M–H overlap may strike a delicate balance leading to an off-axis hydrogen position.

In summary, the assignments favored by us for these single hydride-bridged-carbonyl anions are ca. 1700 cm^{-1} for ν_3, ca. 850 cm^{-1} for ν_2, and <150 cm^{-1} for ν_1.   Alternatively, the ν_3 mode can be near that of ν_2 in the 900–700 cm^{-1} region.   However, we do not favor this assignment because it does not provide a ready explanation for the high frequency modes, which cannot be explained as overtones.   Further confirmation of our preferred assignments can be obtained from normal mode calculations.

A simple valence force field leads to Equation 1 for relationship between a M–H terminal stretch, ν_t, as compared with ν_3 for a linear M–H–M system.

$$\frac{\nu_3}{\nu_t} \approx \sqrt{\frac{2k_b}{k_t}} \tag{1}$$

(Owing to the small degree of bending in the chromium systems, a linear approximation is adequate.) If we take 1370 cm^{-1} as a typical terminal M–D stretching frequency, the observed 1274 cm^{-1} for ν_3 in $[Et_4N][DCr_2(CO)_{10}]$ leads to a force constant for this bridged compound, k_b, that is 42% of the terminal M–H force constant, k_t. This is a reasonable force constant reduction, as shown by the 50% difference between terminal and bridging B–H force constants in diborane. Alternatively, $\nu_3{:}\nu_1$ is predicted to be about 7.2 for a simple valence force field applied to a linear M–H–M system. This ratio, along with the tentatively assigned value of 144 for ν_1, then places ν_3 at 1037 cm^{-1} in $[DCr_2(CO)_{10}^{-}]$. The calculated frequency is lower than the observed 1274 cm^{-1} because the simple valence force field does not include an interaction force constant between the two M–H coordinates in M–H–M, but the magnitude of the calculated frequency substantiates our assignments for ν_1.

Systematics of M–H–M Stretching Modes

In addition to the single hydride bridge discussed above, hydride bridges are found in conjunction with other bridging groups, such as CO (Structure 12). None of these systems have been studied in the detail of the $HM_2(CO)_{10}^{-}$ compounds; however, the M–M stretching modes have been observed in several cases. Since these modes are fairly easily observed in the Raman, their systematics are of some interest for the characterization of hydride-bridged compounds. A bridging structure, such as that shown in Structure 12, is inferred from ir data

12

on $[Et_4N][HFe_2(CO)_8]$ (42). A single M–M stretching mode is observed in the Raman at 255 cm^{-1}, which is comparable to the M–M stretch for other double carbonyl-bridged first-row carbonyls, such as $Co_2(CO)_8$, that displays a M–M stretch at 229 cm^{-1} (43). The indication from this comparison is that, in conjunction with bridging carbonyls, a hydride bridge does not weaken the M–M restoring force, as it does in the case of a single hydride bridge.

The bonding in $H_2W_2(CO)_8^{-2}$, that is depicted in Structure 6, involves two hydride bridges (20), and the Et_4N^{+} salt displays a M–M stretch at 131 cm^{-1}; this is significantly lower than we would estimate for the W–W stretch in a double-bonded tungsten compound (41). These relative frequencies,

$$M = M > M \underset{H}{\overset{H}{\diagdown\diagup}} M > M - M > M-H-M; \quad \underset{M-M}{\overset{CO \ CO}{\underset{|}{\times}|}} \sim \underset{\underset{H}{\diagdown\diagup}}{\overset{CO \ CO}{\underset{M-M}{\overset{|}{\times}|}}} ,$$

can be helpful in the characterization of dinuclear hydride compounds.   In general, the low frequency Raman modes, which involve a large component of M–M stretching, are the easiest features to observe in the skeleton of metal hydride bridged systems.   The higher frequency modes, which involve a large contribution from hydrogen motion, are most clearly seen at low temperatures; but even with low temperature spectroscopy, their observation is not trivial.

Summary

The fundamentals have been assigned for the M–H–M backbone of $HM_2(CO)_{10}^-$, M = Cr, Mo, and W.   When it is observable, the asymmetric M–H–M stretch occurs around $1700 \ cm^{-1}$ in low-temperature ir spectra.   One or possibly two deformation modes occur around $850 \ cm^{-1}$, in conjunction with overtones that are enhanced in intensity by Fermi resonance.   The symmetric stretch, which involves predominantly metal motion, is expected below $150 \ cm^{-1}$. For the molybdenum and tungsten compounds, this band is obscured by other low frequency features.   Vibrational spectroscopic evidence is presented for a bent Cr–H–Cr array in $[PPN][(OC)_5Cr-H-Cr(CO)_5]$.

Experimental

Compounds were prepared by published methods, and their purities were checked by elemental analysis and ir spectroscopy (*40, 44*).   Raman spectra were obtained on pressed pellets in contact with a copper block cooled to $10°K$ in an Air Products closed-cycle helium refrigerator.   A backscattering geometry was used to illuminate the samples with line-focused 514.5 or 676.4 nm radiation. Details of the monochromator and data collection scheme are given elsewhere (*45*).   It is highly likely that local heating by the laser beam raised the sample temperature somewhat above $10°K$.   Ir data were obtained on Perkin–Elmer 180 and Nicolet 7199 FT IR spectrometers.   The samples were deposited from a THF solution as a thin layer on a gold-coated copper block that was then cooled in the helium refrigerator.   Spectra then were recorded with a Harrick Scientific reflectance unit.   This method was used to insure that the temperatures of the samples were controlled accurately.

Acknowledgment

We greatly appreciate many informative discussions with Mark Ratner.   We also appreciate the free exchange of information with J. L. Peterson, L. F. Dahl,

R. Bau, and Jack Williams. This work was supported by the NSF through grant CHE 74-20004 A02. Spectroscopic measurements were performed in the NSF-supported Northwestern University Materials Research Center Optics Facility.

Literature Cited

1. Adams, D. M., Churchill, R. G., *J. Chem. Soc. A* (1970) 697.
2. Ogawa, T., Miyazawa, T., *Spectrochim. Acta* (1964) **20**, 557.
3. Ramaswami, K., Shanmugam, G., *Acta Phys. Pol. A* (1973) **44**, 349.
4. Coe, D. A., Nibler, J. W., *Spectrochim. Acta* (1973) **29A**, 1789.
5. Emery, A. R., Taylor, R. C., *Spectrochim. Acta* (1960) **16**, 1455.
6. Matsui, Y., Taylor, R. C., *J. Am. Chem. Soc.* (1968) **90**, 1363.
7. Nibler, J. W., *J. Am. Chem. Soc.* (1972) **94**, 3349.
8. Lord, R. C., Nielsen, E., *J. Chem. Phys.* (1951) **19**, 1.
9. Davies, N., Wallbridge, M. G. H., Smith, B. E., James, B. D., *J. Chem. Soc., Dalton Trans.* (1973) 162.
10. Keiderling, T. A., Wozniak, W. T., Gray, R. S., Jurkowitz, D., Bernstein, E. R., Lippard, S. J., Spiro, T. G., *Inorg. Chem.* (1975) **14**, 576.
11. Price, W. C., *J. Chem. Phys.* (1949) **17**, 1044.
12. Price, W. C., Longuet-Higgins, H. C., Rice, B., Young, T. F., *J. Chem. Phys.* (1949) **17**, 217.
13. Marks, T. J., Kennelly, W. J., Kolb, J. R., Shimp, L. W., *Inorg. Chem.* (1972) **11**, 2540, and references therein.
14. Kaesz, H. D., Saillant, R. B., *Chem. Rev.* (1972) **72**, 231.
15. Smith, J. M., Fellmann, W., Jones, L. H., *Inorg. Chem.* (1965) **4**, 1361.
16. Petersen, J. L., Dahl, L. F., Williams, J. M., ADV. CHEM. SER. (1978) **167**, 11.
17. Roziere, J., Williams, J. M., Stewart, R. P., Jr., Petersen, J. L., Dahl, L. F., *J. Am. Chem. Soc.* (1977) **99**, 4497.
18. Wilson, R. D., Graham, S. A., Bau, R., *J. Organomet. Chem.* (1975) **91**, C49.
19. Olsen, J. P., Koetzle, T. F., Kirtley, S. W., Andrews, M., Tipton, D. L., Bau, R., *J. Am. Chem. Soc.* (1974) **96**, 6621.
20. Churchill, M. R., Ni Chang, S. W. Y., Berch, M. L., Davison, A., *J. Chem. Soc., Chem. Commun.* (1973) 691.
21. Dapporto, D., Midollini, S., Sacconi, L., *Inorg. Chem.* (1975) **14**, 1643.
22. Bau, R., Carroll, W. E., Hart, D. W., Teller, R. G., Koetzle, T. F., ADV. CHEM. SER. (1978) **167**, 73.
23. Broach, R. W., Dahl, L. F., Longoni, G., Chini, P., Schultz, A. J., Williams, J. M., ADV. CHEM. SER. (1978) **167**, 93.
24. Mays, M. J., Simpson, R. N. F., *J. Chem. Soc. A* (1968) 1444.
25. Müller, J., Dorner, H., *Angew. Chem., Int. Ed. Engl.* (1973) **12**, 843.
26. Kubas, G. J., Shriver, D. F., *J. Am. Chem. Soc.* (1970) **92**, 1949.
27. Kubas, G. J., Shriver, D. F., *Inorg. Chem.* (1970) **9**, 1951.
28. Dapporto, P., Fallani, G., Midollini, S., Sacconi, L., *J. Am. Chem. Soc.* (1973) **96**, 2021.
29. Brintzinger, H. H., Bercaw, J. E., *J. Am. Chem. Soc.* (1970) **92**, 6182.
30. Bercaw, J. E., Brintzinger, H. H., *J. Am. Chem. Soc.* (1969) **91**, 7301.
31. Kirtley, S. W., dissertation, University of California at Los Angeles (1971).
32. Cotton, F. A., Kalbacher, B. J., *Inorg. Chem.* (1976) **15**, 522.
33. Huggins, D. K., Fellmann, W., Smith, J. M., Kaesz, H. D., *J. Am. Chem. Soc.* (1964) **86**, 4841.
34. Shapley, J. R., Keister, J. B., Churchill, M. R., DeBoer, B. G., *J. Am. Chem. Soc.* (1974) **97**, 4145.
35. Johnson, B. F. G., Lewis, J., Williams, J. G., *J. Chem. Soc. A* (1970) 901.
36. Knox, S. A. R., Koepla, J. W., Andrews, M. A., Kaesz, H. D., *J. Am. Chem. Soc.* (1975) **97**, 3942.

37. Churchill, M. R., Baird, P. H., Kaesz, H. D., Bau, R., Fontal, B., *J. Am. Chem. Soc.* (1968) **90**, 7135.
38. Koeploe, J. W., Johnson, J. R., Knox, S. A. R., Kaesz, H. D., *J. Am. Chem. Soc.* (1975) **97**, 3947.
39. This work.
40. Onaka, S., Cooper, C. B., Shriver, D. F., unpublished data.
41. Harris, D. C., Gray, H. B., *J. Am. Chem. Soc.* (1975) **97**, 3073.
42. Furmery, K., Killner, M., Greatrex, R., Greenwood, N. N., *J. Chem. Soc. A* (1969) 2339.
43. Onaka, S., Shriver, D. F., *Inorg. Chem.* (1976) **15**, 915.
44. Hayter, R. G., *J. Am. Chem. Soc.* (1966) **88**, 4376.
45. Shriver, D. F., Dunn, J. B. R., *J. Appl. Spectrosc. (Eng. Transl.)* (1974) **88**, 3191.
46. Bau, R., personal communication.

RECEIVED July 11, 1977.

18

Solid State Nuclear Magnetic Resonance Study of Heavy Metal Hydrides

A. T. NICOL and R. W. VAUGHAN

Division of Chemistry and Chemical Engineering, California Institute of Technology, Pasadena, CA 91125

A group of heavy metal hydrides, Th_4H_{15}, $H_2Os_3(CO)_{10}$, $H_4Os_4(CO)_{12}$, and $H_4Ru_4(CO)_{12}$ have been examined with both conventional-pulsed and high-resolution solid state nuclear magnetic resonance techniques. Results of lineshape analysis, relaxation time measurements, and line shifts are summarized for the solid state binary hydride, Th_4H_{15}. Experimental lineshapes for the carbonyl hydrides are compared with theoretical calculations for small spin systems, and solid state chemical shift tensors are also reported.

This chapter presents results of NMR studies of several heavy metal hydrides, including both a summary of completed work by Lau et al. (1) on a binary hydride, Th_4H_{15}, and preliminary results on several carbonyl hydrides, $H_2Os_3(CO)_{10}$, $H_4Os_4(CO)_{12}$, and $H_4Ru_4(CO)_{12}$. The binary hydride, Th_4H_{15}, has attracted interest recently with the discovery of its super-conducting properties (2) and the carbonyl hydrides are metal cluster hydrides (3) which are of interest as models in the study of catalysis (4).

Efforts to understand the state of hydrogen in metals and metal hydrides have involved the use of NMR for many years. This study combines the conventional solid state NMR techniques with more recently developed high-resolution, solid state NMR techniques (5, 6). Conventional NMR techniques furnish information on dipolar interactions and thus can furnish static geometrical information on hydrogen positions and information on proton motion within such solids. The newer multiple pulse techniques suppress proton–proton dipolar interaction and allow information on other, smaller interactions to be obtained. This chapter reports what the authors believe is the first observation of the powder pattern of the chemical shift tensor of a proton that is directly bonded to a heavy metal.

These materials are all particularly well suited to a NMR investigation since they are simple in structure with no more than one or two chemically different

0-8412-0390-3/78/33-167-248/$05.00/0
© American Chemical Society

protons present. Th_4H_{15} is stoichiometric and has been characterized to a substantial extent (7–14).   It contains two structurally different kinds of protons, while the carbonyl cluster hydrides are thought to contain only a single kind of structural proton (3).

Experimental Details

Proton NMR measurements were made at 56.4 MHz on a spectrometer that was described previously (15). T_1 was measured with a 180°--t--90° pulse sequence, and lineshapes were determined from free-induction decay signals following a 1–2 μsec 90° pulse. $T_{1\rho}$ was measured with a 90° x-pulse followed by an attenuated y-pulse whose length was varied from 10 μsec to 40 msec.

For the multiple pulse studies, an eight-pulse cycle that has been discussed in detail previously (5, 6, 16, 17, 18) was used.   Cycle time, t_c, the time required for a single eight-pulse cycle, was 42–48 μsec for measurements reported here.

The temperature range of 40°–460°K has been achieved with two probes of different constructions.   The probe with a temperature range of 100°–460°K used nitrogen gas as the coolant.   The low temperature probe, connected through a liquid helium transfer line to a liquid helium Dewar flask, used helium as coolant.

For line shift measurements with the eight-pulse cycle between 180°K and room temperature, the reference was acetyl chloride.   Its frequency was measured relative to a spherical tetramethylsilane (TMS) sample at room temperature, and all results are reported relative to this TMS on the τ scale.   ($\tau = \sigma + 10$ ppm, where σ is the signed chemical shift used in solid state NMR.)   At lower temperatures, the reference was a single crystal of $Ca(OH)_2$, oriented in the magnetic field such that the major axis of its proton chemical shift tensor was parallel to the external field (19).   Thus, it is assumed that the proton chemical shift of the $Ca(OH)_2$ remained unchanged as the temperature was varied.

The Th_4H_{15} samples were prepared by Cameron Satterthwaite and co-workers (1, 2) and consisted of two polycrystalline samples that were determined to be within 1% of the stoichiometric composition, $Th_4H_{15\pm0.15}$.   The samples differed primarily in the pressure and temperature used in their synthesis.   The sample hydrided under lower pressure and temperature conditions (1 atm of H_2 and a temperature cycle initiating at 800°K and dropping to 450°K before removing the H_2) is labeled the LP sample, and the one hydrided under higher pressure and temperature (1100°K and ~10,000 psi of H_2) is labeled HP.   The sample preparation and characterization of this stoichiometric compound is apparently critical since the results of the present study differ significantly from the previous NMR studies (12, 13, 14), and some difference was detected between the LP and HP samples themselves.   The carbonyl hydride samples were furnished kindly by John R. Shapley (20).

Results and Discussion

Th_4H_{15}.   Three kinds of information are obtained from the samples of Th_4H_{15}:   information on rigid lattice structure from free induction decays, on proton motion from relaxation time measurements, and on internal fields from peak locations that were found using the multiple pulse techniques.   Figure 1

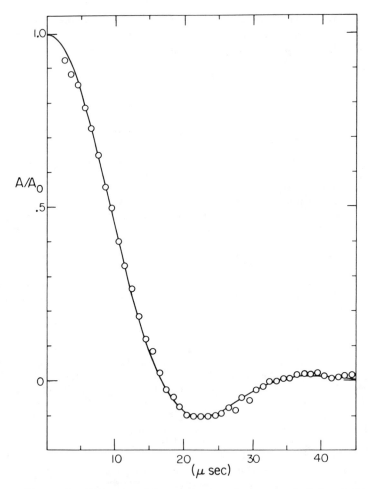

*Figure 1.   Normalized free-induction decay signal of the Th_4H_{15}
powder sample (O).   The solid line is the analytical function, exp
$(-a^2t^2/2)$ sin (bt)/bt, where a and b were determined from the
calculated second and fourth moments for the proposed
structure.*

illustrates the free-induction decay of Th_4H_{15}, and a beat structure is clearly
present.   This decay was temperature independent below room temperature,
and no difference was noted between the two samples.   It is possible to extract
two structural parameters from such a rigid lattice line shape since both the second
and fourth moment of the line were obtained by fitting the experimental free
induction decay (21, 22).   Results of early x-ray studies furnished a basis upon
which one could hypothesize the proton locations within the structure (9), and
more recent neutron diffraction results (10) have confirmed the predicted

structure.   Further confirmation was obtained by comparing theoretical second and fourth moments (*21, 22*) that were calculated from this structure with these experimental values.   The experimental and calculated values are, respectively: 19.5 G^2 and 20.1 G^2 for the second moment and 844 G^4 and 895 G^4 for the fourth moment.   A second way to compare the theoretical values with the experimental line shape is to use an empirical function of the form, $\exp\left(-a^2 t^2/2\right) \sin(bt)/bt$ (*23*), and to calculate values of a and b from the theoretical second and fourth moments.   The curve in Figure 1 was generated in this fashion using the theoretical second and fourth moments, and it fits well with the experimental results.

To extract information on the proton motion from NMR measurements, one can use the following approximate equations appropriate for the various NMR relaxation times measured in this work (*24, 25, 26*).

$$\frac{1}{T_2} \approx \gamma^2 M_2 \tau \quad \text{for} \quad \tau M_2^{1/2} \ll 1 \tag{1}$$

$$\frac{1}{T_1} \approx \frac{4}{3}\gamma^2 M_2 \frac{1}{\omega_0^2 \tau} \quad \text{for} \quad \omega_0\tau \gg 1 \tag{2}$$

$$\frac{1}{T_{1\rho}} \approx \gamma^2 M_2 \frac{\tau}{1 + 4\omega_1^2\tau^2} \tag{3}$$

where ω_0 = Larmor frequency corresponding to external field, H_0; ω_1 = Larmor frequency corresponding to rf locking field, H_1; τ = correlation time for proton motion and M_2 = second moment of proton resonance.   These equations are derived from a single correlation time model for dipolar relaxation of protons, and if one assumes further that τ obeys an Arrhenius relation, the corresponding activation energies, ΔE, can be obtained from the temperature dependence of the relaxation parameters.

$$\frac{1}{\tau} = \frac{1}{\tau_0} e^{-\Delta E/kT} \tag{4}$$

Thus, τ can be estimated by several different measurements, and ΔE is obtained by fitting the slope of a plot of $\ln(1/T_2)$, $\ln(1/T_1)$, or $\ln(1/T_{1\rho})$ as a function of $1/T$.

The relaxation results (*1*) for the LP sample can be summarized.   Line-shape measurements indicated that motional narrowing started at around 60°C, and by 110°C, the lines appeared completely Lorentzian.   The activation energy obtained from T_2 above 120°C was 16.3 ± 1.2 kcal/mol.

The observed T_1 can have two contributions: $1/T_1 = 1/T_{1e} + 1/T_{1d}$. T_{1e} results from relaxation effects caused by conduction electrons, and T_{1d} results from dipolar relaxation effects caused by the motion of nuclear spins.   At temperatures below 110°C, the conduction electron effect appears to dominate T_1 since a plot of T_1 vs. $1/T$ can be fit with a straight line passing through the origin.   At higher temperatures, the relaxation caused by the proton motion becomes dominant and by subtracting out the conduction electron contribution, one obtains an activation energy for proton motion of 18.0 ± 3.0 kcal/mol, which agrees well with the activation energy from the T_2 data.

Two sets of $T_{1\rho}$ data were obtained at substantially different holding fields, $H_1 = 4.7$ G and $H_1 = 20.6$ G. The measured $T_{1\rho}$ values are proportional to $H_1{}^2$ at a fixed temperature over the temperature range studied. This is what the model for proton motion discussed above (26) would predict and is strong evidence that $T_{1\rho}$ is dominated by lattice motion. Activation energies for proton motion of 10.9 ± 0.7 kcal/mol are obtained from the slopes of the lines fitted through both sets of $T_{1\rho}$ data. The difference between the activation energies for proton motion (16.3 ± 1.2 and 18.0 ± 3.0 kcal/mol above 120°C and 10.9 ± 0.7 kcal/mol below 80°C) obtained from the NMR relaxation rates is well outside the limits of experimental error. This suggests a more complex mechanism for proton motion than a simple, single correlation time-activated process and could imply more than a single mechanism for proton motion.

In addition to activation energies, Equations 1, 2, and 3 can be used to obtain estimates of the correlation time, τ, for the proton motion. These results are illustrated in Figure 2. However, the estimation of correlation times depends upon the numerical values for constants relating the relaxation times (T_1, T_2, $T_{1\rho}$) to the correlation time, τ, in Equations 1, 2, and 3, and these constants are defined only within factors of two or three. As shown in Figure 2, the τ values obtained from T_2 data by using Equation 1 are approximately 2.6 times the τ values from the T_1 data using Equation 2. Using Equation 3, the τ values derived from $T_{1\rho}$ data at different H_1 holding fields agree with one another. Furthermore, both τ curves (i.e., above 120°C and below 80°C) extrapolate to the same region in the intermediate temperature range, indicating that the available data give a rather consistent picture of the proton motion in the Th_4H_{15} sample.

Additional evidence for the applicability of the above data interpretation is obtained by using Equation 3 to predict values of $1/(T_{1\rho})$ min. These are 2.8 × 10⁴ sec⁻¹ and 6.4 × 10³ sec⁻¹ for $H_1 = 4.7$ G and 20.6 G, respectively, and are consistent with the $T_{1\rho}$ and T_2 data. The motional properties reported here differ significantly from previous measurements: the activation energy obtained from a temperature above 120°C is two to three times the values obtained in Ref. 13, and motional narrowing of the lineshape occurs at a significantly higher temperature than that in Ref. 12.

Little difference was found between the NMR results on the LP and HP samples at either room temperature or as the temperature was raised, although the T_1 for the HP sample was 10% shorter than that for the LP sample. After maintaining the samples at temperatures near 200°C for 1 hr., a major difference was noted upon cooling, with the HP sample exhibiting a marked time–temperature hysteresis of all of the measured NMR properties while the LP sample exhibited no time–temperature hysteresis in any of the NMR parameters measured. A more dramatic manifestation of this is that after the sample was cooled to room temperature, the proton motion took weeks to return to the original state

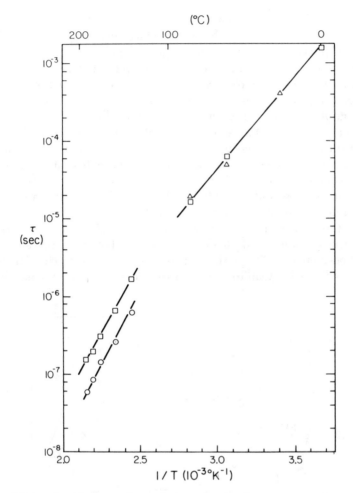

Figure 2.   Estimated correlation times, τ, for proton motion as a function of inverse temperature in Th₄H₁₅: (□) τ values estimated from T_2 data; (O) from T_1 data; (△) from $T_{1\rho}$ with $H_1 = 20.6$ g; and (□) from $T_{1\rho}$ with $H_1 = 4.7$ g below 100° C.

as indicated by the T_2, T_1, and $T_{1\rho}$ data (*see* Figure 3). After the phenomena were observed the first time, the experiment was repeated a month later, and the same effect was observed in T_1, T_2, and $T_{1\rho}$ measurements.

T_1 is dominated at room temperature by conduction electron effects while T_2 and $T_{1\rho}$ are controlled by motional properties of the protons. The large time–temperature hysteresis observed in all of these parameters indicates strongly a phase change in the material on heating. The hysteresis is large enough that it is difficult to understand without moving the thorium atoms to new locations since the mobility of the protons is such that they would relocate in times short compared with these. The fact that samples of the same composition and physical characteristics can behave so differently under mild heating is indicative that there is still much to be learned about this complex material. Detailed x-ray studies as a function of sample preparation and temperature are being conducted by C. B. Satterthwaite's research group at this time.

Multiple-pulse measurements were performed on both the LP and HP samples at 20° and −80°C, and when no differences were noted, lower temperature measurements were performed only on the LP sample. Multiple-pulse spectra for the LP sample are illustrated in Figure 4 together with the eight-pulse spectrum of the reference used for the low-temperature measurements, $Ca(OH)_2$. The lineshapes observed are quite broad, and the line center is a function of temperature. The line width was separated into three contributions by performing three related multiple-pulse measurements (1). These indicated that the main contributions to the linewidth came from both relaxation and second-order dipolar effects. The maximum possible field inhomogeneity Hamiltonian is estimated to be less than 16 ppm by this means, which indicates that the com-

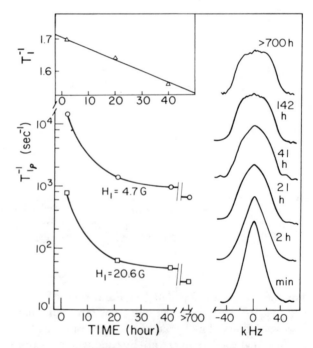

Figure 3.   Log of the inverse of the spin lattice relaxation time (T_1^{-1}), *the spin-lattice relaxation time in the rotating frame* ($T_{1\rho}^{-1}$), *and the line shape as a function of time for the HP sample after the sample was heated to 460°K and brought back to room temperature.*

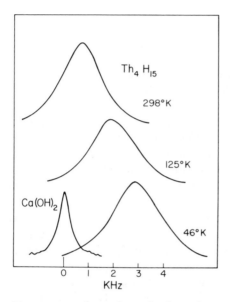

Figure 4.   The eight-pulse line shape and the peak locations of the Th_4H_{15} (LP) powder sample as a function of temperature using a $Ca(OH)_2$ single crystal as reference. The reference is oriented such that the major principal axis of the proton chemical shift tensor is parallel to the external magnetic field. A shift to the left signifies an increase in the value of σ, i.e., the internal magnetic field at the proton site is larger in Th_4H_{15} than in $Ca(OH)_2$.

bination of H_o inhomogeneity, field distortions (caused by susceptibility effects), and chemical and Knight shift anisotropies do not add up to more than this value.

Although the field inhomogeneity Hamiltonian is small, the line center moves by approximately 50 ppm as the temperature is lowered from room temperature to 46°K. This is illustrated in Figure 4. The observed shift was corrected (reduced by up to 20%) for temperature-dependent bulk susceptibility (demagnetization effect) (27) of the sample before constructing Figure 4, and consequently, the experimentally observed shift was larger than that indicated in this figure.

The absolute chemical shifts of protons in diamagnetic solids are typically near 30 ppm (28), and the Knight shift caused by conduction electrons through contact interaction is estimated to be −31.2 ppm, using the experimental T_1T value of 180 K-sec and the Korringa relation (29):

$$T_1 \left(\frac{\Delta H}{H} \right)^2 = \frac{\hbar}{4\pi k T} \frac{\gamma_e^2}{\gamma_n^2}$$

Both of these contributions are temperature independent and are smaller than the temperature-dependent shift observed experimentally. Temperature-dependent shifts have been reported in rare earth intermetallic compounds (30, 31) and attributed to hyperfine couplings with the rare earth electron spin. Additionally, the magnitude of temperature-dependent shifts found in several compounds with the A-15 structure (i.e., ^{51}V and $^{69,71}Ga$ in V_3Ga (32, 33) have been correlated with the super-conducting transition temperature of these materials, and an explanation has been suggested (34) based on the necessity of a sharp peak in the electron density of states at the Fermi level. The temperature-dependent shift reported here for protons in Th_4H_{15} is large when compared with the expected size of the proton chemical or temperature-independent Knight shift and is much smaller than the temperature-dependent shifts seen on other nuclei (32, 33).

Carbonyl Hydrides. This section presents preliminary results obtained from the application of both conventional and multiple-pulse NMR techniques to a series of metal cluster compounds, $H_2Os_3(CO)_{10}$, $H_4Ru_4(CO)_{12}$, and $H_4Os_4(CO)_{12}$. A review by H. D. Kaesz (3), together with more recent articles (35–43), summarizes the present knowledge of the structure of these and closely related molecules. Some information on proton positions has been obtained from NMR studies of similar molecules dissolved in nematic liquid crystals (39, 40) but in general, the proton positions in such symmetric molecules have been inferred from the x-ray-determined geometry of the heavier atoms. X-ray studies in most cases were performed on derivatives containing a large ligand to facilitate crystallization rather than on the parent globular molecule. Thus, efforts to obtain information on the geometry and motional properties of protons within these structures directly with NMR techniques appear worthwhile. Solid state NMR is suited particularly for examining the highly symmetric parent molecules and would, in contrast, be more difficult to apply to the more structurally complex derivatives. These materials present a relatively tractable situation for proton NMR studies since available data indicate that all protons in the structures are equivalent, and the carbonyl groups isolate the protons somewhat within a single molecule from interactions with protons on neighboring molecules. Thus, we are faced with either two- or four-particle problems that must be solved in first order to understand the NMR lineshapes. For solids containing a limited number of spin $\frac{1}{2}$ particles, the dipolar lineshape can be calculated in detail and thus is more informative than in the many-particle case (as the Th_4H_{15} discussed above). G. E. Pake first calculated the two-spin $\frac{1}{2}$ particle lineshape (23, 44), and later the tetrahedral four-spin lineshape was calculated by Bersohn and Gutowsky (45). A single crystal spectra calculation for the rectangular four-spin problem was reported by J. Itoh et al. (46), and more recently, Eichhoff and Zachman (47) discussed powder patterns for four protons on a rectangle. The situation of the molecules to be considered here is, in addition, complicated by the presence of a fraction of the metal nuclei with a nonzero dipole moment. In particular, ^{189}Os has a spin of 3/2 and a large quadrupole moment ($\sim 2 \times 10^{-24}$ cm^2) and is found in a natural abundance of 16.1% while ruthenium has two isotopes with spin 5/2—^{99}Ru (12.8% natural abundance) and ^{101}Ru (17% natural abundance).

$H_2Os_3(CO)_{10}$. Figure 5 presents a proton NMR absorption spectrum for $H_2Os_3(CO)_{10}$ that was obtained by Fourier transformation of the free induction decay. The spectrum was taken at room temperature, but no difference was

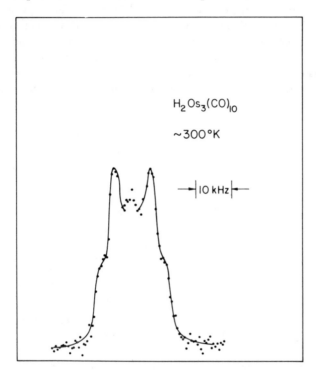

Figure 5.   The proton free-induction decay line shape for $H_2Os_3(CO)_{10}$. The points in the center of the spectrum where no line was drawn are caused partly by probe impurities.

noted as the temperature was lowered to 140°K, thus implying an absence of any change in molecular motions fast enough to average the dipolar interactions. This spectrum is qualitatively similar to that expected from a pair of rigid isolated protons (23, 44); however, it does deviate significantly. For example, the separation between the two inner peaks should be one-half the separation of the two outer peaks (shoulders), but the inner peak separation is somewhat greater than this. Possibly more important, the intensity of the shoulder peaks is greater than would be expected from a single pair interaction. It does not appear possible to account for such deviations with interactions from protons on neighboring molecules, and we presently are investigating the possibility of multiple proton positions (i.e., a double potential well) and that the ^{189}Os might be producing the anomalous effects. Since ^{189}Os has a large quadrupole moment, and the osmium site symmetry is lower than cubic, it is likely that the osmium spin is not in a Zeeman state but primarily is under the influence of a very large quadrupole interaction, with the Zeeman interaction serving only as a perturbation. The ability of non-Zeeman state spins to produce anomalous effects in NMR spectra is well documented (48, 49), and we presently are investigating whether or not the effects observed in Figure 5 can be simulated analytically on this basis. The results are important in attempting to extract proton–proton distances from a spectrum such as that in Figure 5. For example, a second moment calculated from the spectrum in Figure 5 is 1.6 G^2, and if we assume 0.1 G^2 for the combined effects of intermolecular proton interactions and proton–osmium interactions

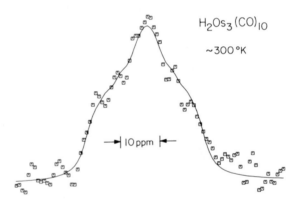

$H_2Os_3(CO)_{10}$

~300 °K

→| 10 ppm |←

Figure 6.   Multiple-pulse proton line shape for
$H_2Os_3(CO)_{10}$ *at room temperature*

(proper for the case where the osmium is in a Zeeman state), we obtain a proton–proton distance of 2.5 Å. If one accepts that the main-peak splittings in Figure 5 were produced by a proton pair alone, one obtains values near 2.55 Å; if one uses the shoulder spacings, one obtains values greater than 2.6 Å. Thus, such distortions must be understood to extract accurate proton–proton distances.

Figure 6 is a plot of the proton NMR spectrum obtained from $H_2Os_3(CO)_{10}$ when using an eight-pulse cycle (5, 6, 16, 17, 18) to suppress the effects of proton–proton dipolar interactions. The curve results from a computer fit that assumes the lineshape is caused by the chemical shift tensor. The center of the spectrum is near $\tau = 19$ ppm, and thus it agrees reasonably with that expected from the solution NMR results ($\tau = 21.7$ ppm (37)). The three principal values of the tensor, according to this fit, are at τ values 5.6 ppm, 19.9 ppm, and 31.6 ppm. Since approximately one-third of the proton pairs interact with a near

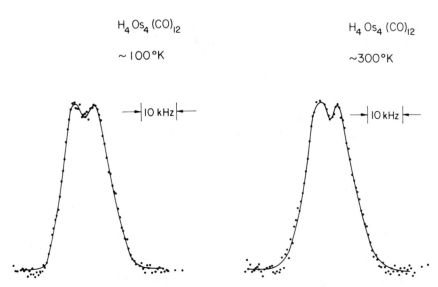

Figure 7. Free-induction decay line shapes for $H_4Os_4(CO)_{12}$ *near* $100°K$ *and room temperature. The lack of complete symmetry in the line shape may be caused by a probe impurity signal located slightly to the left of the spectra's center.*

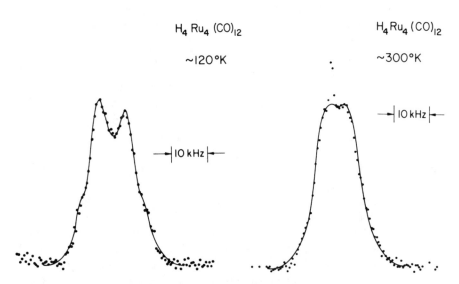

Figure 8. Free-induction decay line shapes for $H_4Ru_4(CO)_{12}$ *near* $120°K$ *and room temperature. The sharp peak located slightly left of the center in the room-temperature spectrum (and the asymmetry in the lower temperature spectrum) is caused partially by probe impurities.*

neighbor osmium with a spin, these values can furnish an upper limit to the proton chemical shift anisotropy ($\sim$26 ppm) in this environment. This is, to our knowledge, the first direct observation of a chemical shift tensor for a proton believed to be bonded directly to a transition metal. The size of the anisotropy of the chemical shift tensor is similar to that found in several proton tensors (5, 6) and in particular, is similar to the O-H$\cdots$O hydrogen-bonded proton tensor (5, 6, 28). Thus, the phenomena responsible for producing the large τ chemical shifts that are unique to the direct bonding of the proton to a transition metal ion does not necessarily produce a large chemical shift tensor anisotropy, as has been speculated for the case of terminally bonded protons (50).

$H_4Os_4(CO)_{12}$ AND $H_4Ru_4(CO)_{12}$. Figures 7 and 8 illustrate the wide-line spectra of these two materials. As indicated above, these cluster hydrides consist of nearly tetrahedral units of four metal atoms and are thought to have bridging hydrogens on four of the six metal–metal bonds (3, 35, 36, 37). The two $H_4Os_4(CO)_{12}$ spectra are similar and indicate that the protons in this molecule are not involved in a motional process sufficient to average the dipolar interaction differently at 100° and at 300°K. The proton spectrum taken of $H_4Os_4(CO)_{12}$ with the eight-pulse cycle (5, 6, 16, 17, 18) was centered at τ = 30 ppm but is otherwise similar to Figure 6 with approximately the same chemical shift anisotropy. Thus, the same remarks can be made about the chemical shift spectra of $H_4Os_4(CO)_{12}$ as were made for $H_2Os_3(CO)_{10}$. In particular, there are no indications that molecular motions are rapid enough to average either the chemical shift or dipolar spectra. One cannot eliminate the possibility of a highly restricted motion with a small activation energy (51, 52) that could produce similar motional averaging at 100° and 300°K. The high molecular symmetry of $H_4Os_4(CO)_{12}$ makes it difficult to envision such a motional process that would not average the chemical shift or dipolar spectra, except possibly for a double potential well in the osmium–hydrogen osmium bond.

The two $H_4Ru_4(CO)_{12}$ spectra in Figure 8 present a different picture. The room temperature spectrum is similar to, although slightly narrower than, the two $H_4Os_4(CO)_{12}$ spectra in Figure 7. The 120°K $H_4Ru_4(CO)_{12}$ spectrum is, however, quite different in shape. A change in shape of a wideline NMR spectrum as a function of temperature normally indicates that thermally activated nuclear motions can average partially the dipolar interaction at the higher temperature; one suspects that the protons are moving sufficiently rapidly at room temperature to narrow partially the dipolar line width. Figure 9, which illustrates the eight-pulse spectrum of $H_4Ru_4(CO)_{12}$ taken at two different temperatures (-45° and 22°C), confirms the presence of molecular motion as the spectrum changes significantly over this temperature range. Also note that the center of this motionally averaged tensor is at τ = 25 ppm compared with 28 ppm in solution (32, 48). Since the chemical shift tensor can be averaged with a motion slower than that required to affect the proton–proton dipolar interaction, a marked effect occurs in the chemical shift pattern. Thus, we have confirmed qualitatively the results obtained by J. R. Shapley (54) which indicated that motional narrowing occurs in $H_4Ru_4(CO)_{12}$ at room temperature, and we find no evidence for similar motional processes in $H_4Os_4(CO)_{12}$ in the same temperature range.

As indicated in the introduction to this section, these materials are particularly interesting since the number of interacting spins is limited by the separation between molecules that is created by the carbonyl groups; thus, one can compare results with detailed lineshape calculations. For these two materials, $H_4Os_4(CO)_{12}$ and $H_4Ru_4(CO)_{12}$, one must deal with a four-proton problem possibly modified by the presence of quadrupolar nuclei with small dipole moments. The four-spin $\frac{1}{2}$ problem has been the subject of a number of papers, but there is only one, the work of Eichhoff and Zachmann (47), dealing with the spin $\frac{1}{2}$ particles arranged on a rectangle (the case here if one assumes the sug-

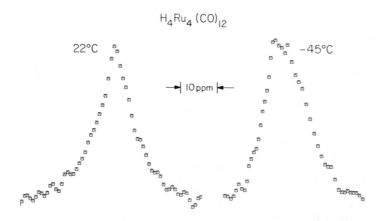

Figure 9. Multiple-pulse proton spectra for $H_4Ru_4(CO)_{12}$ at 22° and −45°C. The increased width of the lower temperature spectrum indicates less motional averaging of the chemical-shift powder pattern as the temperature is lowered.

gested edge-bridging structures), and it is not in a form that can be used without more detailed calculations. Thus, at this early point in our study we cannot make the same comparisons with detailed lineshapes as we did for $H_2Os_3(CO)_{10}$. However, such calculations are in progress and we will hopefully be able to make such detailed comparisons in the future.

Acknowledgment

This work was supported by the Energy Research and Development Administration, and laboratory equipment purchased with funds from the National Science Foundation was used. A. T. Nicol expresses appreciation for partial support from an I.B.M. predoctoral fellowship.

Literature Cited

1. Lau, K. F., Vaughan, R. W., Satterthwaite, C. B., *Phys. Rev.* (1977) **B15**, 2449.
2. Satterthwaite, C. B., Toepke, I. J., *Phys. Rev. Lett.* (1970) **25**, 741.
3. Kaesz, H. D., *Chem. in Britain* (1973) **9**, 344.
4. Robinson, A. L., *Science* (1976) **194**, 1150.
5. Mehring, M., "High Resolution NMR Spectroscopy in Solids," *NMR Basic Principles and Progress*, P. Kiehl, E. Fluck, R. Kosfeld, Eds., Vol. 11, Springer–Verlag, New York, 1976.
6. Haeberlen, U., "High Resolution NMR in Solids, Selective Averaging," *Advances in Magnetic Resonance*, J. S. Waugh, Ed., Supp. 1, Academic, New York, 1976.
7. Mueller, W. M., Blackledge, J. P., Libowitz, G. G., "Metal Hydrides," Academic, New York, 1968.
8. Miller, J. F., Caton, R., Satterthwaite, C. B., *Phys. Rev.* (1976) **B14**, 2795.
9. Zachariasen, W. H., *Acta Cryst.* (1953) **6**, 393.
10. Carpenter, J. M., Mueller, M. H., Beyerlein, R. A., Worlton, T. G., Jorgensen, J. D., Brun, T. O., Skold, K., Pelizzari, C. A., Peterson, S. W., Watonabe, N., Kimura, M., Gunning, J. E., *Proc. Neutron Diffraction Conf. Netherlands, 1975.*
11. Schmidt, H.-G., Wolf, G., *Solid State Comm.* (1975) **16**, 1085.
12. Spalthoff, W., *Z. fuer Phys. Chem. Neue Folge* (1961) **29**, 258.
13. Will, J. D., Barnes, R. G., *J. Less Common Met.* (1967) **13**, 131.
14. Schreiber, D. S., *Solid State Comm.* (1974) **14**, 177.
15. Vaughan, R. W., Elleman, D. D., Stacey, L. M., Rhim, W.-K., Lee, J. W., *Rev. Sci. Instrum.* (1973) **43**, 1356.
16. Rhim, W.-K., Elleman, D. D., Vaughan, R. W., *J. Chem. Phys.* (1973) **58**, 1772.
17. Rhim, W.-K., Elleman, D. D., Vaughan, R. W., *J. Chem. Phys.* (1974) **59**, 3740.
18. Rhim, W.-K., Elleman, D. D., Schreiber, L. B., Vaughan, R. W., *J. Chem. Phys.* (1974) **60**, 4595.
19. Schreiber, L. B., Vaughan, R. W., *Chem. Phys. Lett.* (1974) **28**, 586.
20. Shapley, J. R., Dept. of Chemistry, University of Illinois, Urbana, IL 61801.
21. Lowe, I. J., Norberg, R. W., *Phys. Rev.* (1957) **107**, 46.
22. Van Vleck, J. H., *Phys. Rev.* (1948) **74**, 1168.
23. Abragam, A., "The Principle of Nuclear Magnetic Resonance," Ch. 4, Oxford University, London, 1961.
24. Bloembergen, N., Purcell, E. M., Pound, R. V., *Phys. Rev.* (1948) **73**, 679.
25. Barnaal, D. E., Kopp, M., Lowe, I. J., *J. Chem. Phys.* (1976) **65**, 5495.
26. Look, D. C., Lowe, I. J., *J. Chem. Phys.* (1966) **44**, 2995.
27. Miller, Jim, unpublished data.
28. Lau, K. F., Vaughan, R. W., *Chem. Phys. Lett.* (1975) **33**, 550.
29. Korringa, J., *Physica* (1950) **16**, 601.
30. Jaccarino, V., Matthias, B. T., Peter, M., Suhnl, H., Wernick, J. H., *Phys. Rev. Lett.* (1960) **5**, 251.
31. Jones, E. D., *Phys. Rev.* (1969) **180**, 455.
32. Blumberg, W. E., Eisinger, J., Jaccarino, V., Matthias, B. T., *Phys. Rev. Lett.* (1960) **5**, 149.
33. Weger, M., Goldberg, I. B., "Solid State Physics," *Advances in Research and Applications*, Ehrenreich, H., Seitz, F., Turnbull, D., Eds., Vol. 28, 1973.
34. Clogston, A. M., Jaccarino, V., *Phys. Rev.* (1961) **121**, 1357.
35. Wilson, R. D., Bau, R., *J. Am. Chem. Soc.* (1976) **98**, 4687.
36. Koepke, J. W., Johnson, J. R., Knox, S. A. R., Kaesz, H. D., *J. Am. Chem. Soc.* (1975) **97**, 3947.
37. Knox, S. A. R., Koepke, J. W., Andrews, M. A., Kaesz, H. D., *J. Am. Chem. Soc.* (1975) **97**, 3942.
38. Shapley, J. R., Keister, J. B., Churchill, M. R., DeBoer, B. G., *J. Am. Chem. Soc.* (1975) **97**, 4145.
39. Yesinowski, J. P., Bailey, D., *J. Organometal. Chem.* (1974) **65**, C27.
40. Buckingham, A. D., Yesinowski, J. P., Canty, A. J., Rest, A. J., *J. Am. Chem. Soc.* (1973) **95**, 2732.

41. Shapley, J. R., Richter, S. I., Churchill, M. R., Lashewycz, R. A., *Conf. Chem. Inst. Canada and Am. Chem. Soc., 2nd, Montreal, 1977.*
42. Churchill, M. R., Hollander, F. J., Hutchinson, J. P., *Conf. Chem. Inst. Canada and Am. Chem. Soc., 2nd, Montreal, 1977.*
43. Wilson, R. D., Wu, S. M., Love, R. A., Bau, R., unpublished data.
44. Pake, G. E., *J. Chem. Phys.* (1948) **16**, 327.
45. Bersohn, R., Gutowsky, H. S., *J. Chem. Phys.* (1954) **33**, 651.
46. Itoh, J., Kusaka, R., Yamagata, Y., Kiriyama, R., Ibamoto, H., *J. Physiol. Soc. Jpn.* (1953) **8**, 293.
47. Eichhoff, V. U., Zachmann, H. G., *Kolloid Z. Z. Polym.* (1970) **241**, 928.
48. VanderHart, D. L., Gutowsky, H. S., Farrar, T. C., *J. Am. Chem. Soc.* (1967) **89**, 5056.
49. Stoll, M. E., Vaughan, R. W., Saillant, R. B., Cole, T., *J. Chem. Phys.* (1974) **61**, 2896.
50. Buckingham, A. D., Stephens, P. J., *J. Chem. Soc.* (1964) Part **III**, 2747.
51. Watton, A., Sharp, A. R., Petch, H. E., Pintar, M. M., *Phys. Rev.* (1972) **B5**, 4281.
52. Peternelj, J., Hallsworth, R. S., Pintar, M. M., *J. Mag. Res.* (1977) **25**, 413.
53. Kaesz, H. D., Saillant, R. B., *Chem. Revs.* (1972) **72**, 231.
54. Shapley, J. R., et al., unpublished data.

RECEIVED July 11, 1977.

19

Electronic Structure of Transition Metal Hydrides

ALFRED C. SWITENDICK

Sandia Laboratories, Albuquerque, NM 87115

Metal hydrides form crystalline solids of various types ranging from ionic crystals through semiconductors to metals. They also can be considered as perhaps the simplest example of a class of solid state systems called interstitial alloys wherein a nonmetallic element (hydrogen) is incorporated into an empty hole in a simple metal lattice. The effect of this incorporation on the structure and electronic properties of the metal has been investigated theoretically using the solid state counterpart of molecular orbital theory-band structure theory. Results of these calculations are given for a prototype dihydride system TiH₂ and the well-studied monohydride PdH. In particular, the nature of the metal-hydrogen bond and hydrogen-hydrogen interactions are elucidated.

The incorporation of hydrogen into metal lattices leads to formation of metal–hydrogen compounds called metal hydrides. The properties of these systems have been discussed in numerous monographs and review articles (*1, 2, 3*). Similarly a number of theoretical papers have been written about the change in electronic properties induced by hydrogen in various metal lattices (*4–11*). In this chapter, some of these theoretical results will be summarized. In particular, the nature of the bonding between the hydrogen and the metal (hydrogen) will be examined to assess the factors leading to the stability of these systems.

The occurrence of transition and rare earth metal hydrides is shown in Figure 1. Not shown are the alkali hydrides and alkaline earth dihydrides that bear a close resemblence to their halide counterparts. Also not shown are compounds formed from elements to the right that are the topic of much of the rest of this symposium volume.

We shall address ourselves to the systems shown in Figure 1 and discuss two particular systems in detail, showing how many of the regularities of these materials can be understood. Generally speaking, these materials are metals with

0-8412-0390-3/78/33-167-264/$05.00/0

the exception of the rare earth trihydrides, that become semiconducting. The major feature of Figure 1 is the preponderance of dihydrides on the left, which become increasingly unstable and display competing phases along the line from chromium to tantalum. On the right, the monohydride of nickel and palladium form, but platinum does not hydride. These factors are of interest in terms of an electronic structure model.

A distinct dihydride phase is the strongest confirmation of the magnitude of the influence of hydrogen on the metal systems. As the metal is exposed to hydrogen gas, spontaneous uptake occurs within the metal lattice. This concentration is usually very small but can have strong influence on the mechanical properties, a phenomenon known as embrittlement. With further increase of hydrogen, the metal lattice transforms to a cubic close packed lattice with

IIIB	IVB	VB	VIB	VIIB	VIIIB		
ScH_2	TiH_2	VH VH_2	CrH --	(Mn) --	(Fe) --	(Co) --	NiH
YH_2 YH_3	ZrH_2	NbH NbH_2	(Mo) --	(Tc) --	(Ru) --	(Rh) --	PdH
See Rare Earth Series	HfH_2	TaH	(W) -	(Re) --	(Os) --	(Ir) --	(Pt) --

LaH_{2-3}	CeH_{2-3}	PrH_{2-3}	NdH_{2-3}	Pm ?	SmH_2 SmH_3	EuH_2	GdH_2 GdH_3	TbH_2 TbH_3	DyH_2 DyH_3	HoH_2 HoH_3	ErH_2 ErH_3	TmH_2 TmH_3	YbH_2 $YbH_3(?)$	LuH_2 LuH_3

Figure 1.   Occurrence of binary transition and rare earth metal hydrides

nominally two hydrogens per metal. In the rare earth systems, trihydride phases are even possible, but we shall limit this discussion primarily to these dihydride phases. Very high hydrogen densities can be obtained in these metals, which leads to their application for hydrogen storage and for reactor moderator materials.

As we indicated earlier and as is pointed out in the recent review article by McClellan and Harkins (12), hydrogen plays a somewhat ambivalent role, as it is both the alkali and the halide in its period; this leads to alternative models for the electronic structure of these systems. In the anion model, the hydrogen states are presumed to lie below the metal states. Each state can be filled with two electrons, and since the hydrogen only contributes one electron, electrons from the transition metal fill the low-lying hydrogen states, lowering the Fermi level in the metal manifold. In the proton model, the hydrogen states are higher in energy than the metal states, and the electrons from the hydrogen fill up the metal states.   There is experimental evidence for both viewpoints, and the questions for a model become: where do the states associated with the hydrogen lie and which states are filled by the added electrons?

Our results will be based on the one electron energy band theory of solids (13) that forms the basis for the present-day understanding of metal and semi-conductor physics.   It is the counterpart of the chemist's molecular orbital theory, and we shall try to relate our results back to the underlying atomic structure.

Results and Discussion

We now turn to the formation of some of these hydride structures.   The majority of them are based on a fcc array of metal atoms, as shown by the open circles in Figure 2.   The dihydride structure comes from filling the tetrahedral interstice (large solid circles) in the lattice with hydrogens and gives the well known CaF_2 or calcite structure.   Similarly, if one fills the octahedral interstice (small solid circles), one gets the NaCl or rocksalt structure found in nickel hydride and palladium hydride, which we will discuss near the end of this chapter.

Titanium Dihydride.   The first candidate for study is TiH_2 which forms a prototype dihydride system.   Figure 3 indicates schematically what the cal-culations (14) show, which we will examine further in some detail.   Titanium atoms are put together to form a metal, starting on the left with the atomic con-

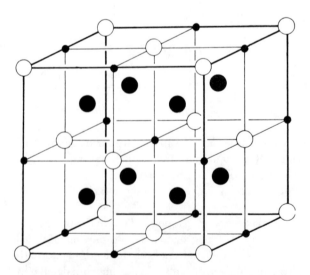

Figure 2.   Fcc hydride structures.   Large open circles represent metal atom positions, solid large circles represent hydrogen atom positions in dihy-dride structure, small solid circles represent hydrogen atom positions in NiH and PdH.

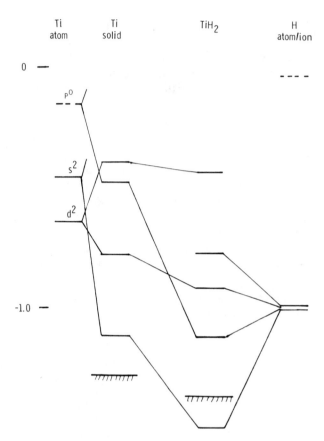

Figure 3.   *Energy level diagram indicating modification of energy levels of titanium and hydrogen upon formation of titanium dihydride.   The diagonally shaded line indicates the potential between the atoms.*

figuration $d^2 s^2$ and with unoccupied p levels.   The atomic potentials overlap in the metal; this leads to a lowering of the energy between the atoms.   As in the molecular case, the degeneracy of the isolated atom levels is broken by the formation of a variety of bonding and antibonding states from the 10^{23} atomic s, p, and d levels.   As in the molecular case, the accommodation of electrons in low lying electronic states leads to bonding and cohesion.   The states broaden about their atomic values, and we have a broad s band now partly below the d states and a fairly well-defined d band.   The relative numbers of low-lying s and d states and the number of electrons determine the highest filled energy, the Fermi level.

On the right, we have the levels associated with the hydrogen atom (ion): the $1s$ level at -1 Rydberg and the H^- ion at -0.72 eV.   It seems clear that the lowest valence energy of the combined system is associated with the hydrogen

$1s$, that the highest is $1s^2$, and that some compromise between these extremes will be met in the solid. The hydrogen level shown is doubled since there are two hydrogens in each unit cell. As the hydride forms, the hydrogen $1s$ levels interact with each other to form a bonding and antibonding pair. Also, the overall potential is lowered by the presence of the hydrogens. Interactions with transition metal d, s, and p electrons are indicated in the second and third columns. The magnitudes of the lowerings are sizeable ($\sim$5 eV). The d band manifold largely is maintained. The major result is that the hydrogen-derived s states are low and can be filled with electrons both from the transition metal and from the hydrogen. The results of the band structure calculations show in detail how this comes about. Instead of labels like Π, σ, a_{1g}, B_{1u}, etc., the three-dimensional periodic nature of the solid defines symmetry labels that are vectors. These vectors are defined in the Brillouin zone that is complementary to the real space unit cell of Figure 2. Instead of plotting 10^{23} levels one above the other, we spread them out, giving quasi-continuous points (lines) E vs. $\overline{k}$. The important fact is that each line (band) can hold two electrons per unit cell, and the lowest bands are filled first.

Figure 4 shows our results for fcc titanium. We see the low-lying states derived from the titanium s states. We also see the fairly narrow d band structure

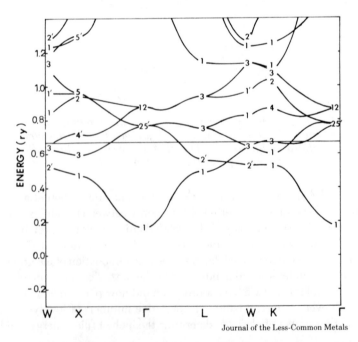

Figure 4. Energy levels (bands) for fcc titanium. The symbols at the bottom designate specific values of k *(see Figure 2, Ref. 11). The Fermi energy is indicated by the horizontal line (14).*

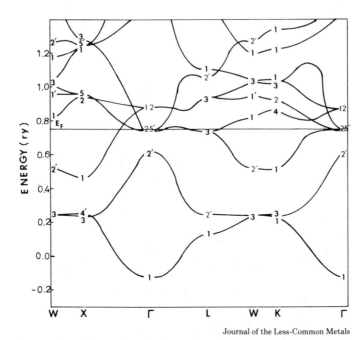

Figure 5.   Energy bands for titanium dihydride, TiH$_2$ with same remarks as Figure 4 (14)

derived from the atomic d states W2′–W1′, L1–L3, K1–K2, and X1–X5. The states at zone center $k = 0$, Γ, of t_{2g} and e_g symmetry are those familiar in ligand field theory, only they are now called Γ25′ and Γ12, respectively. Filling the states of Figure 4 with the four electrons of titanium gives a Fermi energy in the lower part of the d complex. Figure 5 shows what happens when we add hydrogen in the tetrahedral site to form the dihydride structure. As already indicated, the lowest levels drop. But most importantly, a new band (Γ2′ at zone center) appears mostly below the d complex; this band is formed largely from the antibonding combination of hydrogen s levels with some d admixture. This band can be filled with the electrons added by the hydrogen. It is this result that accounts for the occurrence and stability of the dihydride phase.

We can look at this result another way by plotting line density as a function of energy, the so called density of states as shown in Figure 6. The major feature is the occurrence of the low energy peak that is associated largely with the hydrogens. This peak is observed by Eastman (15) in photo-emission from a sample of titanium exposed to hydrogen where a peak develops below the main d band peak of titanium. However, we can go even further and decompose the density of states according to wave function character around the titanium and hydrogen sites, as shown in Figures 7–12. Thus, Figure 7 confirms our earlier statement that the low-lying peak is associated with the hydrogens while Figures 8 and 9 display the bonding and antibonding parts of the hydrogen s character, respec-

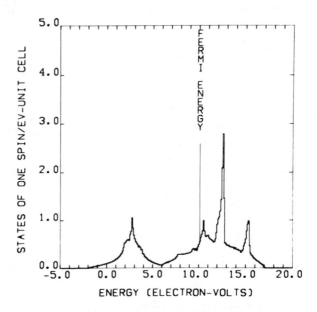

*Figure 6.   Density of states for TiH$_2$.   There are two
electrons per spin state.*

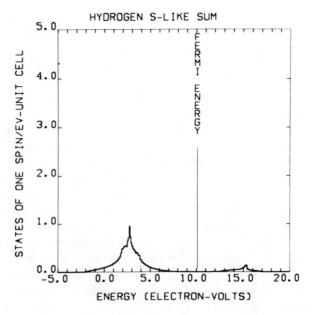

*Figure 7.   Density of states associated with s-like
character about the hydrogens*

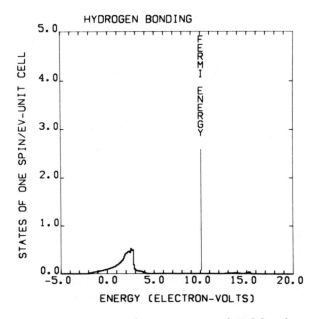

Figure 8.   Density of states associated with bonding
s-like character about the hydrogens

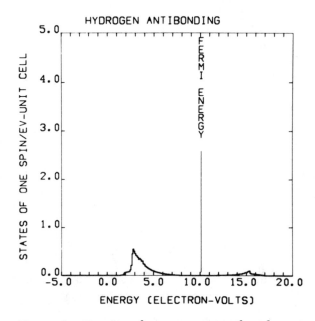

Figure 9.   Density of states associated with antibonding s-like character about the hydrogens

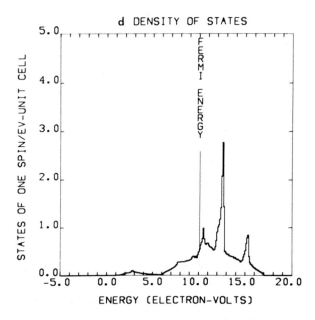

Figure 10.   *Density of states associated with titanium*
d-like character

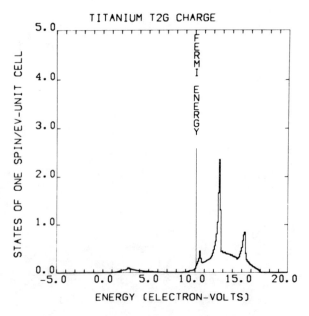

Figure 11.   *Density of states of titanium* t_{2g}
character

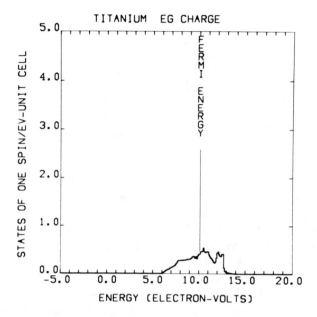

Figure 12.   Density of states of titanium e_g
character

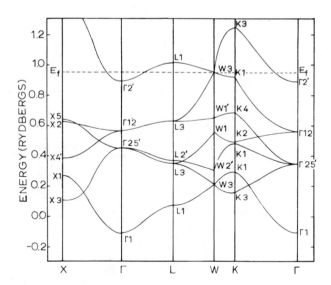

Figure 13.   Energy bands for palladium dihydride,
PdH_2

tively. Figure 10 shows that the upper three peaks of Figure 6 largely are associated with d character around the titanium. Figure 10 also shows some evidence of d bonding with the hydrogen. Figures 11 and 12 show the t_{2g} (xy, yz, zx) and e_g ($x^2 - y^2$, $3z^2 - r^2$) decomposition of the titanium d character, respectively. Figure 11 shows that the d bonding with the hydrogen is exclusively of t_{2g} character, as is most of the structure in Figure 10, while most of the occupied states are of e_g character. These site and character decompositions are useful and serve as further check on the theory when used in conjunction with spectral probes that are site and character sensitive such as soft x-ray emission and absorption and Auger spectra.

We see that these calculations can account for the stability of the dihydride structure in terms of low-lying hydrogen states that can be filled and are considerably detailed in the explanation of the nature of the interactions between the atomic system. Can this approach also explain why palladium doesn't form a dihydride but forms a monohydride instead (8)?

Palladium Hydride. The energy band structure for PdH_2 is shown in Figure 13. Comparison with Figure 5 shows the states associated with the antibonding hydrogen band ($\Gamma 2'$) now are well above the d manifold, and the added two electrons must be accommodated in empty d states of palladium and in these higher antibonding states above $\Gamma 2'$. The reason these states are high in palladium but low in titanium becomes clear when we recall the energy dependence of the bonding Σ_g and antibonding Σ_u states of the hydrogen molecule shown in Figure 14. The antibonding states rise more rapidly as the hydrogens come

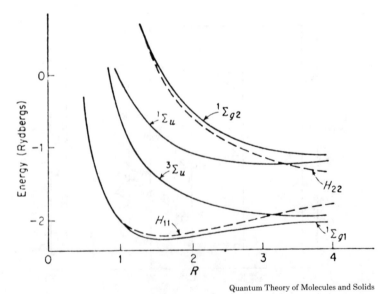

Quantum Theory of Molecules and Solids

Figure 14.   Total energy curves for hydrogen molecule according to Ref. 16

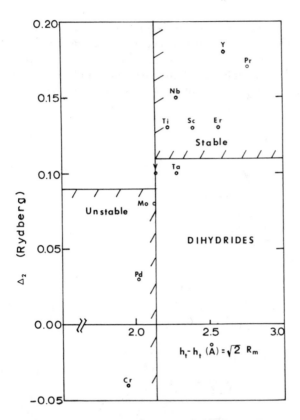

Figure 15.   *Position of state* Γ2′ *relative to Fermi energy* $\Delta_2 = E_F = \Gamma 2'$ *as a function of tetrahedral hydrogen–hydrogen spacing in dihydride structures*

closer together.   This behavior of the antibonding states in dihydrides also has been observed where Γ2′ is low when the hydrogens are far apart, and it is energetically favorable to fill this band—large atoms ≈ left side of periodic table ≈ larger hydrogen–hydrogen spacing.   This behavior is summarized in Figure 15 where the $\Delta_2 = E_F - \Gamma 2'$ is plotted vs. the tetrahedral–tetrahedral hydrogen spacing or equivalently, the metal atoms size R_m.   We observe that all stable dihydrides have hydrogen–hydrogen spacing > 2.14Å and $\Delta_2 > .10$, while the unstable dihydrides have hydrogen–hydrogen < 2.14Å and $\Delta_2 < .10$.   Of the two borderline cases, vanadium and tantalum, Figure 1 shows VH_2 exists while TaH_2 does not.   The validity of various approximations, including the lack of self consistency, prevents these results from being even relatively quantitative.

To introduce our picture of PdH, we again show the energetics as we go from isolated palladium and hydrogen atoms to palladium and palladium hydride in

Figure 16. The overall trend of levels going down in energy is favorable to stability, as the s level drops to become occupied in palladium metal and d states empty (holes). Figure 16 also shows significant H–s and H–p lowerings such that unoccupied p states in palladium are lowered and will become occupied in PdH. The detailed bandstructures and densities of states are given in Figures

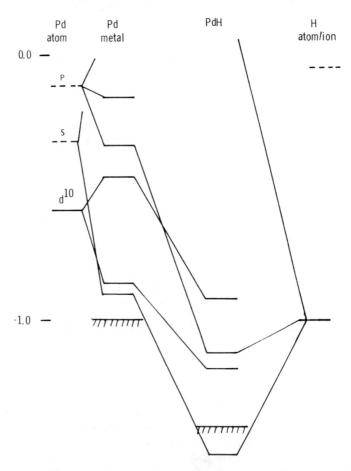

Figure 16. Energy level diagram indicating modification of energy levels of palladium and hydrogen upon formation of palladium hydride. Same remarks as those in Figure 3.

17–25. As indicated in Figure 17, the bandstructure of palladium consists of an almost-full d band X1–X5, L1–L3, K1–K2, overlapped by the s–p band Γ1—X4′, Γ1–L2′, Γ1–K1, with the highest filled state determining the Fermi energy, E_F, falling just below the top of the d bands. The total density of states for palladium is shown in Figure 18, where it is seen that the Fermi energy falls in a region of high density of states that is rapidly decreasing above it. The results of band-

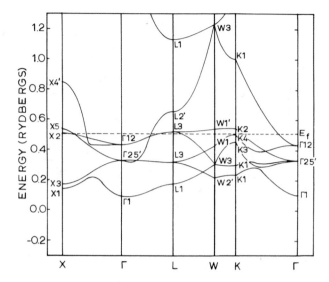

*Figure 17.   Energy bands for palladium metal.   The
Fermi energy* E_F *is indicated by the dashed horizontal
line.*

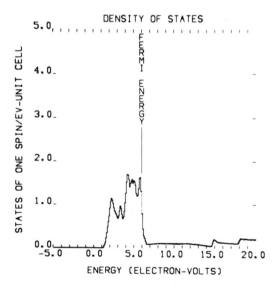

*Figure 18.   Total density of states for palladium
metal*

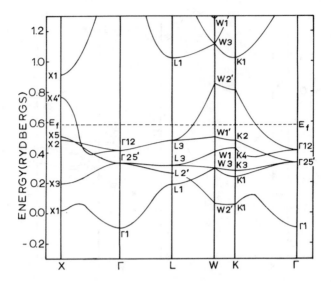

Figure 19.   Energy band structure for palladium hy-
dride

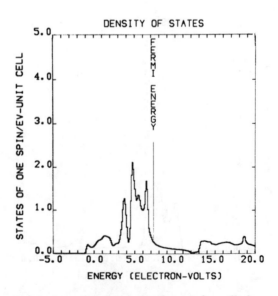

Figure 20.   Total density of states for palladium
hydride

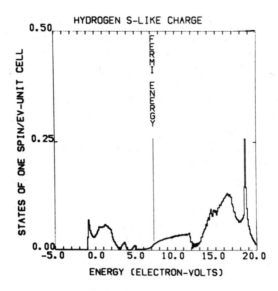

Figure 21.   Density of states associated with
s-like character about the hydrogen

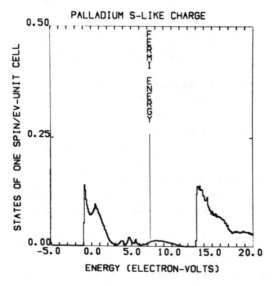

Figure 22.   Density of states associated with
palladium s-like character

structure calculations for PdH are shown in Figure 19. The d-like states (X5, Γ12, Γ25′; L3 K2) are affected little by the presence of the hydrogen. States having s-like character (h) around the hydrogen site Γ1, L2′, W2′ are lowered significantly. This lowering reflects metal s–h, p–h, d–h type interactions as outlined in Figure 16. In particular, the empty states L2′ of palladium are lowered 5 eV and become filled in PdH. This lowering and filling of empty states (as in the dihydrides) contributes to the formation of PdH and NiH. These states already are filled in platinum and presages the lack of PtH formation indicated in Figure 1. The total density of states for PdH is shown in Figure 20. Again, the low-lying states below the d band are characteristic of hydrogen influence on the host electronic states and have been observed in photoemission (7).

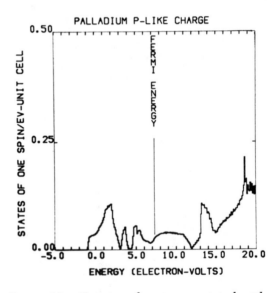

*Figure 23.   Density of states associated with
palladium p-like character*

However, since these states were largely filled in palladium and the states associated with L2′ contain only about 0.4 electrons, the remaining 0.6 electron is accommodated: 0.36 electron in the d band holes and 0.24 electron go to increasing the Fermi energy such that the state density is much lower in PdH than in Pd but not as low as Pd + 1 electron. Comparison of Figures 17 and 19 and Figures 18 and 20 indicates several similarities as well as considerable differences in detail.

As before, the total density of states can be broken down into site and wavefunction character to better reveal the details of the hydrogen–palladium interaction. Figure 21 shows hydrogen s-like character contributing to the lowest peak as well as being present and increasing (while the total is decreasing) at the Fermi energy. These latter facts have explained the occurrence of supercon-

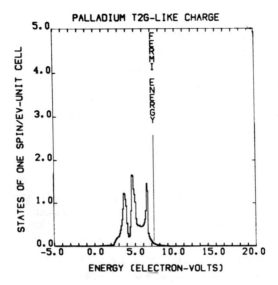

Figure 24. *Density of states associated with palladium t_{2g}-like character*

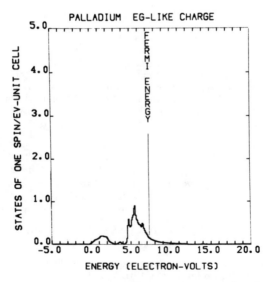

Figure 25. *Density of states associated with palladium e_g-like charge in palladium hydride*

ductivity in PdH (17) while their antibonding nature and arguments, similar to those in discussing Figures 14 and 15, can account for the pressure dependence of T_c (18). Figure 22 shows states with s-like character around the palladium showing sizeable evidence of s–h interaction associated with the lowest peak. This is also seen in Figure 23 for the palladium p character. Finally, Figures 24 and 25 show the t_{2g} and e_g d character associated with the palladium, respectively. This time, the hydrogen interacts primarily with the e_g d-character overlap, which the octahedral site occupancy in PdH favors.

Conclusions

In conclusion, calculations for metal hydride systems have shown evidence of sizeable metal–hydrogen and hydrogen–hydrogen interactions. Stable hydride formation is favored by filling energetically low states. The more of these that are empty in the metal, the better. The added electrons should go into states associated with the hydrogen and not in states associated with the metal (at E_F—high and empty). We also have seen that detailed wavefunction information can be derived from such calculations, which can help put the results on a firmer basis.

Literature Cited

1. Lewis, F. A., "The Palladium Hydrogen System," Academic, London, 1967.
2. Libowitz, G. G., "Solid State Chemistry of Binary Metal Hydrides," W. A. Benjamin, Inc., New York, 1965.
3. Mueller, W., Blackledge, J. P., Libowitz, G. G. Eds. "Metal Hydrides," Academic, 1968.
4. Switendick, A. C., Solid State Commun. (1970) 8, 1463.
5. Ibid. (1970) 34.
6. Switendick, A. C., Int. J. Quantum Chem. (1971) 5, 459.
7. Eastman, D. E., Cashion, J. K., Switendick, A. C., Phys. Rev. Lett. (1971) 27, 35.
8. Switendick, A. C., Ber. Bunseges. Phys. Chem. (1972) 76, 535.
9. Switendick, A. C., "Hydrogen in Metals—A New Theoretical Model," Hydrogen Energy, Part B, T. N. Veziroglu, Ed., Plenum, New York, 1975.
10. Gelatt, C. D. Jr., Weiss, J. A., Ehrenreich, H., Solid State Commun. (1975) 17, 663.
11. Zbasnik, J., Mahnig, M., Z. Phys. (1976) 23, 15.
12. McClellan, R. B., Harkins, C. G., Mater. Sci. Eng. (1975) 18, 5.
13. Callaway, J., "Energy Band Theory," Academic, New York, 1964.
14. Switendick, A. C., J. Less Common Met. (1976) 49, 283.
15. Eastman, D. E., Solid State Commun. (1972) 10, 933.
16. Slater, J. C., "Quantum Theory of Molecules and Solids," Vol. 1, McGraw-Hill, New York, 1963.
17. Papaconstantopoulos, D. A., Klein, B. M., Phys. Rev. Lett (1975) 35, 110
18. Switendick, A. C., Bull. Am. Phys. Soc. (1975) 20, 420.

RECEIVED October 5, 1977. This work was supported by the United States Energy Research and Development Administration, ERDA, under Contract AT(29-1)789.

Palladium–Hydrogen: The Classical Metal–Hydrogen System

TED B. FLANAGAN

Department of Chemistry, University of Vermont, Burlington, VT 05401

W. A. OATES

Department of Metallurgy, University of Newcastle, Newcastle, New South Wales, Australia

The current status of the palladium–hydrogen system is discussed with emphasis on thermodynamics of hydrogen solution. New pressure–composition–temperature data have been obtained in the dilute region of primary solubility, i.e., before hydride-phase formation. From these data, partial excess entropies of solution have been obtained and are shown to be independent of hydrogen content. These data suggest that the hydrogen–hydrogen interaction mainly is enthalpic in origin; the interaction enthalpies have been obtained over a wide temperature range. These new data are discussed relative to existing models for this system.

Palladium–hydrogen occupies a unique position among metal–hydrogen systems. There have been hundreds of publications on this system since the discovery by Graham in 1866 (1) that palladium can absorb large amounts of hydrogen. Interest in palladium–hydrogen has continued partly because of its great experimental convenience. There are no oxidation problems with palladium, and the hydride phase can be formed without mechanical disintegration of the metal, although extensive plastic deformation occurs (2). By contrast, many of the other hydrogen-absorbing metals are disrupted massively upon hydride formation. Recently, interest in palladium–hydrogen has intensified because techniques of neutron physics were applied to this system (3) and superconductivity in palladium–hydrogen with very high H:Pd ratios (4) was discovered.

0-8412-0390-3/78/33-167-283/$05.00/0
© American Chemical Society

Review of Pressure–Composition–Temperature Relationships

One of the most useful features of metal–hydrogen systems are their pressure–composition–temperature data, P–C–T. Such relationships for palladium–hydrogen are shown in Figure 1. For compositions and temperatures within the envelope, two solid phases coexist, as required by the phase rule. The lower hydrogen-content α-phase represents solution of hydrogen into the metal, and the higher hydrogen-content β-phase is the hydride. Both α and β are

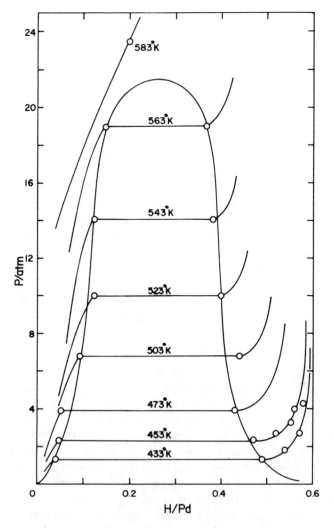

Figure 1.   Pressure–composition–temperature relationships for palladium–hydrogen (43)

nonstoichiometric and, as pointed out by Anderson (5), this was the first non-stoichiometric system for which detailed thermodynamic data were available. The phase diagram is known at much lower temperatures (2) than shown in Figure 1, but it cannot be complete since the third law requires that nonsto-ichiometry cannot exist at $0°$ K; however, reasonable extrapolation of the β-phase boundary does not meet this requirement. There is probably a eutectoid reaction at quite low temperatures; some indications of an anomalous transformation have been observed (6, 7).

An early application (1952) of neutron diffraction (8) revealed that the hydrogen atoms occupy the octahedral interstices within the fcc palladium lattice in the β-phase. More recent neutron scattering data have indicated that these same interstices are occupied in the α phase (9). Thus, the α and β phases differ only in their hydrogen content and in the lattice parameters of the containing palladium matrix. For palladium–hydrogen within the two-phase envelope, two distinct fcc lattice parameters can be measured. Baranowski et al. (10) have shown that the volume expansion of palladium per added hydrogen atom is the same for both α and β phases, i.e., the partial molar volume of hydrogen, V_H, is independent of hydrogen content up to H:Pd $= n \simeq 0.75$ ($V_H = 1.65$ cm^3 mol H^{-1}). Indeed, the same value of V_H is found for fcc palladium alloys and for other fcc systems that absorb hydrogen into octahedral interstices (10).

In the α-phase, it is found that at very low hydrogen concentrations $n \propto P_{H_2}^{1/2}$. This square-root dependence demonstrates that molecular hydrogen dissociates upon entering the metal (2). The dependence of solubility upon $P_{H_2}^{1/2}$, which represents Henry's Law for a dissociating gas, is called Sieverts' Law when discussing metal–hydrogen systems. Deviations from this ideal, dilute solution behavior commence as n increases. The observed equilibrium pressures then are progressively smaller with an increase of n than predicted by Sieverts' Law. This nonideal behavior is attributed to attractive interactions among dissolved hydrogen atoms, and it is this attraction that ultimately leads to hydrogen condensation, i.e., hydride formation.

While the hydrogen-saturated metal is being transformed to the hydride phase, the equilibrium hydrogen pressures remain constant or nearly so, but once completely transformed, the chemical potential of dissolved hydrogen increases rapidly as hydrogen dissolves further. When hydrogen is desorbed from a hy-drided sample, the invariant hydrogen pressures over the two-phase coexistence region are lower than those during the absorption cycle when the hydride is being formed (11). Such hysteresis occurs frequently for metal–hydrogen systems in which hydride formation occurs. Hysteresis can limit efficiencies in the practical use of hydrides as energy (hydrogen) storers (12). It is likely that hysteresis originates in the plastic deformation that occurs because of volume increases associated with hydride-phase formation, which is 11% for palladium hydride (2). Birnbaum et al. (13) have shown recently, for the niobium–hydrogen system, that plastic deformation also occurs during the reverse process of α-phase formation from β-phase decomposition. As a result of these plastic deformations, there is a net energy loss while traversing the hysteresis loop. From the values of the absorption and desorption pressures during hysteresis, $\frac{1}{2} RT \ln(P_{\alpha \to \beta}/P_{\beta \to \alpha})$, this loss can be estimated as 800 J/mol H (300°K). The abrupt lattice expansion or contraction accompanying the formation or decomposition of β phase generates large dislocation densities. These densities are approximately 10^{12} cm^{-2} (14), which is the same order of magnitude as the dislocation density that results from heavy cold-work (14). If it is assumed that fresh dislocations are generated both during absorption and desorption, the elastic energy stored in the dislocations can be estimated as ~300 J/mol H. The remainder of the energy lost during the hysteresis loop must be lost as thermal energy from dislocation motion and annihilation.

Flanagan and co-workers have shown (15) that hydrogen segregates to the tensile stress fields around the edge components of dislocations in heavily cold-worked palladium. This leads to a net increase of solubility of hydrogen in the α phase. It was shown also (16) that a comparable increase of solubility occurs in the α phase following the formation and subsequent decomposition of β phase. It follows that accurate $P-C-T$ data for the α phase of the palladium–hydrogen system can be obtained only with samples that previously have not formed β phase unless a high temperature anneal has been carried out prior to the solubility studies.

Models for Dissolved Hydrogen in Palladium

Many years ago, Lacher (17) explained the $P-C-T$ data shown in Figure 1 with a statistical mechanical model, a model that has formed the basis of subsequent treatments of nonstoichiometry in other systems as well. With the assumption that the hydrogen atoms are localized to each site, the resulting partial configurational entropy is given by

$$S_H^c = +R \ln (n_s - n)/n \qquad (1)$$

where n and n_s are the H:Pd atom ratios at any hydrogen concentration and at the limiting concentration, respectively. To describe the increase of chemical potential, $\mu_H \propto RT \ln P_{H_2}^{1/2}$, for the dissolved hydrogen in the β phase (see Figure 1), Lacher chose $n_s = 0.59$ and predicted most of the system's features known at that time. However, the choice of $n_s = 0.59$ was incorrect; neutron diffraction results have shown that the octahedral interstices are occupied, and, therefore, the limiting value of n should be 1. Since Lacher's theoretical treatment of this system, values of n very close to 1 have been obtained experimentally. For example, by charging palladium with hydrogen via reaction with hydrogen atoms generated in the gas phase, Flanagan and Oates (18) have charged palladium samples to $n = 0.99 \pm 0.01$. Low-temperature, electrochemical techniques also have been successful in producing hydrogen-saturated β phase with values of n close to unity (*19*), and direct high-pressure techniques have been used by Baranowski and co-workers (*20, 21, 22*).

The similarity in the appearance of the phase diagram for palladium–hydrogen, and other metal–hydrogen systems such as niobium–hydrogen, to the P–C–T relationships for free gases has been noted and exploited (*23, 24*). Alefeld and Buck (*25*) and de Ribaupierre and Manchester (*26*) have shown, from the critical behavior of the palladium–silver ($X_{Ag} = 0.10$)–H(*25*) and palladium–hydrogen (*26*) systems, that the observed critical-point exponents agree with those predicted by mean-field theory. The hydrogen–hydrogen attractive interaction must be long range for mean-field theory to obtain, and for metal–hydrogen systems, it is believed that the interaction is elastic in origin. The hard-core repulsion part of the interaction is satisfied since only a single hydrogen atom can occupy an interstice. The recent verifications of mean-field theory explain the relatively successful application of the simple Bragg–Williams model to the palladium–hydrogen system when suitable modified for $n_s = 1$ (*27*).

The partial molar volume of hydrogen in palladium and its alloys is 1.65 cm^3 (mol H)$^{-1}$ up to $n = 0.75$ (*10*); at greater hydrogen contents, it appears that there is a smaller partial molar volume of ~ 0.4 cm^3 (mol H)$^{-1}$. The decrease of V_H corresponds to the region of hydrogen concentrations where superconductivity first appears in palladium–hydrogen (*4*). Baranowski and his co-workers (*28*) have suggested tantalizingly that in this region of large hydrogen concentrations, metallic-like hydrogen obtains since, for example, the measured partial molar volume of hydrogen in this region corresponds to predicted volumes of metallic hydrogen. The remarkable feature of superconductivity in palladium–hydrogen is that pure palladium is not superconducting, and palladium–hydrogen does not become superconductive until very large hydrogen concentrations are reached, i.e., $n > 0.75$ (*4*). A maximum superconducting, transition temperature of $8.5°K$ is reached at $n \sim 1$. Surprisingly, a higher transition temperature, $11°K$, is found for the deuteride (*29*), which is contrary to the expected isotope effect for superconductivity. Miller and Satterthwaite (*30*) have suggested that because of its greater zero-point motion, the binding of hydrogen to the interstice might be greater than that of deuterium. Subsequently, Rahman et al. (*31*) have found that the palladium–hydrogen force constant must be 20% greater than the palladium–deuterium force constant to agree with the phonon spectra of $PdD_{0.63}$ and $PdH_{0.63}$, as determined by neutron scattering. Oates and Flanagan (*32*) independently arrived at a similar difference in force constants from analyzing the isotope effect for hydrogen and deuterium solubility in palladium. Ganguly (*33*) has rationalized the isotope effect for superconductivity in terms of the $\sim 20\%$ difference in force constants for hydrogen and deuterium. By using ion implantation for the introduction of hydrogen, a hydrogen-saturated palladium–copper ($X_{Cu} = 0.40$) alloy has achieved a superconducting transition temperature of $16.6°K$ (*34*), and again, as for palladium, the hydrogen-free alloy is not a superconductor.

Hydrogen Solution in α-Phase

Until about 10 years ago, there were no reliable thermodynamic data for hydrogen dissolved in bulk palladium below about 400°K because of the slowness of the $H_2 \rightleftarrows 2H$(chemisorbed) step at the surface. In the earlier studies, vacuum systems were used in which mercury and grease were possible contaminants. Wicke and Nernst (11) surmounted this problem with the aid of hydrogen-transfer catalysts such as finely divided copper powder in intimate contact with the metal surface. The exact mechanism whereby these catalysts function still remains somewhat mysterious, but, in any case, their use enabled Wicke and Nernst to obtain the first accurate P-C-T data from 194.5 to 348°K for the palladium–hydrogen system. In 1973, Clewley and co-workers (35) found that such transfer catalysts were not necessary for low-temperature equilibration of hydrogen with bulk palladium provided that the sample was free from contamination by grease and mercury. Subsequent investigations in the low-hydrogen content α phase have yielded data that substantially agree (36) with the results of Clewley et al. (35) and Wicke and Nernst (11).

Apart from an early investigation by Gillespie (37) and a study by Nace and Aston (38), calorimetric studies of hydrogen absorption by metals have been lacking inexplicably until recently. Lynch and Flanagan (39) have obtained heats of solution in both the α and β phases of palladium at room temperature in substantial agreement, in the α phase at least, with data based on the application of the Clausius–Clapeyron equation to the P-C-T data (11, 35, 36). Using a more accurate high-temperature calorimeter, Boureau, Kleppa, and co-workers (40, 41) since have obtained P-C-T and calorimetric data at low hydrogen concentrations and at elevated temperatures for palladium–hydrogen. They found good agreement with data derived from P-C-T results by Clewley et al. (35) for the variation of the partial molar entropies and enthalpies with temperature, for the infinitely dilute solutions. Both quantities increase significantly with temperature.

The chemical potential of hydrogen in the gas phase, at equilibrium with the dissolved hydrogen in the solid, is given by Equation 2 where $\mu_{H_2}^0$ is the standard (1 atm) chemical potential of gaseous hydrogen. At small hydrogen concentrations in the metal, μ_H is given by Equation 3, (11, 25) where the next-

$$\mu_H = \tfrac{1}{2}\mu_{H_2}^0 + RT \ln P_{H_2}^{1/2} \tag{2}$$

$$\mu_H = H_{H\rightarrow 0} - TS_{H\rightarrow 0}^{xs} + RT \ln (n/1 - n) + W_{HH}^f \, n \tag{3}$$

to-last term on the right is the negative of temperature times the partial ideal configuration entropy; W_{HH}^f is the partial-interaction free energy of hydrogen. Generally this latter term has been labeled $2\,W_{HH}$ (17, 27), but it seems more convenient to omit the factor of 2. The superscript f refers to a free-surface solid, and $H \rightarrow 0$ indicates conditions of infinite dilution of hydrogen. The excess terminology refers to all partial contributions to μ_H and S_H, save that arising from the ideal partial configurational entropy term. Equating the chemical potentials of hydrogen atoms in the gas (*see* Equation 2) and metal (*see* Equation 3), Equation 4 is obtained where, e.g., the relative partial molar enthalpy $\Delta H_{H\to 0}$ = $H_{H\to 0} - \frac{1}{2} H_{H_2}^0$. From the slopes of plots of $\Delta\mu_H^{xs}$ against n, values of W_{HH}^f can be obtained, and at $300°K$, $W_{HH}^f = -47.6$ kJ(mol H)$^{-1}$ (*11*). Wicke and Nernst (*11*) have expressed the temperature dependence of W_{HH}^f by Equation 5.

$$\Delta\mu_H^{xs} = \Delta H_{H\to 0} - T\Delta S_{H\to 0}^{xs} + W_{HH}^f\, n = RT \ln P_{H_2}^{1/2}(1-n)/n \qquad (4)$$

$$W_{HH}^f = -18,830(1 + 445/T)\text{J}(\text{mol H})^{-1} \qquad (5)$$

Their measurements extended over the temperature range $273°-348°K$, but data of extrapolation to higher temperatures reasonably agreed with earlier data of Gillespie and Gaulstaun (*42*) and Brüning and Sieverts (*43*). Moon (*44*) analyzed all of the data available in 1956, which was mainly higher temperature data, and found that W_{HH}^f increased with temperature. $P-C-T$ data also can be used to determine H_{HH}^f over a narrow temperature range from the relationship

$$H_{HH}^f = (\partial\Delta H_H/\partial n)_{P,T} = \frac{\partial}{\partial n}\left[-T^2\partial\left(\frac{\Delta\mu_H}{T}\right)\bigg/\partial T\right]_{P,n} \qquad (6)$$

where $H^f{}_{HH}$, the partial interaction enthalpy, is defined by

$$\Delta H_H = H_{H\to 0} + H_{HH}^f\, n \qquad (7)$$

Thus, determination of the hydrogen–hydrogen interaction from Equation 4 can give values which include both entropic and enthalpic contributions, whereas in Equation 7, there are only enthalpic contributions. Boureau et al. (*40, 41*) have determined values of H_{HH}^f from calorimetric measurements. They found $H_{HH}^f = -125.5$ kJ (mol H)$^{-1}$ ($555°K$), and from their $P-C-T$ data, W_{HH}^f, -52.7 kJ (mol H)$^{-1}$ ($555°K$). Burch and Francis (*36*), on the other hand, found, from their $P-C-T$ data ($\sim600°K$), $H_{HH}^f = -54.2$ kJ(mol H)$^{-1}$.

Because of the question raised by the recent calorimetric investigations (*40, 41*) as to whether the hydrogen–hydrogen interaction contains a significant entropic contribution, it was decided to determine $P-C-T$ over a temperature range similar to that used when the calorimetric data were determined.

Experimental Results at Low Hydrogen Contents

$P-C-T$ data were obtained with a palladium sample (~20 g) in the form of 1-mm rods (99.9% purity). Thus, the surface-to-volume ratio was quite small. The sample was not coated with palladium black but merely was cleaned mechanically after being annealed at $\sim1100°K$ for several days. The sample had never been subjected to the hydride-phase transformation.

Two absorption runs at elevated temperatures (556.7° and 558°K) are shown in Figure 2, and these data agree substantially with data of Boureau et al. (40), as derived from the empirical representation of their data for 555°K. Values of $\Delta\mu_{H\rightarrow 0}^{xs}$ are 19.79 and 19.85 kJ (mol H)$^{-1}$ at 557° and 558°K, respectively, compared with 19.75 kJ(mol H)$^{-1}$ derived by Boureau et al. (40) at 555°K; hence, the values differ by only 30 J(mol H)$^{-1}$ when corrected for the small temperature differences. The values of the slopes of the lines, W_{HH}^{f}, shown in Figure 3 differ relatively more than the values of $\Delta\mu_{H\rightarrow 0}^{xs}$ do. Thus, we obtain -41.4 kJ(mol H)$^{-1}$ at 557°K and -38.6 kJ(mol H)$^{-1}$ at 558°K, compared with -52.7 kJ(mol H)$^{-1}$ at 555°K (40), although our reanalysis of their data by using Equation 4 gives -46 kJ(mol H)$^{-1}$. This illustrates that the slopes of $\Delta\mu_{H\rightarrow 0}^{xs}$ values plotted against n have considerably greater error than the values do at infinite dilution. This is particularly true at elevated temperatures where the range of n values is quite small. A trace of residual hydrogen in the sample or the surface contributions to hydrogen uptake can affect the slopes markedly.

Absorption isotherms were measured in the dilute range of the α phase from $n = 0$ to $n = 0.012$ and from 418° to 586°K. From four isotherms in the range

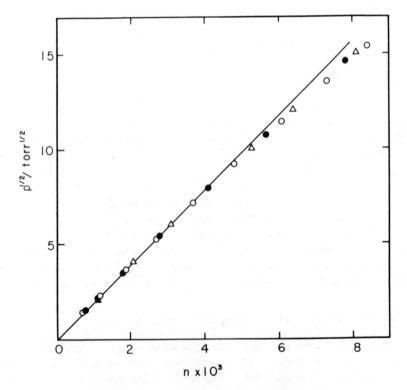

Figure 2. Plot of $P_{H_2}^{1/2}$ against H–Pd (= n). ($\triangle$) Ref. 40 data, 555°K; (O) present data, 556.7°K; (●) present data, 558°K.

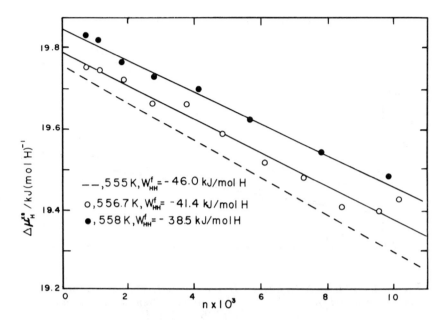

Figure 3. Plot of RT ln $P_{H_2}^{1/2}(1-n)/n\,(= \Delta\mu_H^{xs})$ against H–Pd. (--) Ref. 40 data; (O) present data, 556.7°K; (●) present data, 558°K.

515°–586°K, the variations of ΔH_H with n were determined directly from variations of $\ln P_{H_2}^{1/2}$ with T^{-1}, i.e., $\Delta\mu_H = RT \ln P_{H_2}^{1/2} = \Delta H_H - T\Delta S_H$. By using Equation 7, H_{HH}^f was evaluated as -32 kJ(mol H)$^{-1}$ (Figure 4), which is close to the average value of -40 kJ(mol H)$^{-1}$ found for W_{HH}^f (*see* Figure 3) but quite different than the value of -125.5 kJ(mol H)$^{-1}$ found for H_{HH}^f calorimetrically (41). The intercept at $n \to 0$, $\Delta H_{H\to 0}$, from the P–C–T data of the present study (Figure 4), agrees quite well with the corresponding calorimetric value (41), i.e., within 100 J(mol H)$^{-1}$. The calorimetric determination of H_{HH}^f values are based on extremely small differences of heat evolution (40, 41). Therefore, on the basis of the present results which are consistent with earlier results (e.g., Equation 11), it will be assumed that, for palladium–hydrogen, $W_{HH}^f \sim H_{HH}^f$ in the temperature range 515°–586°K.

Since it has been shown that $W_{HH}^f \sim H_{HH}^f$, Equation 4 can be rewritten as

$$S_H^{xs} = \tfrac{1}{2}S_{H_2}^0 + \Delta H_H/T - R \ln P_{H_2}^{1/2}\left(\frac{1-n}{n}\right) \tag{8}$$

where $S_H^{xs} = \Delta S_H^{xs} + \tfrac{1}{2}S_{H_2}^0$ and $\Delta H_H = \Delta H_{H\to 0} + W_{HH}^f n$ were used to obtain Equation 8 from Equation 4. The subscript H → 0 on the excess entropy has been omitted intentionally since the purpose of this section is to show that $S_{H\to 0}^{xs} = S_H^{xs}$, at least over the range of the dilute hydrogen contents studied here. Values of S_H^{xs}, determined directly from the absorption data by using Equation 8, are

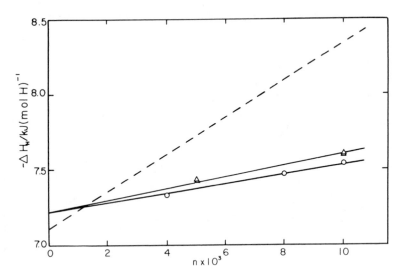

*Figure 4.   Plot of ΔH_H against H-Pd. (--) Ref. 41 data; ($\triangle$) present
data determined from $\Delta H_{H \rightarrow 0} + W^f_{HH}\, n;$ ($\bigcirc$) present data determined
from P–C–I data 515° to 568°K.*

shown in Table I.   Table I shows that values of S^{xs}_H are remarkably independent
of n.   Alternatively, S^{xs}_H values can be determined from plots of $\Delta\mu^{xs}_H$ against T,
and the results again lead to values closely independent of n.   Since the excess
partial entropy must be ideal at infinite dilution, and it was observed here that
$S^{xs}_{H \rightarrow 0} = S^{xs}_H$, it follows that the partial entropy is ideal over this range of n values.
The absolute reliabilities of these S^{xs}_H values are not as good as those suggested by
the values reported in Table I, but negligible changes with n definitely are indicated.   The absolute errors are probably $\pm\ 1\ J(\text{mol H})^{-1}\ K^{-1}$.   These results
confirm that the partial configurational entropy for these dilute ranges of hydrogen contents is indeed the ideal value.

Values of W^f_{HH} are shown as a function of temperature in Figure 5, as determined in this research and elsewhere.   Empirical equations given in the literature $(11, 44)$ suggest that the dependence of W^f_{HH} upon temperature is non-
linear, i.e., the dependence upon temperature is small at elevated temperatures.

Table I.   Partial Excess Entropies at Selected Temperatures

$10^3 n$	$461°K$ S^{xs}_H/R	$515°K$ S^{xs}_H/R	$557°K$ S^{xs}_H/R
0	2.34	2.80	3.11
2	2.34	2.80	3.11
4	2.34	2.80	3.12
6	2.34	2.78	3.12
8	2.25	2.80	3.12
10	2.34	2.80	3.12

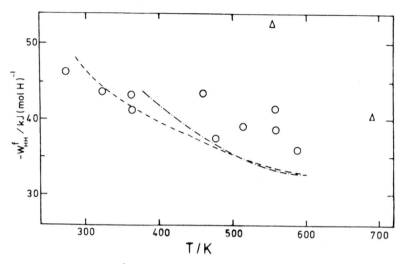

Figure 5. Plot of W^f_{HH} against temperature. (O) data of present study; (— · —) from equation given in Ref. 44: (△) from equation given in Ref. 11: (- -) data from Ref. 4.

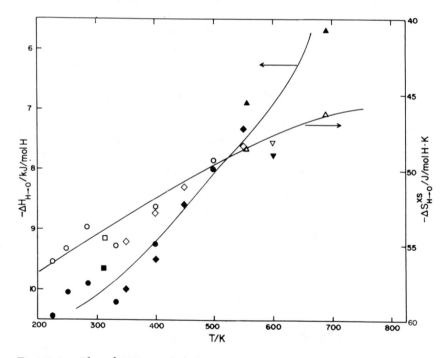

Figure 6. Plot of $\Delta H_{H \to 0}$ and $\Delta S^{xs}_{H \to 0}$ against temperature. (△) Ref. 41 data; (▽) Ref. 36 data; (□) Ref. 11 data; (O) Ref. 35 data; (◇) data of present study. Filled and open symbols refer to $\Delta H_{H \to 0}$ and $\Delta S^{xs}_{H \to 0}$, respectively.

Our finding that $H^f_{HH} \sim W^f_{HH}$ in the temperature range $515°-586°K$ supports this. The small temperature dependence in this range (Figure 5) can reflect changes of elastic constants with temperature (*see* below). It would be useful to determine whether $H^f_{HH} \sim W^f_{HH}$ at lower temperatures, but this would be difficult since large samples with small surface-to-bulk ratios are needed and the times for equilibrium would be long.

Figure 6 shows values of $\Delta H_{H \to 0}$ and $\Delta S^{xs}_{H \to 0}$ as a function of temperature, and it can be seen that the present data agree sufficiently with earlier data and with the data of Boureau et al. (*40, 41*).

Volume and Elastic Effects

Wagner (*45*) has shown that

$$(\partial \Delta \mu^{xs}_H / \partial n)_{V,T} = (\partial \Delta \mu^{xs}_H / \partial n)_{P,T} + V^2_H / \kappa_T V_s \tag{9}$$

where V_s, V_H, and κ_T are the molar volume of the metal (palladium) solvent, the partial molar volume of hydrogen, and the compressibility of the metal, respectively. From Equation 4, $(\partial \Delta \mu^{xs}_H / \partial n)_{P,T} = W^f_{HH}$.

Alefeld and co-workers (*24, 25*) have discussed the hydrogen–hydrogen attractive interaction by using the elasticity theory developed for defects in solids by Eshelby and others (*46*). The strength of the elastic dipole moment is related to the volume expansion resulting from the interstitial hydrogen (*25*) (Equation 10), where P is the strength of the elastic dipole caused by the interstitial species

$$\Delta V_s(n)/V_s = 4\kappa_T Pn/a^3 \tag{10}$$

and a is the lattice parameter of the metal. Alefeld and Buck (*25*) have shown that the partial hydrogen–hydrogen interaction for a free surface solid, i.e., a solid with unconstrained surfaces, which is the usual experimental condition for solubility studies, is given by Equation 11, where the γ_G factor is introduced to

$$W^f_{HH} = -4\kappa_T \gamma_G P^2 / a^3 \tag{11}$$

account for local field effects. Substitution of the expression for P derived from Equation 10 into Equation 12 gives, for W^f_{HH},

$$W^f_{HH} = -\frac{\gamma_G a^3}{\kappa_T 4n^2}(\Delta V_s(n)/V_s)^2 = -\gamma_G V^2_H / \kappa_T V_s \tag{12}$$

It can be shown for a constant volume solid that

$$W^v_{HH} = (1 - \gamma_G)V^2_H / \kappa_T V_s \tag{13}$$

which follows from Equations 9 and 12.

Wagner (45) has shown from experimental data for palladium–hydrogen that $W_{HH}^{\upsilon} \simeq 0$. However, elasticity theory predicts, using $\gamma_G = 0.3$, that $W_{HH}^{\upsilon} > 0$ and that the hydrogen–hydrogen interaction under free surface conditions is less negative than that found experimentally. The γ_G value is somewhat uncertain but, nonetheless, is expected to be in the range from 0.3 to less than 1, and therefore, the presence of short-range interactions between hydrogen atoms of a non-elastic origin is indicated. In any case, the similar functional dependence of the interaction energy among hydrogen atoms in a free surface solid upon V_H, V_s, and κ_T, predicted from thermodynamics (*see* Equation 9) and from elasticity theory (*see* Equation 12), supports the view that an appreciable fraction of the hydrogen–hydrogen interaction must have its origin in the image displacement forces (23, 24, 45). The elastic interaction is long-ranged and therefore justifies the application of mean-field theory to this part of the interaction (25, 26). Equation 12 predicts that values of W_{HH}^f should become less negative with an increase of temperature, as observed (*see* Figure 5), because V_s and κ_T increase with an increase of temperature while V_H does not change much with temperature. The predicted decrease in the magnitude of $|W_{HH}^f|$ is, however, smaller than that observed; this may be caused by the unknown temperature dependence of the short-range interaction.

Wagner (45) also has derived the relationship given by Equation 14, where α is the coefficient of expansion of the metal; the first term on the left is $-S_{H,P}^{xs}$, i.e., the partial excess entropy at constant pressure (free surface conditions); and the first term on the right is the corresponding partial excess entropy under constant volume. By combining Equation 9 with Equation 14 and integrating, Oates and Flanagan (47) have obtained Equation 15, where the reference con-

$$(\partial \Delta \mu_H^{xs}/\partial T)_{V,n} = (\partial \Delta \mu_H^{xs}/\partial T)_{P,n} + \alpha V_H/\kappa_T \tag{14}$$

$$\Delta \mu_{H,v}^{xs}(T,n) = \Delta \mu_{H,P}^{xs}(\text{ref}) + \int_0^n (\partial \Delta \mu_H^{xs}/\partial n)_{P,T}\,dn + \int_{300}^T (\partial \Delta \mu_H^{xs}/\partial T)_{P,n}\,dT$$

$$+ \int_0^n \frac{V_H^2}{\kappa_T V_s}\,dn + \int_{300}^T (\alpha V_H/\kappa_T)\,dT \tag{15}$$

ditions are pure palladium at 300°K. With the aid of Equation 15, Oates and Flanagan (47) have converted experimental data obtained at constant pressure (free surface conditions) to data at constant volume. The first two terms on the right were evaluated from the P–C–T data obtained under experimental conditions of constant pressure on the solid palladium (free surface conditions), and the remaining integrals were evaluated from data relating elastic moduli and expansion coefficient changes with hydrogen concentration and temperature. A comparison of values of $\Delta \mu_{H,P}^{xs}$ and $\Delta \mu_{H,v}^{xs}$ and their variations with n at 300°K is shown in Figure 7. A phase transition occurs under constant pressure conditions since the excess chemical potential passes through a minimum but does not occur under conditions of constant volume.

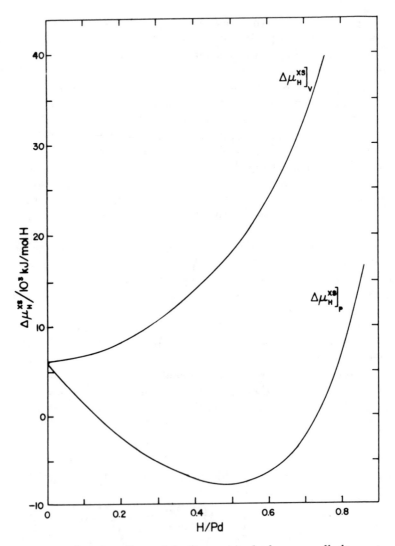

Figure 7. Plot of $\Delta\mu_{H,P}^{xs}$ and $\Delta\mu_{H,V}^{xs}$ against hydrogen–palladium at 300° K from results of Ref. 47

Partial Excess Entropy of Dissolved Hydrogen

In many studies of metal–hydrogen systems, attention has focused on the partial excess entropy of dissolved hydrogen (*see* Table I) and the evaluation of the various contributions to it. The partial excess entropy of hydrogen can be written

$$S_H^{xs} = S_H + R \ln (n/(1-n)) = S_H^c + S_H^v + S_H^{el} \qquad (16)$$

where S_H^c is the nonideal configurational entropy of hydrogen and has been shown here to be zero, and S_H^v and S_H^{el} refer to the partial vibrational entropy and the partial electronic entropy, respectively. Since measurements of the separate contributions are performed on samples with free surfaces and not constant volumes, S_H^{xs} values obtained from the usual solubility measurements are the quantities for comparison; the constant volume–constant pressure correction (*see* Equation 14) is not needed. Neutron-scattering techniques (*3*) allow the evaluation of $\nu v 3_H$ for use with the Einstein model of localized oscillators for S_H^v evaluation. However, as Fast (*48*) has pointed out, the introduction of interstitial hydrogen into the lattice affects the surrounding metallic matrix, and this should be reflected in the partial vibrational entropy term. This has been recognized recently by Boureau et al. (*40, 41*), and they have attempted to allow for this effect by using the results of Rowe et al. (*49*) who find that the acoustic modes of palladium are 20% lower for $PdD_{0.63}$ than for pure palladium. Magerl et al. (*50*) also have evaluated this contribution to the excess partial entropy and found a value of 9.97 J(mol H)$^{-1}$ K^{-1} (400°K). Using a more recent value for ν_H(*51*) and a slightly different estimate of S_H^v (acoustic), we obtain $S_H^{xs} = 11.55$ (H-vib) + 11.22 (Pd-acous) − 4.1 (elect) = 18.7 J(mol H)$^{-1}$ K^{-1} (400°K) compared with the experimental value of 17.5 ± 1.0 J(mol H)$^{-1}$ K^{-1}, which can be considered in good agreement. Magerl et al. (*50*) have pointed out that the lack of agreement in previous analyses of the partial excess entropy for dissolved hydrogen (*35, 36, 40*) resulted from the failure to include the S_H^v (acous) and/or the unnecessary use of Equation 14.

Two-Phase and Critical Phenomena

According to Equation 3, under conditions of constant pressure, the critical temperature for the phase transition is related to W_{HH}^f. However, in regions of hydrogen concentration other than dilute regions, Equation 3 is not complete because another n-dependent term contributes to μ_H. This term will be represented by $\mu_e(n)$. Attempts to interpret $\mu_e(n)$ in terms of electronic effects (e.g., *27*) have been made, but the correctness of this interpretation will not be discussed here. Regardless of the origin of this term, it can be evaluated from differences between $\Delta\mu_H$(exp) and the contributions to $\Delta\mu_H$ determined from data in the dilute region where Equation 3 is valid. The critical point can be related to W_{HH}^f and $\mu_e(n)$ since, at that point, $(\partial\Delta\mu_H/\partial n) = 0$ and $(\partial^2\Delta\mu_H/\partial^2 n) = 0$, where $\Delta\mu_H$ is given by

$$\Delta\mu_H = \Delta\mu_{H\rightarrow0}^{xs} + RT \ln n/(1-n) + W_{HH}^f n + \mu_e(n) \qquad (17)$$

The n_c value for palladium–hydrogen is 0.25 from magnetic susceptibility measurements (*52*) with $T_c = 564°$K (*52*), and from P–C–T data, de Ribaupierre and Manchester (*26*) estimate $n_c = 0.29$ and $T_c = 566°$K. The Bragg–Williams approximation gives a reasonable W_{HH}^f value by using an average value of n_c, the critical temperature, and an analytical expression for $\mu_e(n)$ determined elsewhere (53). The equation used by Lacher (17), $|W^f_{HH}|/4R = T_c$, gives incorrect values for W^f_{HH} because $n_c \neq 0.5$, as was assumed by Lacher; in addition, he omitted the contribution from the important $\mu_e(n)$ term.

Flanagan and Lynch (53) have drawn attention to the remarkable linearity found in plots of $\ln P_{H_2}$ (two-phase) against T^{-1} for palladium–hydrogen and other metal–hydrogen systems. It is clear that the compositions of the coexisting hydrogen-saturated metal (α) and the nonstoichiometric hydride phase (β) vary markedly with temperature (see Figure 1), and it is known that the partial enthalpy and entropy for solution of hydrogen vary markedly with composition (53). In view of these considerations, it is surprising that plots of $\ln P_{H_2}$ (two-phase) vs. T^{-1} have constant slopes over wide temperature ranges; the derived thermodynamic parameters, ΔS (two-phase) and ΔH (two-phase), of course, also must be temperature independent. Flanagan and Lynch (53) have shown from the variation of $\Delta\mu_H$ with n and temperature that for palladium–hydrogen, plots of $\ln P_{H_2}$ (two-phase) against T^{-1} should be linear over an extended temperature range, and that the slopes and intersection of these plots should give thermodynamic parameters for the phase change, i.e., ΔH (two-phase) $= \Delta H_{\alpha\to\beta}$ and ΔS (two-phase) $= \Delta S_{\alpha\to\beta}$.

The treatment by Flanagan and Lynch (53) will be expanded in the following: Equation 18 is valid in single-phase regions that are infinitely small close to the phase boundaries, and where n_α and n_β represent $n_{\alpha max}$ and $n_{\beta min}$, respectively, the compositions at the phase boundaries. It can be shown analytically from Equation 4 that, for the α-phase boundary, Equation 19, which is a required condition for Equation 18, must be true. However, it is not necessary to use Equation 4, that is based on a specific model, to arrive at the equality given in Equation 19. It is clear that Equation 19 must be generally valid for both phase boundaries because the first term within the brackets of Equation 18 differs for the α- and β-phase boundaries, i.e., $(\partial\Delta\mu_H/\partial n)_{n_\alpha} \neq (\partial\Delta\mu_H/\partial n)_{n_\beta}$; therefore, for Equation 18 to be valid at all temperatures, the term in brackets must be identical to $-\Delta S$(two-phase). This leads to the result that Equation 19 must be generally true. Evaluation of the term in brackets (Equation 18) for palladium–hydrogen, by using the known dependence of $\Delta\mu_H$ on n and T, confirms that for palladium–hydrogen this term is equal indeed to $-\Delta S$(two-phase) and is independent of temperature (53).

$$(d\Delta\mu_H/dT)_{n_\alpha} = [(\partial\Delta\mu_H/\partial T)_n + (\partial\Delta\mu_H/\partial n)_T(dn/dT)]_{n_\alpha \text{ or } n_\beta}$$
$$= (d\Delta\mu_H/dT)_{n_\beta} = -\Delta S(\text{two-phase}) \quad (18)$$

$$(\partial\Delta\mu_H/\partial n)_T(dn/dT) = \Delta S_H - \Delta S(\text{two-phase}) \quad (19)$$

It is quite clear that the equilibrium two-phase pressure at any temperature must be a measure of $\Delta\mu_H$ for the reaction

$$\tfrac{1}{2}H_2(g, 1\ atm.) + \frac{PdH_{n_\alpha}}{(n_\beta - n_\alpha)} \to \frac{PdH_{n_\beta}}{(n_\beta - n_\alpha)} \quad (20)$$

As discussed by Flanagan and Lynch (53), the correspondence between the equilibrium two-phase pressures and the $\Delta\mu_H$ values can be appreciated most easily by considering standard electrode potentials ($f_{H_2} = 1$ atm) for Equation 20. The standard cell potential gives a direct measure of the standard free-energy change, which in turn, can be related to the two-phase equilibrium pressure, i.e., $\Delta G^0 = -FE^0 = RT \ln P_{H_2}^{1/2}$(two-phase). The temperature dependence of E^0 (the standard cell potential), or $\ln P_{H_2}^{1/2}$(two-phase), then yields the enthalpy and entropy changes corresponding to Equation 20; therefore, it follows that ΔS(two-phase) = $\Delta S_{\alpha \to \beta}$ and ΔH(two-phase) = $\Delta H_{\alpha \to \beta}$. The only conceptual difficulty in equating thermodynamic parameters derived from the equilibrium two-phase pressures with the $\alpha \to \beta$ phase change is related to the possible effect of temperature on $\Delta S_{\alpha \to \beta}$ and $\Delta H_{\alpha \to \beta}$. Since it has been shown for palladium–hydrogen (53) that thermodynamic parameters for the phase transition essentially are independent of temperature, this difficulty is resolved.

A more general explanation of the temperature independence of the phase-transition quantities will be illustrated with $\Delta S_{\alpha \to \beta}$, although it could be illustrated equally as well with $\Delta H_{\alpha \to \beta}$. The integral entropy change for the phase transition is related to the partial entropy changes by Equation 21 (53).

$$\Delta S_{\alpha \to \beta} = \frac{1}{(n_\beta - n_\alpha)} \int_{n_\alpha}^{n_\beta} \Delta S_H dn \tag{21}$$

The analogy between gas–liquid and metal–hydrogen systems has been mentioned elsewhere in this chapter, and therefore the rule of rectilinear diameter, that holds for gas–liquid systems, might apply to metal–hydrogen systems. This rule suggests that the following relations should hold at any temperature:

$$n_m = \frac{n_\alpha + n_\beta}{2}, \; n_\alpha = n_m - \Delta n, \text{ and } n_\beta = n_m + \Delta n \tag{22}$$

Judging from experimental data for palladium–hydrogen (11, 52) the relations in 22 hold very well. The integral in Equation 21 now may be rewritten, using the relations in 22, as:

$$\Delta S_{\alpha \to \beta} = \frac{1}{2 \, \Delta n} \int_{(n_m - \Delta n)}^{(n_m + \Delta n)} \Delta S_H dn \cong \frac{1}{6} [\Delta S_H(n_m - \Delta n) + \Delta S_H(n_m + \Delta n)$$
$$+ 4 \, \Delta S_H(n_m)] \cong \Delta S_H(n_m) \cong \Delta S_H(n_c) \tag{23}$$

where Simpson's approximation has been used for evaluating the integral. The approximation that $\Delta S_{\alpha \to \beta} \simeq \Delta S(n_c)$ is quite good, judging from results for the palladium–hydrogen system. $\Delta S_H(n_c) = -44.1$ J(mol H)$^{-1}$ K^{-1}, using the ideal partial configurational entropy with $\Delta S_{H \to 0}^{xs} = -53.6$ J(mol H)$^{-1}$ K^{-1} (11), and the experimental value evaluated from the variation of $\ln P_{H_2}^{1/2}$(two-phase) with T^{-1} is -43.2 J(mol H)$^{-1}$ K^{-1} (53); the experimental $\Delta H_{\alpha \to \beta}$ value is -40.1 kJ(mol H)$^{-1}$ (53) and the ΔH_H value at n_c is -42.8 kJ (mol H)$^{-1}$ (53). The approximation is a good one since $\Delta S_H(n_m - \Delta n) + \Delta S_H(n_m + \Delta n) \cong 2 \, \Delta S_H(n_m) \cong 2$

ΔS_H (n_c), i.e., values at α_{max} are offset by values at β_{min} such that their sum is approximately twice that at the critical composition. It can be shown that the approximation given in Equation 23 is good even though the partial entropy is not ideal, for example, as in niobium–hydrogen.

It has been shown that the two-phase pressure variation for palladium–hydrogen yields values of $\Delta S_{\alpha \to \beta}$ and $\Delta H_{\alpha \to \beta}$, and that these values are closely temperature independent. The temperature independence results because the value of the integral in Equation 21 can be approximated closely by the corresponding relative partial value at the critical composition. Variations of $\Delta S_{H \to 0}^{xs}$ and $\Delta H_{H \to 0}$ with temperature (see Figure 6) are apparently too small to be detected in the plot of $\ln P_{H_2}^{1/2}$(two-phase) against T^{-1}, but can be detected by the more sensitive plot of $RT \ln P_{H_2}^{1/2}$ against T or possibly could be detected by calorimetric determinations of $\Delta H_{\alpha \to \beta}$ over a wide temperature range.

Acknowledgment

T.B.F. wishes to thank the National Science Foundation for financial support of his research on solid-hydrogen systems.

Literature Cited

1. Graham, T., *Philos. Trans. R. Soc. London* (1866) **156**, 415.
2. Lewis, F. A., "The Palladium Hydrogen System," Academic, New York, 1967.
3. Hunt, D. G., Ross, D. K., *J. Less Common Met.* (1976) **49**, 169.
4. Skoskiewicz, T., *Phys. Status Solidi A* (1972) **11**, K123.
5. Anderson, J. S., "Non-Stoichiometric Compounds," ADV. CHEM. SER. (1963) **39**, 1.
6. Jacobs, J. K., Manchester, F. D., *J. Less Common Met.* (1976) **49**, 67.
7. Manchester, F. D., *J. Less Common Met.* (1976) **49**, 1.
8. Worsham, J. E., Wilkinson, M. G., Shull, C. G., *J. Phys. Chem. Solids* (1957) **3**, 303.
9. Sköld, K., Nelin, G., *J. Phys. Chem. Solids* (1967) **28**, 2369.
10. Baranowski, B., Majchrzak, S., Flanagan, T. B., *J. Phys. F* (1971) **1**, 258.
11. Wicke, E., Nernst, G., *Ber. Bunsenges Phys. Chem.* (1964) **68**, 224.
12. Reilly, J. J., Johnson, J. R., *World Hydrogen Energy Conf., 1st, Florida, 1976*.
13. Birnbaum, H., Grossbeck, M., Amano, M., *J. Less Common Met.* (1976) **49**, 357.
14. Wise, M. L., Farr, J. P., Harris, I., Hirst, H. R., "L'Hydrogene dans les Metaux," Tome 1, Ed. Sci. et Ind., Paris, 1972.
15. Flanagan, T. B., Lynch, J. F., Clewley, J. D., von Turkovich, B., *J. Less Common Met.* (1976) **49**, 13.
16. Lynch, J. F., Clewley, J. D., Curran, T., Flanagan, T. B., *J. Less Common Met.* (1977) **55**, 153.
17. Lacher, J. R., *Proc. Roy. Soc., London* (1937) **161A**, 525.
18. Flanagan, T. B., Oates, W. A., *Can. J. Chem.* (1975) **53**, 694.
19. Burger, J. P., Senoussi, S., Soufache, B., *J. Less Common Met.* (1976) **49**, 213.
20. Baranowski, B., *Ber. Bunsenges Phys. Chem.* (1972) **76**, 714.
21. Baranowski, B., Majchrzak, S., Flanagan, T. B., *J. Phys. Chem.* (1970) **74**, 4299.
22. Baranowski, B., Majchrzak, S., Flanagan, T. B., *J. Phys. Chem.* (1973) **77**, 35.
23. Alefeld, G., *Comments on Solid State Physics* (1975) **6**, 53.
24. Alefeld, G., *Ber. Bunsenges Phys. Chem.* (1972) **76**, 746.
25. Buck, H., Alefeld, G., *Phys. Status Solidi B* (1972) **49**, 317.

26. deRibaupierre, Y., Manchester, F. D., *J. Phys. C* (1974) **7**, 2126, 2140.
27. Simons, J. W., Flanagan, T. B., *Can. J. Chem.* (1965) **43**, 1665.
28. Baranowski, B., Skoskiewicz, T., Szafranski, A. W., Phys. Low Temperatures," *Akad. Nauk SSSR* (1975) **1**, 616.
29. Stritzker, B., Buckel, W., *Z. Phys.* (1972) **257**, 1.
30. Miller, R. J. Satterthwaite, C. B., *Phys. Rev. Lett.* (1975) **34**, 144.
31. Rahman, A., Sköld, K., Pelizzari, K., Sinha, S. K., *Phys. Rev. B* (1976) **14**, 3630.
32. Oates, W. A., Flanagan, T. B., *J. Chem. Soc., Faraday Trans. 1* (1977) **73**, 407.
33. Ganguly, B. N., *Z. Phys.* (1975) **22**, 127.
34. Stritzker, B., *Z. Phys.* (1974) **268**, 261.
35. Clewley, J. D., Curran, T., Flanagan, T. B., Oates, W. A., *J. Chem. Soc., Faraday Trans. 1* (1973) **69**, 449.
36. Burch, R., Francis, N. B., *J. Chem. Soc., Faraday Trans. 1* (1973) **69**, 1978.
37. Gillespie, L. J., Ambrose, H. A., *J. Phys. Chem.* (1931) **35**, 3105.
38. Nace, D. M., Aston, J. G., *J. Am. Chem. Soc.* (1957) **79**, 3619, 3623.
39. Lynch, J. F., Flanagan, T. B., *J. Chem. Soc., Faraday Trans. 1* (1974) **70**, 814.
40. Boureau, G., Kleppa, O. J., Dantzer, P., *J. Chem. Phys.* (1976) **64**, 5247.
41. Boureau, G., Kleppa, O. J., *J. Chem. Phys.* (1976) **65**, 3915.
42. Gillespie, L., Gaulstaun, L. S., *J. Am. Chem. Soc.* (1936) **58**, 2565.
43. Brüning, H., Sieverts, A., *Z. Phys. Chem., Abt. A* (1933) **163**, 409.
44. Moon, K. A., *J. Phys. Chem.* (1956) **60**, 502.
45. Wagner, C., *Acta Metall.* (1971) **19**, 843.
46. Eshelby, J. D., "Solid State Physics," 3, p. 79, Academic, New York, 1956.
47. Oates, W. A., Flanagan, T. B., *J. Chem. Soc., Faraday Trans. 1* (1977) **73**, 993.
48. Fast, J. D., "Interaction of Gases and Metals," Barnes and Noble, New York, 1972.
49. Rowe, J. M., Rush, J. J., Smith, H. G., Mostoller, M., Flotow, H. E., *Phys. Rev. Lett.* (1974) **33**, 1297.
50. Magerl, A., Stump, N., Wipf, H., Alefeld, G., *J. Phys. Chem. Solids* (1977) **38**, 683.
51. Drexel, W., Murani, A., Tocchetti, D., Kley, W., Sosnowska, I., Ross, D. K., *J. Phys. Chem. Solids* (1976) **37**, 1135.
52. Frieske, H., Wicke, E., *Ber. Bunsenges Phys. Chem.* (1973) **77**, 48.
53. Flanagan, T. B., Lynch, J. F., *J. Phys. Chem.* (1975) **79**, 444.

RECEIVED July 19, 1977.

21

Preparation and Properties of TiCuH

ARNULF J. MAELAND

Materials Research Center, Allied Chemical Corp., Morristown, NJ 07960

γ-TiCu reacts with hydrogen below ~200°C to form a hydride with the approximate composition TiCuH. The enthalpy of the reaction is −75 kJ/mol H_2 and the entropy change is −113 J/deg mol H_2. The tetragonal unit cell undergoes a 9% vol increase during hydriding. TiCuH is thermodynamically unstable and decomposes into TiH_2 + Cu when heated above ~200°C. The ease of decomposition can be explained in terms of the structure. The stability of TiCuH with respect to dissociation is discussed in relation to Miedema's model.

M any metals react exothermally and reversibly with hydrogen according to Reaction 1. In many cases, the resulting metal hydrides contain more

$$M + \frac{x}{2} H_2 \rightleftarrows MH_x \qquad (1)$$

hydrogen per unit volume than liquid or even solid hydrogen. In titanium hydride, for example, the hydrogen concentration is 9.2×10^{22} atoms per cm^3 while the corresponding concentrations for liquid (20°K) and solid (4.2°K) hydrogen are 4.2×10^{22} and 5.3×10^{22} atoms per cm^3, respectively. This high density of hydrogen atoms in metal hydrides, coupled with the ease of recovery of the hydrogen simply by heating, have created an interest in these materials as energy-storage media. For practical storage applications, however, additional properties are required. These and the above mentioned properties recently have been discussed by Libowitz (1) and are summarized in Table I. Since none of the binary metal hydrides possesses all or even most of the properties listed in Table I, current research efforts are directed toward developing new alloy hydrides with the required properties, either through modification of known hydrides by appropriate alloying or by synthesis of new intermetallic-compound hydrides.

A hydride formed in the reaction of a binary solid-solution alloy with hydrogen can be considered as a solid solution of two binary hydrides and will have properties related to the properties of the constituent binary hydrides. An intermetallic-compound hydride, however, formed in accordance with Reaction

0-8412-0390-3/78/33-167-302/$05.00/0

© American Chemical Society

Table I.   Properties Required of Metal Hydrides for Energy Storage

1. Large hydrogen density, i.e., high hydrogen-to-metal ratio (H:M).
2. Ease of hydrogen recovery, i.e., low temperature of dissociation ($\leq 100°C$).
3. High rates of hydrogen absorption and desorption.
4. Low enthalpies of formation.
5. Stability with respect to oxygen and moisture.
6. Low material cost.
7. Light weight of material.

2 has, in general, properties that bear little or no resemblance to those of the constituent metal hydrides (*1,2*).   The intermetallic-compound hydrides, in fact, conveniently can be looked upon as pseudobinary metal hydrides.   Unfortunately, current understanding of metal hydrides does not permit one to predict with a high degree of certainty which intermetallic compounds will form hydrides or the approximate properties of such new hydrides although efforts along these lines have been made by Miedema and co-workers (*3,4,5*).

$$MM_a' + xH_2 \rightleftarrows MM_a'H_{2x} \tag{2}$$

In some cases, absorption of hydrogen by intermetallic compounds causes decomposition as indicated by Reaction 3.   Decomposition also can lead to the formation of a new intermetallic compound, as is observed when Mg_2Cu reacts with hydrogen (*6*) in Reaction 4.   Reactions 3 and 4 take place at elevated temperatures where diffusion of the metal atoms becomes possible.

$$MM_a' + xH_2 \rightleftarrows MH_{2x} + M_a' \tag{3}$$

$$2Mg_2Cu + 3H_2 \rightleftarrows 3MgH_2 + MgCu_2 \tag{4}$$

Titanium iron hydrides are among the materials which, at the present time, appear to have potential for practical applications as an energy-storage medium (*7*).   The formation and properties of titanium iron hydride have been studied by Reilly and Wiswall (*8*), who found that the reaction proceeds in two steps as indicated by Reactions 5 and 6.   Both hydrides have dissociation pressures above 1 atm at room temperature in contrast to TiH_2 which is very stable.   Titanium iron is representative of intermetallic compounds that consist of an element (titanium) capable of forming a stable hydride and another element (iron) that is not a hydride former or at best, forms a hydride with great difficulty.   Iron presumably plays a role in destabilizing the hydrides.   Titanium also forms a 1:1 compound with copper (there are other intermetallic compounds in the titanium–copper system) and this fact, coupled with the observation that copper

$$2.13TiFeH_{0.10} + H_2 \rightleftarrows 2.13TiFeH_{1.04} \quad \Delta H = -28 \text{ kJ/molH}_2 \tag{5}$$

$$2.20TiFeH_{1.04} + H_2 \rightleftarrows 2.20TiFeH_{1.95} \quad \Delta H = -31 \text{ KJ/molH}_2 \tag{6}$$

does not form a stable hydride (ΔH = 65.9 kJ/mol H_2 (9)), prompted an investigation of the TiCu–H_2 system to establish whether any intermetallic hydride phases formed and, if so, to determine their properties.

TiCu reportedly crystallizes in two forms: γ-phase, which is bc tetragonal of the B11 type; and δ-phase, which is fc tetragonal of the $L1_0$ type (10,11). The δ-phase has a homogeneity range of 45–50 atom % titanium while the homogeneity range for the γ-phase is 50–53 atom % titanium (10). The present work deals with the γ-phase.

Beck (12), in a survey of hydriding characteristics of numerous intermetallic compounds, reported that γ-TiCu reacted exothermally with hydrogen to form $TiCuH_{0.42}$. An x-ray pattern of the product indicated that a 0.5% lattice expansion had taken place. In a recent paper that appeared while this study was in progress, Yamanaka et al. (13) observed that TiCu decomposed according to Reaction 3 when exposed to 130 atm H_2 at 200°C; the intermetallic compound could be regenerated by degassing at 850°C. Our investigation has shown that γ-TiCu does form an intermetallic hydride under appropriate conditions but that the intermetallic hydride is unstable with respect to TiH_2 + Cu and will decompose at elevated temperatures, as was reported by Yamanaka et al. (13).

Experimental

The γ-TiCu samples were prepared by arc melting equimolar amounts of 99.99% titanium and 99.9% copper under an Ar atm. The resulting buttons were remelted several times to ensure homogeneity. Lattice parameters, determined from x-ray powder patterns, were in good agreement with published values for the 50 atom % alloy (10). All of the samples contained a trace of the δ phase in agreement with previous observations (10), but the presence of this minor impurity, which can have some effect on the details of the isotherm, does not affect our general conclusions. Prior to exposure to hydrogen, the samples were broken up into small chunks and degassed by heating to 400°–500°C in vacuum (10^{-5} Torr). After cooling to room temperature, approximately 1 atm of pure hydrogen, generated by decomposition of TiH_2, was admitted. Absorption started slowly at room temperature but proceeded rapidly when the sample was heated to ~150°C. When absorption was complete, the sample was cooled slowly to room temperature and removed for x-ray examination. The composition was calculated from the initial and final pressures, measured to ±0.01 Torr on a Texas Instrument, Model 145, Precision Pressure Gage, the known volume of the system (82.3 cm^3 plus storage bulbs of 500 and 1000 cm^3), and the weight of the sample. A representative isotherm was determined by immersing the sample in a constant temperature oil bath (±2°C), admitting known amounts of hydrogen, and recording the equilibrium pressures after each addition of hydrogen. Sample compositions were determined as before from the pressure change, the volume of the system, and the sample weight. The enthalpy of the reaction

$$\frac{2}{x-y} TiCuH_y \text{ (α phase)} + H_2 \leftrightarrows \frac{2}{x-y} TiCuH_x \text{ (β phase)} \qquad (7)$$

was determined from the van't Hoff relationship by plotting the log of the plateau pressures in the isotherms as a function of the reciprocal temperature. The relationship is linear of the form

$$\ln P = (A/T) + B$$

where A and B are constants and T is the absolute temperature; ΔH is obtained from the slope and ΔS from the intercept. Thermal analysis data were obtained on a Mettler TA-1 apparatus.

Results and Discussion

γ-TiCu reacted with hydrogen at room temperature and slightly higher temperatures to form an intermetallic hydride. The composition of the hydride was determined to be $TiCuH_{0.97\pm0.03}$ at room temperature under 1 atm H_2. An x-ray pattern of the hydride showed that the *a* axis of the tetragonal unit cell had contracted slightly while the *c* axis had expanded greatly for a net volume increase of about 9%. The x-ray data are summarized in Table II.

Table II. X-Ray Data on TiCu and TiCuH

	Lattice Parameters ($\mathring{A}$)	*Vol* ($\mathring{A}^3$)	*% Vol Increase*
TiCu	$a = 3.13 \pm 0.01; c = 5.91 \pm 0.01$	57.9	—
$TiCuH_{0.97}$	$a = 3.04 \pm 0.01; c = 6.85 \pm 0.01$	63.3	9

A pressure–composition absorption isotherm ($215°C$) is shown in Figure 1. Hydrogen first dissolves in TiCu to form a solid solution (α phase), whose composition depends on the hydrogen pressure. The solid solution region is represented by the steeply rising portion on the left side of the isotherm in Figure 1. When the alloy becomes saturated with hydrogen, the nonstoichiometric hydride, β phase, is formed; with further addition of hydrogen, more alloy is converted to hydride, and the pressure remains invariant across the two-phase region as shown by the horizontal portion of the isotherm. An x-ray examination of a separate sample with a composition near the middle of the plateau pressure region ($TiCuH_{0.47}$) confirmed the presence of two phases. After complete conversion to the hydride phase, the hydrogen pressure increases again (right side of Figure 1) as the nonstoichiometric hydride absorbs hydrogen. The sample was removed for x-ray examination after completion of the run that took a total of 18 days. The x-ray pattern showed the presence of the intermetallic hydride with lattice parameters very close to those given in Table II. In addition, traces of titanium hydride and metallic copper also were indicated in the pattern. Evidently, some decomposition of the intermetallic hydride had occurred in the long time period it took to obtain the isotherm. The decomposition does not affect the general shape of the isotherm, nor the general conclusions regarding it, but is expected to have some affect on the compositions of the phase boundaries. The phase boundaries indicated by the isotherm in Figure 1, should, therefore,

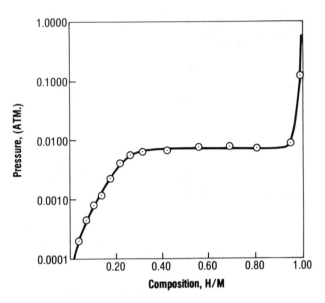

*Figure 1. Pressure-composition absorption isotherm
(215°C) for the titanium-copper-hydrogen system*

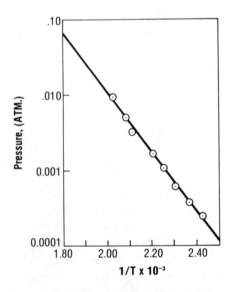

*Figure 2. Equilibrium dissociation
pressure vs. the reciprocal temperature
for TiCuH$_{1-x}$*

be observed with caution.   We will return to the problem of decomposition of
the intermetallic hydride below.

Figure 2 shows a plot of equilibrium dissociation pressure vs. the reciprocal
temperature for $TiCuH_{1-x}$.   The plot was obtained by selecting a composition
near the middle of the plateau in Figure 1 and measuring the dissociation pressure
as a function of temperature.   Since the volume of the system was very small and
a large sample was used, the sample composition was nearly constant during the
measurements.   The enthalpy of Reaction 7 determined by this method is -75
kJ/mol H_2, and the entropy change is -113 J/deg mol H_2.

To account for the decomposition of TiCuH, we proceed as follows: although both x and y in Reaction 7 vary with temperature, ΔH remains reasonably
constant with temperature and composition, as seen by the straight line in Figure
2.   Such behavior is, in fact, observed experimentally in most metal hydrides (*14*).
This being the case, we approximate the enthalpy of formation of TiCuH by
rewriting Reaction 7 (Reaction 8) and assume ΔH for this reaction to be -75

$$2TiCu + H_2 = 2TiCuH \tag{8}$$

kJ/mol H_2.   If the enthalpy of Reaction 9 is -100 kJ, as estimated by the method
given in Ref. *15*, the enthalpy of formation of TiCuH, $\Delta H_{f(298)}$, given by Reaction
10 is -175 kJ/mol H_2; the corresponding entropy of formation, $\Delta S_{f(298)}$, becomes

$$2Ti + 2Cu = 2TiCu \tag{9}$$

$$2Ti + 2Cu + H_2 = 2TiCuH \tag{10}$$

-113 J/mol deg, neglecting the small entropy change associated with Reaction
9.   From $\Delta G = \Delta H - T\Delta S$, we calculate the free energy of formation, $\Delta G_{f(298)}$,
to be -141 kJ/mol H_2.   For Reaction 11, ΔG is given by Equation 12.

$$2TiCuH + H_2 = 2TiH_2 + 2Cu \tag{11}$$

$$\Delta G = 2\Delta G_{f(298),TiH_2} - 2\Delta G_{f(298),TiCuH} \tag{12}$$

Substituting -86 kJ/mol H_2 for $\Delta G_{f(298),TiH_2}$ (*14*) and -141 kJ/mol H_2 for
$2\Delta G_{f(298),TiCuH}$, the free energy change for Reaction 11 is -31 kJ/mol H_2; this
shows that TiCuH + H_2 is thermodynamically unstable with respect to TiH_2 +
Cu.   At room temperatures below $\sim$200°C, diffusion of the metal atoms is very
slow and Reaction 11 is kinetically unfavorable.   Therefore, hydriding TiCu at
temperatures below $\sim$200°C will proceed according to Reaction 7, with the
formation of TiCuH rather than TiH_2 + Cu.   If TiCuH is heated at $\sim$200°C in
hydrogen for an extended period of time, however, some TiH_2 + Cu will be
formed, as was observed in the isothermal experiment.   In the study by Yamanaka
et al. (*13*), the conditions were such (200°C, 130 atm H_2) that effective sample
temperatures, caused by the exothermic reaction, well above 200°C were reached,
making Reaction 11 kinetically favorable.   Notice that high hydrogen pressures

also favor formation of TiH_2. The structure of TiCuH recently has been determined (16), and it provides a clue to the reason why titanium hydride precipitates so readily when the intermetallic hydride is heated to relatively low temperatures ($>200°C$); it also points out the importance of the intermetallic hydride structure in considering stability with respect to decomposition. Hydrogen is located in tetrahedral sites in the TiCuH structure (16), surrounded by four titanium atoms; the titanium–hydrogen distances are very close to those in TiH_2 (17) where hydrogen is also in tetrahedral sites. Very little thermal energy is necessary to cause precipitation of TiH_2. In contrast, TiFeH also is unstable thermodynamically with respect to TiH_2 + Fe, but decomposition does not occur until much higher temperatures are reached because hydrogen is presumably in octahedral sites surrounded both by titanium and iron atoms (18), and considerable atomic motion is necessary to precipitate TiH_2.

Miedema and co-workers (3,4,5) have used Equation 13

$$\Delta H(MM_a'H_{2x}) = \Delta H(MH_{2x-y}) + \Delta H(M_a'H_y) - \Delta H(MM_a') \qquad (13)$$

to estimate the enthalpy of Reaction 2 and thus to predict the stability of ternary metal hydrides with respect to dissociation. The entropy change in Reaction 2 is predominantly caused by the entropy loss of hydrogen gas, -131 J/deg mol H_2, as it enters the metal. Since this is approximately constant for all metal–hydrogen systems (3), stability of metal hydrides can be discussed in terms of the enthalpy of Reaction 2 rather than the customary free energy change. In Equation 13, $\Delta H(MM_a'H_{2x})$ is the enthalpy of Reaction 2; $\Delta H(MH_{2x-y})$ and $\Delta H(M_a'H_y)$ are the enthalpies of formation of the constituent metal hydrides, MH_{2x-y} and $M_a'H_y$, respectively; $\Delta H(MM_a')$ is the enthalpy of formation of the intermetallic compound, MM_a'. Equation 13 applies to compounds richer in metal M' than M, and the boundary case is that of equal concentrations of M and M' (5). Thus, $a \geq 1$. This approach has resulted in the Rule of Reversed Stability (3) which states that within a given series of intermetallic compounds, the greater the stability of the intermetallic compound, the less stable is its hydride. Although successful in many cases, numerous instances are known in which the rule fails completely. In the series TiFe, TiCu, TiNi, and TiCo, for example, the stability of the intermetallic compound, as measured by the enthalpy of formation, increases from TiFe to TiCo [ΔH_f = -42 kJ/mol (19), -50 kJ/mol (15), -68 kJ/mol (20), and -92 kJ/mol (5), respectively]. The Rule of Reversed Stability predicts decreasing stability of the intermetallic hydride in the same order. Experimental data are not in agreement with this prediction. The enthalpy of Reaction 2 is -75 kJ/mol H_2 for TiCuH, -57 kJ/mol H_2 for TiCoH (13), and -28 kJ/mol H_2 for TiFeH (8). The value for TiNi hydride is not available, but this hydride is also much more stable than TiFeH (13), in direct opposition to the prediction of the Rule of Reversed Stability. Yamanaka et al. (13) suggested that in considering the stability of hydrides of intermetallic compounds, the electronic structure is of major importance and must be considered in addition to the thermodynamic arguments. Another reason for the failure of the Rule of Reversed Stability in this series of compounds (and in other cases as well) can be that hydrogen positions in the intermetallic hydride structure are completely ignored in Miedema's approach.   The assumption made in deriving Equation 13 is that as hydrogen is introduced into the intermetallic compound, M–H bonds are formed at the expense of M–M′ bonds.   Although this is true in many structures, i.e., when hydrogen goes into positions between M and M′, it is not always the case.   In TiCuH, for example, hydrogen enters sites that are surrounded tetrahedrally by titanium atoms without changing the metal lattice structure and without interfering with the titanium–copper bonds.   In such cases, subtracting the term $\Delta H(M-M_a{}')$ from the right of Equation 13 therefore may lead to an incorrect estimation of the enthalpy of Reaction 2.

Continuing with TiCuH as our example, Equation 13 can be written as

$$\Delta H(\text{TiCuH}) = \Delta H(\text{TiH}_{1-y}) + \Delta H(\text{CuH}_y) - \Delta H(\text{TiCu}) \qquad (14)$$

If we assume $y \ll 1$, which makes $\Delta H(\text{CuH}_y) = 0$, take $\Delta H(\text{TiCu}) = -50 \text{ kJ/mol}$ and $\Delta H(\text{TiH}) = -54 \text{ kJ/mol}$ (estimated from data in Ref. *21*), $\Delta H(\text{TiCuH})$ is calculated to be -4 kJ/mol TiCuH or -8 kJ/mol H_2.   Agreement with the measured value, -75 kJ/mol H_2, is very poor.   Somewhat better agreement is obtained if, as suggested by the structural arguments, $\Delta H(\text{TiCu})$ is not subtracted from the right of Equation 14.   In this case, $\Delta H(\text{TiCuH})$ is estimated to be -108 kJ/mol H_2.

The $\text{TiCuH}_{0.97\pm0.03}$ sample was analyzed by thermogravimetric analysis (TGA) and the results are shown in Figure 3.   Hydrogen evolution started very

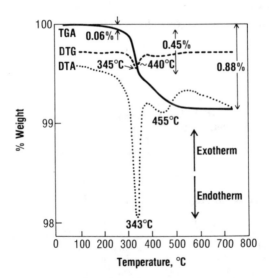

Figure 3. TGA, DTA, and DTG curves for titanium copper hydride.   The heating rate was 25°C/min in purified helium.   Sample 32.1 mg, contained in an Al_2O_3 crucible.

slowly around 100°C, began increasing near 300°C, and reached its maximum rate at 345°C (minimum in the DTG curve), as seen by the superimposed differential thermogravimetric analysis (DTG) curve. Another, much smaller increase in the hydrogen evolution rate is indicated by the second, very broad, minimum in the DTG curve near 440°C. The total weight loss when hydrogen evolution had been completed was 0.88%, which corresponds to the composition $TiCuH_{0.98}$ in excellent agreement with the composition calculated from the pressure–volume measurement. Differential thermal analysis (DTA) of the sample also was obtained, and the results are shown by the dotted curve in Figure 3. Two endothermic processes are indicated by the DTA curve, at 343° and 455°C. These processes are correlated with the minima in the DTG curve and can be represented by the following reactions (15 at 343°C and 16 at 455°C).

$$2TiCuH \rightarrow \frac{1}{2} TiH_2 + \frac{3}{2} TiCu + \frac{1}{2} Cu + \frac{1}{2} H_2 \qquad (15)$$

$$TiH_2 + Cu \rightarrow TiCu + H_2 \qquad (16)$$

Reactions 15 and 16 depend on experimental conditions such as heating rate, helium flow rate, etc., and both reactions occur simultaneously to some extent. Therefore, it was not possible to determine the enthalpies of these reactions individually. However, by using a differential scanning calorimeter, we determined the combined enthalpy of Reactions 15 and 16 which is, of course, the enthalpy of Reaction 8 with the sign reversed. The value obtained was 100 ± 10 kJ/mol H_2, which is in fair agreement with the value obtained from the van't Hoff relationship.

Literature Cited

1. Libowitz, G. G., "Metal Hydrides for Energy Storage," *Critical Materials Problems in Energy Production*, C. Stein, Ed., pp. 825–852, Academic, New York, 1976.
2. van Mal, H. H., *Philips Res. Rep. Suppl.* (1976) **1**.
3. van Mal, H. H., Buschow, K. H. J., Miedema, A. R., *J. Less Common Met.* (1974) **35**, 65.
4. Buschow, K. H. J., van Mal, H. H., Miedema, A. R., *J. Less Common. Met.* (1975) **42**, 163.
5. Miedema, A. R., Buschow, K. H. J., van Mal, H. H., *J. Less Common Met.* (1976) **49**, 463.
6. Reilly, J. J., Wiswall, R. H., Jr., *Inorg. Chem.* (1967) **6**, 2220.
7. Reilly, J. J., Hoffman, K. C., Strickland, G., Wiswall, R. H., "Iron Titanium Hydrides as a Source of Hydrogen Fuel for Stationary and Automotive Applications," *Proc. Ann. Power Sources Conf., 26th, New Jersey, 1974*.
8. Reilly, J. J., Wiswall, R. H., Jr., *Inorg. Chem.* (1974) **13**, 218.
9. Warf, J. C., *J. Inorg. Nucl. Chem.* (1966) **28**, 1031.
10. Karlson, N., *J. Inst. Met.* (1951) **79**, 391.
11. Raub, E., Walter, P., Engel, M., *Z. Metallkd.* (1952) **43**, 112.
12. Beck, R. L., LAR-55, Summary Report, 1961.
13. Yamanaka, K., Saito, H., Someno, M., *J. Chem. Soc. Jpn.* (1975) **8**, 1267.

14. Libowitz, G. G.,"The Solid-State Chemistry of Binary Metal Hydrides," p. 55, W. A. Benjamin, Inc., New York, 1965.
15. Miedema, A. R., *J. Less Common Met.* (1976) **46**, 67.
16. Santoro, A., Maeland, A. J., Rush, J. J., "Neutron Diffraction Determination of the' Crystal Structure of TiCuD," *Winter Meeting Am. Crystallogr. Assoc., California, 1977.*
17. Sidhu, S. S., Heaton, L., Zauberis, D. D., *Acta Crystallogr.* (1956) **9**, 612.
18. Adkins, C. M., University of Virginia, Charlottesville, Virginia, private communication.
19. Kubaschewski, O., Dench, W. A., *Acta Metalla.* (1955) **3**, 339.
20. Kubaschewski, O., *Trans. Faraday Soc.* (1958) **54**, 814.
21. Muller, W. M., "Titanium Hydrides," *Metal Hydrides*, W. M. Muller, J. P. Blackledge, G. G. Libowitz, Eds., pp. 336–383, Academic, New York, 1968.

RECEIVED August 3, 1977.

22

Thermodynamics and Kinetics of Hydrogen Absorption in Rare Earth–Cobalt (R_2Co_7 and RCo_3) and Rare Earth–Iron (RFe_3) Compounds

A. GOUDY, W. E. WALLACE, R. S. CRAIG, and T. TAKESHITA

Department of Chemistry, University of Pittsburgh, Pittsburgh, PA 15260

Pressure–composition isotherms were obtained for some rare earth intermetallic hydrides of the type R_2Co_7 (R = a rare earth). The compounds absorbed large quantities of hydrogen, giving the limiting composition $R_2Co_7H_9$ at 100 atm. This solubility exceeds that of the RCo_5 compounds but is less than that of the RCo_3 compounds. Pr_2Ni_7 and $ErNi_3$ were included in this study for comparison. Kinetic studies of hydrogen desorption were conducted on the following representative compounds: Dy_2Co_7, Gd_2Co_7, $DyCo_3$, $ErCo_3$, $DyFe_3$, and $ErFe_3$. Desorption was second order, indicating the recombination of hydrogen atoms on the metal surface to form molecular hydrogen as the likely rate-determining step. Activation energies for this process range from 15–36 kcal/mol. Hydride stability varies systematically with atomic number of the constituents.

In 1969 Zijlstra and Westendorp (1) reported that the rare earth intermetallic $SmCo_5$ extensively absorbed hydrogen. They subjected this material, the most powerful permanent magnet known, to metallographic examination, polishing and acid-etching it to reveal its grain structure. They found that the magnetic properties of $SmCo_5$ that had been processed this way were degraded sharply. The effect was traced to hydrogen that was absorbed into the lattice during the acid etch. Almost simultaneously, Neumann (2) observed that $LaNi_5$, which has a $CaCu_5$ structure like $SmCo_5$, absorbs hydrogen. Shortly thereafter, Van Vucht, Kuijpers, and Bruning (3, 4) found that $LaNi_5$ and some structurally related rare earth intermetallics (RCo_5, etc., where R represents a rare earth) absorb hydrogen extensively, and the absorption (or desorption) occurs

0-8412-0390-3/78/33-167-312/$05.00/0

extremely rapidly. (For information about the stoichiometries, structures, and properties of this extensive class of materials, *see* Ref. 5.) The solubility is such that the proton density is $\sim6 \times 10^{22}$ cm^{-3} or roughly 50% greater than the proton density of liquid hydrogen (for an intermetallic with H$_2$ gas at ~2 atm at 25°C).

As impressive as the hydrogen solvent capacity of LaNi$_5$ and similar materials is, the remarkable feature of these materials is their rapid exchange of hydrogen between the hydride and gaseous phases under mild temperature and pressure conditions. For LaNi$_5$, 95% of the dissolved hydrogen is released within 5 min at 50°C if the restraining pressure is dropped from 3 to 1 atm (3). In contrast, UH$_3$ or YH$_3$, which have higher proton densities, readily release hydrogen only at elevated temperatures. Since hydrogen is absorbed dissociatively, the gas can exist, at least fleetingly, as monatomic hydrogen on the metal's surface. This suggests that the surfaces of the rare earth intermetallics are quite active, which indicates their possible use as heterogeneous catalysts. Some rare earth intermetallics (6, 7, 8, 9) were shown recently to be very effective as synthetic NH$_3$ and as methanation catalysts.

Since the rare earth intermetallics can contain large amounts of hydrogen at pressures <10 atm and can readily release this hydrogen, they are being considered in efforts to exploit hydrogen as a fuel. In mobile applications, the containment of hydrogen is a problem. The energy density of gaseous H$_2$ is too low, and liquid H$_2$ has an acceptable energy density but inconvenient cryogenic temperatures. Metallic hydrides are promising in this regard although the low weight percentage of hydrogen and the cost of the metal represent significant disadvantages. Many rare earth intermetallics were studied from a hydrogen storage viewpoint. Van Mal et al. (10) summarized the findings up to early 1976.

In earlier studies from this laboratory, results were presented for the thermodynamics (pressure–composition isotherms and heats of vaporization) of the RCo$_3$–H and RFe$_3$–H (R = Gd, Tb, Dy, Ho, and Er) systems (11) and for YCo$_5$ (12). Interestingly, these results showed that three phases, α, β, and γ, formed in the RCo$_3$–H systems, but only two phases developed in the RFe$_3$–H systems. The phase rule requires that pressure is constant on an isotherm for compositions with two solid phases. These pressures are usually termed plateau pressures. The plateau pressures for the RFe$_3$–H and RCo$_3$–H systems increased monotonically with an increase in the atomic number of R. Also, the plateau pressure is lower for RFe$_3$–H than it is for its cobalt counterpart. These interesting systematics were observed by Kuijpers for the RCo$_5$–H series (4).

Presently, R$_2$Co$_7$ compounds are being studied as hosts for hydrogen. In addition to thermodynamic data, some crystallographic and kinetic information is provided for the R$_2$Co$_7$–H ternaries. Results are included for Pr$_2$Ni$_7$ and for ErNi$_3$ to determine if there is a systematic variation in plateau pressures with changing atomic number of the d-transition element. For practical applications, kinetic information is equally as important as hydrogen capacity information, but little kinetic information has been published, and much of the published material is inappropriate for establishing the kinetic parameters of interest—the reaction order and the activation energy. $SmCo_5$ and $LaNi_5$ were studied by Raichlen and Doremus (13) and Boser (14), respectively. All reported first-order absorption kinetics. Boser found an activation energy of 7.6 kcal/mol whereas Lundin and Lynch (15) reported an activation energy of 21.6 kcal/mol for the absorption of H_2 into $LaNi_5$ and stipulated that it is a second order process. (The discrepancy perhaps exists since the escape of hydrogen (an endothermal process) is controlled by the rate of heat supplied (16).) These supposedly are the only instances where the reaction order and activation energy have been established, and the results for $LaNi_5$ are in disagreement.

Kinetic information is provided not only for the release of hydrogen from the R_2Co_7 compounds but also for several related materials, i.e., $ErCo_3$, $DyCo_3$, $ErFe_3$, and $DyFe_3$.

Experimental

The intermetallic compounds were prepared from the best grade of metals commercially obtainable. The rare earth metals, obtained from Research Chemicals, Inc., were 99.9% pure, and cobalt and nickel, obtained from the United Mineral Corp., were 99.999% pure. After each element was weighed to obtain the correct stoichiometric amounts, the compounds were formed by induction melting in a water-cooled copper boat under an argon atmosphere, purified by passage through a titanium-gettering furnace.

After a sample melted, it was turned over and remelted four times to ensure adequate mixing. This was followed by quenching on the cold boat to reduce the likelihood of the compound separating into two or more phases of higher stability. X-ray diffraction patterns confirmed the formation of the desired compounds and the absence of extraneous phases. When x-ray analysis showed that extraneous phases were present, the samples were subjected to an homogenization treatment. They were sealed in fused quartz tubes under argon at reduced pressure and held in an annealing furnace at an appropriate temperature for as long as four weeks.

The solubility experiments were carried out in a stainless steel system. It had a manifold with ports for adding hydrogen, for evacuating, and for attaching metal chambers of a known volume (a few hundred cubic centimeters). Also connected to the ports were Ashcroft bourdon gages for measuring pressures of 1–100 atm and a mercury manometer for measuring pressures below 1 atm. The ports were equipped with valves, and the volumes of all spaces accessible to the manifold were known accurately.

The compound (8–10 g) was sealed in a stainless steel sample container with a Teflon gasket. The unit extended down from the manifold and, for temperatures up to 100°C, was immersed in a water bath regulated to ±0.1°C. When higher temperatures were required, the sample container was inserted into a tube furnace regulated to ±0.25°C.

Before equilibrium measurements could be obtained, the sample was activated. It was crushed to a particle size of 1 mm or less, placed in the sample container, and outgassed at 250°C under a vacuum (10^{-3} Torr) for at least 8 hr. It was cooled to about 200°C and exposed to hydrogen at pressures up to 100 atm.

Hydrogen was removed by pumping (at 250°C), and the hydrogenation–dehydrogenation process was repeated several times.   Through this treatment, the sample was reduced to a fine powder with a relatively large surface area.

After a sample was loaded with hydrogen, desorption isotherms were obtained by removing measured quantities of hydrogen and determining the pressure after the system had come to equilibrium.   At pressures above 1 atm, the hydrogen flowed out underwater, and the amount removed was measured in a gas buret.   The hydrogen remaining in the dead spaces of the apparatus was accounted for.   At pressures below 1 atm, the sample section was isolated, and the remaining portions of the system were evacuated.   The sample section was reopened and the entire system was allowed to come to a new and lower equilibrium pressure.   The amount of hydrogen removed was determined from the pressure and temperature of the system's known volume.   The time required for the system to reach equilibrium usually varied from 0.5 to 2.0 hr at pressures above 1 atm to 12 hr at lower pressures.

Absorption pressure–composition isotherms were established by metering fixed amounts of hydrogen into the system and determining the pressure after equilibrium.   In each case, the sample chamber was isolated before hydrogen

Table I.   Plateau Pressures of R_2Co_7–H Systems

System	Temperature (°C)	Plateau Pressure (atm) $\alpha + \beta$	Plateau Pressure (atm) $\beta + \alpha$
Ce_2Co_7–H	50	0.10	—
	100	0.80	—
	150	4.40	—
Pr_2Co_7–H	150	0.10	1.47
	175	0.29	3.50
	200	0.70	7.65
Nd_2Co_7–H	125	0.027	0.84
	150	0.101	2.45
	175	0.300	5.80
Gd_2Co_7–H	75	0.031	2.20
	100	0.122	5.35
	125	0.405	10.26
Tb_2Co_7–H	50	0.0265	1.28
	75	0.090	3.80
	101	0.300	9.95
Dy_2Co_7–H	50	0.078	2.25
	74.6	0.25	5.90
	101	0.78	14.0
Ho_2Co_7–H	50	0.225	4.1
	75	0.69	11.5
	101	1.86	26.0
Er_2Co_7–H	50	0.44	8.0
	75	1.20	18.5
	101	3.20	36.0

was added to the remaining portions of the system, then reopened so the entire system could equilibrate.

Thermodynamic Results

Pressure–composition isotherms for the Dy_2Co_7–H system, which are typical (17) of the R_2Co_7–H systems, are shown in Figure 1. Except for Ce_2Co_7–H, two plateau regions were found in all systems studied. This indicates that three crystallographically distinguishable hydrides exist and can be designated as α, β, and γ. The α phase is the solid solution phase based upon R_2Co_7, and γ is the

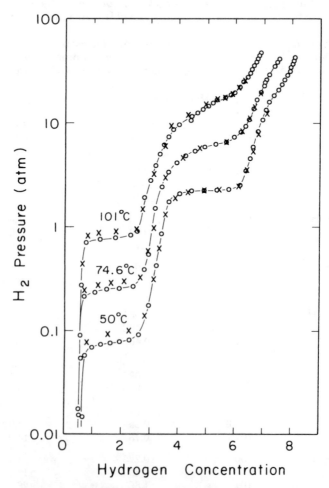

Figure 1. Pressure–composition isotherms for the Dy_2Co_7–H system. Hydrogen concentration is expressed as atoms of H per formula unit of Dy_2Co_7. × designates absorption; O designates desorption.

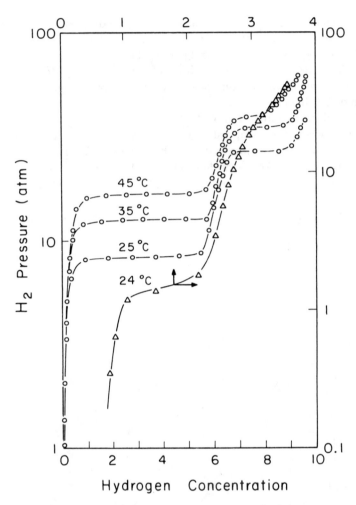

Figure 2.   Desorption pressure–composition isotherms for the Pr₂Ni₇–H system (O) and ErNi₃–H system (Δ) at 24°C. Hydrogen concentration is expressed as atoms of H per formula unit of intermetallic.

most hydrogen-rich phase.   Absorption and desorption isotherms were determined for Dy_2Co_7, Nd_2Co_7, and Ce_2Co_7, but only desorption isotherms were determined for the other R_2Co_7 compounds.   Plateau pressures (for desorption) for the R_2Co_7–H systems studied are given in Table I.   Desorption isotherms for the structurally related R_2Ni_7 compound, Pr_2Ni_7, were determined and are shown in Figure 2.   The R_2Co_7 and R_2Ni_7 systems are similar because they show the existence of three hydride phases.   A desorption isotherm for $ErNi_3$, which illustrates the RNi_3–H systems, is also shown in Figure 2.   The characteristics of this isotherm are similar to those for the RFe_3 compounds (11) in that only a single phase transition occurs within the pressure range studied.

Plots of log P vs $1/T$ for several R_2Co_7–H systems are shown in Figures 3 and 4. The ΔH values, giving the heats of desorption of hydrogen from the compound, were obtained from the slopes of the curves in Figures 3 and 4 and are listed in Table II.

The limiting composition of the R_2Co_7–H systems at 100 atm, from the pressure–concentration–temperature curves, is $R_2Co_7H_9$. On a per-atom basis, these compounds absorb more hydrogen than the RCo_5H_4 systems but less than the RCo_3H_5 systems. Also, hydride stability is inversely proportional to the Co–R ratio; hydride stability is in the order, $RCo_3 > R_2Co_7 > RCo_5$. Thus, the capacity

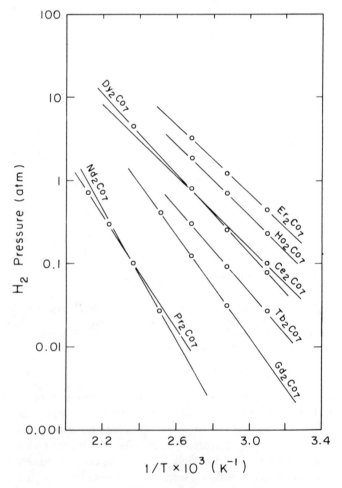

Figure 3. Temperature dependence on the plateau pressure ($\alpha + \beta$) for the R_2Co_7–H systems

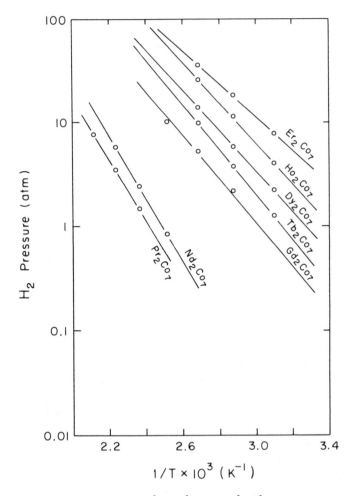

Figure 4. Temperature dependence on the plateau pressures
($\beta + \gamma$) for the R_2Co_7–H systems

for hydrogen absorption and hydride stability increases with increasing rare earth content of the hydride. Lundin et al. (*18*) noted a correlation between stability and the size of the interstitial hole where hydrogen presumably is located.

Absorption experiments for Nd_2Co_7 and Dy_2Co_7 show that the quantity $\Delta P/P$ is small (<0.20) for the $\alpha \rightarrow \beta$ transition in each system. P is the desorption plateau pressure (in atm), and ΔP is the difference between the absorption and desorption plateau pressures. $\Delta P/P$ at different temperatures for the same hydride essentially was constant. For the $\beta \rightarrow \gamma$ transition, surprisingly, no hysteresis was observed. The hysteretic behavior of the RCo_3 hydrides differed somewhat from the behavior of the R_2Co_7 hydrides in that $\Delta P/P$ was consistently larger for the $\beta \rightarrow \gamma$ transition than for the $\alpha \rightarrow \beta$ transition. For Ce_2Co_7, $\Delta P/P$

Table II.   Heats of Desorption of Hydrogen from R_2Co_7–H Systems

System	Phase	ΔH (kcal/mol of H_2)	Temperature Range (°C)
Ce_2Co_7–H	$\alpha + \beta$	10.4	50–150
Pr_2Co_7–H	$\alpha + \beta$	15.5	150–200
	$\beta + \gamma$	13.1	
Nd_2Co_7–H	$\alpha + \beta$	17.2	125–175
	$\beta + \gamma$	13.6	
Gd_2Co_7–H	$\alpha + \beta$	13.8	75–125
	$\beta + \gamma$	9.6	
Tb_2Co_7–H	$\alpha + \beta$	11.5	50–101
	$\beta + \gamma$	9.7	
Dy_2Co_7–H	$\alpha + \beta$	11.0	50–101
	$\beta + \gamma$	8.6	
Ho_2Co_7–H	$\alpha + \beta$	10.0	50–101
	$\beta + \gamma$	8.8	
Er_2Co_7–H	$\alpha + \beta$	9.4	50–101
	$\beta + \gamma$		

($\sim$2) was much larger than for the other R_2Co_7 hydrides; however, it essentially remained constant at the various temperatures.

Two systematic trends are observed from the experimental data. First, the plateau pressures (in atm) for the $ErFe_3$–, $ErCo_3$–, and $ErNi_3$–H systems at 25°C are 0.0067, 0.48, and 1.35, respectively. Hydride stability (stability is measured as the inverse of the hydrogen pressure) for a given rare earth metal in the RT_3 systems (T = Fe, Co, Ni) is in the order, $RFe_3 > RCo_3 > RNi_3$. Since the unit cell size of ErT_3 diminishes in the order, $ErFe_3 > ErCo_3 > ErNi_3$, one might attribute the systematics to the varying size of the interstitial sites occupied by hydrogen. However, electronic factors are also involved. Second, the hydride stabilities of the R_2Co_7 compounds are greater than those for the corresponding R_2Ni_7 compounds since the plateau pressures of Pr_2Ni_7 are much greater than those for Pr_2Co_7.

Effect of Hydrogenation on Lattice Parameters

X-ray diffraction patterns of the metal hydrides were obtained to determine the effect of dissolved hydrogen on the lattice dimensions and to confirm that the sample had not undergone decomposition. Results are presented in Table III. The R_2Co_7 materials are similar crystallographically to the RCo_3 materials in the following respects:

(1) Both form three hydrides (α, β, and γ) upon hydrogenation.

(2) Both systems retain their rhombohedral or hexagonal symmetry upon hydriding. The RCo_5–H systems, on the other hand, were found to be degraded

from the hexagonal CaCu$_5$ structure to an orthorhombic structure upon hydriding.

(3) Expansion takes place mainly along the c-direction upon formation of the β-hydride, but mainly in the basal plane upon formation of the γ-hydride. For the RCo$_5$ hydrides, expansion is in the basal plane with no appreciable change in the c-parameter for either the β- or γ-hydride.

Kinetic Features

Desorption kinetics were measured in the two-phase region and, in a few cases, in the single-phase region. The rate of hydrogen evolution from the hydride was determined gasometrically as a function of time. The compound ($\sim$2–4 g) was sealed in a copper sample chamber and immersed in a Hoskins tube furnace controlled to within $\pm 0.25\,^{\circ}$C by a Paktronics, Inc. temperature controller. The temperature of the sample was measured with a copper–constantan thermocouple placed in contact with the sample chamber. Temperatures were chosen so that the equilibrium vapor pressure in the two-phase region was in excess of 1 atm ($\sim$5–10 atm). All samples were activated before the experiment by sequentially absorbing and desorbing hydrogen at least three times to ensure

Table III. Lattice Parameters and Proton Densities in R$_2$Co$_7$–H Systems

Compound	a (Å)	c (Å)	c/a	V (Å^3)	%ΔV	H/cm^3 $\times 10^{22}$ [a]
Ce$_2$Co$_7$	4.497	24.496	4.952	519.2	—	—
Ce$_2$Co$_7$H$_{5.3}$	5.007	24.986	4.990	542.5	4.5	5.9
Pr$_2$Co$_7$	5.058	24.508	4.845	543.0	—	—
Pr$_2$Co$_7$H$_{2.5}$	5.081	26.300	5.176	588.0	8.3	1.7
Pr$_2$Co$_7$H$_{5.8}$	5.312	26.014	4.897	635.7	17.1	3.7
Nd$_2$Co$_7$	5.053	24.427	4.834	540.1	—	—
Nd$_2$Co$_7$H$_{2.7}$	5.069	26.286	5.186	584.9	8.3	1.9
Nd$_2$Co$_7$H$_{6.2}$	5.268	25.919	4.920	622.9	15.3	4.2
Gd$_2$Co$_7$	5.017	36.309	7.237	791.4	—	—
Gd$_2$Co$_7$H$_{2.6}$	5.012	39.043	7.790	849.3	7.3	1.8
Gd$_2$Co$_7$H$_{5.9}$	5.199	38.624	7.429	904.1	14.2	3.9
Tb$_2$Co$_7$	5.007	36.269	7.244	787.4	—	—
Tb$_2$Co$_7$H$_{2.7}$	5.011	38.964	7.776	847.3	7.6	1.9
Tb$_2$Co$_7$H$_{6.6}$	5.175	38.482	7.436	892.5	13.3	4.4
Dy$_2$Co$_7$	4.988	36.151	7.248	778.9	—	—
Dy$_2$Co$_7$H$_{2.6}$	4.984	38.699	7.765	832.5	6.9	1.9
Dy$_2$Co$_7$H$_{6.4}$	5.169	38.314	7.412	886.5	13.8	4.3
Ho$_2$Co$_7$	4.989	36.172	7.250	779.7	—	—
Er$_2$Co$_7$	4.963	35.886	7.231	765.5	—	—

[a] These can be compared with the proton density in liquid hydrogen, 4.2×10^{22} L/cm^2.

a constant active surface. Hydrogen was then allowed to flow freely from the sample into a gas-measuring buret. Since the sample size was small compared with the copper container, and was bathed in H_2 gas (a good heat conductor), it appeared that there was an adequate rate of heat transfer to the sample which minimized the temperature gradient between the sample's interior and the container wall. (This was inferred ·om the general performance of the equipment and the internal consistency of the data.)

The reaction order for decomposition of metal hydride was determined by fitting the experimental data to a rate equation of the general form:

$$-\frac{dC}{dt} - kC^n \tag{1}$$

where C is the amount of hydrogen in the hydride phase, k is the desorption rate constant, and n is the order of the reaction. In this rate equation, the reverse reaction, i.e., absorption, was not included. The desorption reaction was conducted against a pressure of 1 atm, which is much below the plateau pressures, so absorption can be neglected. For $n = 1$ and $n = 2$, Equation 1 becomes

$$\ln C_0 - \ln C = kt \quad \text{for } n = 1 \tag{2}$$

$$1/C - 1/C_0 = kt \quad \text{for } n = 2, \tag{3}$$

where C_0 is the initial quantity of hydrogen in the desorbing hydride phase. The experimental data were examined using these expressions. The second-order plots fit the data more satisfactorily than the first-order plots.

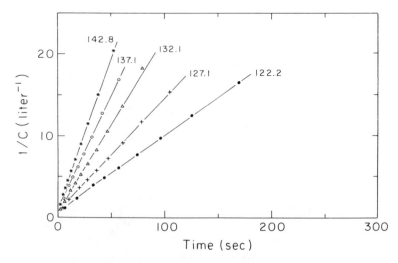

Figure 5.   Rate of hydrogen release from Dy_2Co_7–H on the $\alpha + \beta$ plateau.   C is a measure of the hydrogen concentration remaining in the intermetallic (in liters at $25°C$ and 1 atm).

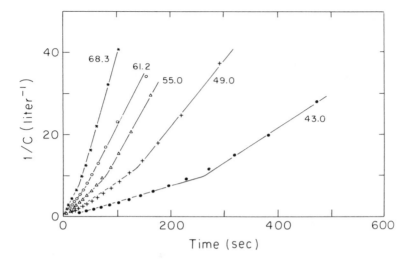

Figure 6. *Rate of hydrogen release from* Dy_2Co_7–H. *For short time intervals, the process converts the* γ *form into the* β *form. For longer times, the material only consists of the* β *form. The break in the curve corresponds to the disappearance of the* γ *form.*

Desorption kinetics are determined for Dy_2Co_7, Gd_2Co_7, $ErCo_3$, $DyCo_3$, $ErFe_3$, and $DyFe_3$. The results are illustrated by the Dy_2Co_7–H system (*see* Figures 5 and 6). Quantity C is the volume of hydrogen, measured at 1 atm and at $25°C$, that is retained in the sample at time t. C is proportional to the hydrogen concentration in the hydride. The data shown in Figure 5 are for the system in the $\alpha + \beta$ two-phase region, whereas those in Figure 6 involve the two-phase $\beta + \gamma$ region in the early reaction stages and the single-phase β region in the latter stages. The break in the curve corresponds to movement from the two-phase to the single-phase region. The curve plotted in Figure 5 is continuous since the variation in hydrogen content cannot sufficiently remove all of the β phase. Log k is linear with $1/T$ (*see* Figure 7). The slope indicates an activation energy of 23 kcal/mol for the Dy_2Co_7–H ($\beta \rightarrow \alpha$) system.

Kinetic data for all the R_2Co_7–H systems studied are summarized in Table IV. The observed behavior in all cases indicated a second-order process. The activation energies exceed the ΔH values given in Table II, as expected of the endothermal nature of the process. In most cases, E_a is two or more times as large as ΔH. More work is necessary to establish full details of the dehydrogenation process since a sequence of steps is involved. Hydrogen migration in the lattice, conversion of one phase into another, recombination of atomic hydrogen at the surface of the metal, and release of molecular hydrogen into the gas phase should be considered. Diffusion of hydrogen most likely is not rate controlling. The second-order nature of the process and the magnitude of the activation energy indicate that the process is not diffusion controlled. Hydrogen diffusion in metals

Table IV. Kinetic Data for the Desorption of Hydrogen from Selected Rare Earth Intermetallics

System	$E_a(kcal/mol)$	Temperature(°C)	$k(net) \times 10^5$ $(l^{-1}sec^{-1})$
Gd_2Co_7–H ($\alpha + \beta$)	36	160.0	5.61
		165.0	12.04
		172.1	23.4
		178.0	43.2
		186.0	59.5
Gd_2Co_7–H ($\beta + \gamma$)	18	68.2	4.15
		75.0	12.8
		81.1	19.6
		87.8	29.1
		99.2	47.4
Dy_2Co_7–H ($\alpha + \beta$)	23	122.2	9.4
		127.1	11.9
		132.1	22.1
		137.1	28.4
		142.8	37.2
Dy_2Co_7–H ($\beta + \gamma$)	15	43.0	3.87
		49.0	7.84
		55.0	11.3
		61.2	16.9
		68.3	26.5
$ErCo_3$ ($\alpha + \beta$)	20	103.5	1.10
		109.0	2.10
		115.4	3.28
		120.7	4.38
		126.2	5.21
$ErCo_3$ ($\beta + \gamma$)	18	53.0	.519
		58.2	.949
		65.2	1.68
		69.1	2.13
		73.0	2.56
$DyCo_3$ ($\alpha + \beta$)	25	139.9	1.18
		145.6	2.45
		151.6	3.77
		156.1	4.80
		161.2	5.50
$DyCo_3$ ($\beta + \gamma$)	16	79.1	0.57
		86.1	1.28
		93.6	2.00
		100.0	2.47
		107.0	3.49
$ErFe_3$	23	159.4	7.02
		163.9	11.63
		169.4	14.85

Table IV. Continued

System	$E_a(kcal/mol)$	Temperature(°C)	$k(net) \times 10^5$ $(l^{-1}sec^{-1})$
		175.0	20.32
		180.1	24.23
		186.5	35.38
DyFe$_3$	25	194.2	2.57
		200.0	4.44
		204.7	5.76
		209.4	7.39
		216.8	7.90

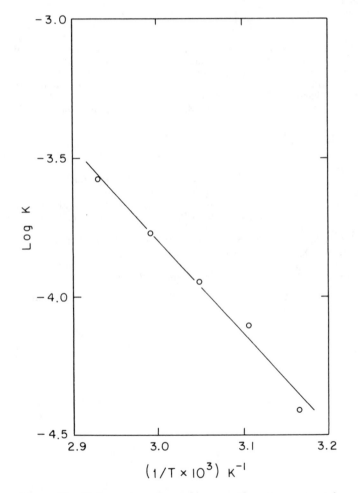

Figure 7. Temperature dependence on the rate constant for hydrogen release from the Dy$_2$Co$_7$-H system on the β + γ plateau

involves an activation energy from 3-12 kcal/mol (19). The second-order nature of the desorption process suggests that atomic hydrogen recombination at the surface is rate controlling.

Literature Cited

1. Zijlstra, H., Westendorp, F. F., *Solid State Commun.* (1969) 7, 857.
2. Neumann, H. H., Ph.D. thesis, Technische Hochschule, Darmstadt, Germany, 1969.
3. Van Vucht, J. H. N., Kuijpers, F. A., Bruning, H. C. A. M., *Philips Res. Rep.* (1970) 25, 133.
4. Kuijpers, F. A., Ph.D. thesis, University of Delft, Holland, 1973.
5. Wallace, W. E., "Rare Earth Intermetallics," Academic, New York, 1973.
6. Takeshita, T., Wallace, W. E., Craig, R. S., *J. Catal.* (1976) 44, 236.
7. Coon, V. T., Takeshita, T., Wallace, W. E., Craig, R. S., *J. Phys. Chem.* (1976) 80, 1878.
8. Elattar, A., Takeshita, T., Wallace, W. E., Craig, R. S., *Science* (1977) 196, 1093.
9. Wallace, W. E., Elattar, A., Takeshita, T., Coon, V., Bechman, C. A., Craig, R. S., "Proceedings of the 2nd International Conference on the Electronic Structure of the Actinides," J. Mulak, W. Suski, R. Troć, Ed., p. 357, Polish Academy of Sciences, Warsaw, 1977.
10. Van Mal, H. H., Buschow, K. H. J., Miedema, A. R., *J. Less-Common Met.* (1976) 49, 473.
11. Bechman, C. A., Goudy, A., Takeshita, T., Wallace, W. E., Craig, R. S., *Inorg. Chem.* (1976) 15, 2184.
12. Takeshita, T., Wallace, W. E., Craig, R. S., *Inorg. Chem.* (1974) 13, 2282 and 2283.
13. Raichlen, J. S., Doremus, R. H., *J. Appl. Phys.* (1971) 42, 3166.
14. Boser, O., *J. Less-Common Met.* (1976) 46, 91.
15. Lundin, C. E., Lynch, F. E., Denver Research Institute, University of Denver, AFOSR Contract No. F44620-74-C-002 (ARPA Order 2552), January 1975, First Annual Report; May 1976, Final Report.
16. Gruen, D. M., Mendelsohn, M., ADV. CHEM. SER. (1978) 167, 327.
17. Goudy, A., Ph.D. thesis, University of Pittsburgh (1976).
18. Lundin, C. E., Lynch, F. E., Magee, C. B., *J. Less-Common Met.*
19. Birnbaum, H. K., Wert, C. A., Ber. Busenges, *Phys. Chem.* (1972) 76, 806.

RECEIVED July 19, 1977. Work was partially assisted by a contract with the Energy Research and Development Administration.

Stability Considerations of AB$_5$ Hydrides in Chemical Heat Pump Applications with Reference to the New LaNi$_{5-x}$Al$_x$ Ternary System

DIETER M. GRUEN and MARSHALL H. MENDELSOHN

Chemistry Division, Argonne National Laboratory, Argonne, IL 60439

AUSTIN E. DWIGHT[1]

Materials Science Division, Argonne National Laboratory, Argonne, IL 60439

Alloys of the general composition AB$_5$, that react rapidly and reversibly with large quantities of hydrogen gas, can be used in chemical heat pump systems. The theoretical operating temperatures of these heat pumps are determined by the thermodynamic values of the enthalpy (ΔH) and the entropy (ΔS) of the reactions of hydrogen with metal alloys. A configurational entropy model was developed to account for experimentally observed differences in ΔS. Because the exact theory of dissociation pressures is not fully understood, an empirical correlation relating the alloy cell volumes to the plateau pressures is useful. A new ternary alloy system (composition LaNi$_{5-x}$Al$_x$), that spans a wide range of pressures without greatly impairing the desirable properties of the alloy LaNi$_5$, was developed.

The discovery of the ability of AB$_5$ compounds (A = rare earth, B = transition metal) to form hydrides with unique properties stimulated research and development in several areas (*1, 2, 3, 4*). Although the physicochemical properties of some AB$_5$ hydrides were measured, the factors that determine the hydrogen dissociation pressures of these materials are not completely understood.

The rapid kinetics of Reaction 1, the high volumetric hydrogen storage densities, and the wide range of hydrogen decomposition pressures of the AB$_5$ hydrides initiated proposals to use them as chemical compressors, cryogenic

---

[1] Present address: Department of Physics, Northern Illinois University, Dekalb, IL 60115

0-8412-0390-3/78/33-167-327/$05.00/0

devices, and as media for thermal energy storage, space heating and cooling, and conversion of low grade heat to work (5).

$$AB_5 + nH_2 \leftrightarrow AB_5H_{2n} \tag{1}$$

The use of AB_5 hydrides as chemically or thermally driven heat pumps is intriguing (6). Since two different AB_5 alloys are involved, the relationship between their respective hydrogen decomposition pressures, as a function of temperature, is the key parameter that determines the thermodynamics of heat pump action.

It is crucial to discover the relationship between chemical compositions and hydrogen decomposition pressures of the AB_5 compounds. Intermetallic compounds of lanthanide and transition metals form an interesting class of structures. The AB_5 series crystallize in the hexagonal $CaCu_5$ (P6/mmm) structure (*see* Figure 1). Generally, radius ratios (r_A/r_B) greater than 1.30 form the $CaCu_5$-type

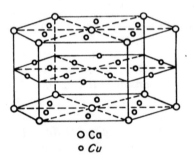

Figure 1. $CaCu_5$ type of   o Ca
structure   o Cu

configuration and compounds with ratios less than 1.30 form the cubic UNi_5 structure. Compounds formed by rare earths to the right of lanthanum in the periodic table display the lanthanide contraction that decreases the AB_5 unit cell volume. The AB_5 phase is generally stable over the composition range ($AB_{4.8-5.2}$) (5, 7).

Hexagonal AB_5 compounds form orthorhombic hydrides. Basal plane expansion, caused by hydrogen occupation of interstitial sites (4), change the compound's structure and can be ordered or disordered. For example, Kuijpers and Loopstra (4) found, by neutron diffraction, that the deuterium atoms in $PrCo_5D_4$ were ordered on certain interstitial octahedral and tetrahedral sites. The sites occupied by hydrogen in various AB_5 hydrides are not fully understood because of insufficient neutron diffraction data. The available information is considered in the section on configurational entropies.

Thermodynamic Properties of AB_5 Hydrides

The equation

$$RT \ln P = \Delta G = \Delta H - T\Delta S \tag{2}$$

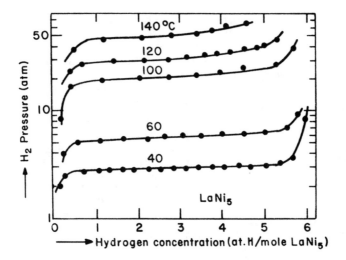

Figure 2.   Pressure vs. composition for LaNi₅

expresses the relationships among the dissociation pressure (P), the free energy of formation (ΔG), the enthalpy change (ΔH), and the entropy change (ΔS) of Reaction 1.

Phase diagrams consisting of pressure–composition curves are available for many intermetallic hydrides (5). In some cases ($CeCo_5$, $SmCo_5$, and YCo_5), single plateaus approach the composition AB_5H_3. In other cases ($PrCo_5$ and $NdCo_5$), two plateaus exist with a maximum composition approaching AB_5H_4. Certain AB_5 intermetallics possessing one ($LaNi_5$), two ($NdNi_5$), or three ($LaCo_5$) plateaus absorb more than four atoms of hydrogen. $LaCo_5$ absorbs nine atoms at pressures approaching 1250 atm (8).

Data are available on quaternary hydrides, where another element is partially substituted at the A or B sites (9). Desorption isotherms were measured for $La_{1-x}Y_xNi_5$, where x = 0.3, 0.4, 0.5 (10), and for $Mm_{1-x}Ca_xNi_5$, where x = 0–1 at room temperature (11). Several isotherms were reported for $LaNi_4Cu$ and one for $LaNi_3Cu_2$ (12), while van Mal et al. have shown 40°C isotherms in the series $LaCo_{5-x}Ni_x$ for x = 0–5 (13).

Typical pressure–composition isotherms for the $LaNi_5$–H_2 system are shown in Figure 2 (14). From plots of $\ln P_{plateau}$ vs. $1/T$, several workers determined the experimental heats and entropies of Reaction 1. The enthalpies are the heats of formation for the β-phase hydride from the α-phase hydrogen-saturated alloy (*see* Table I).

Ternary Metal Hydrides as Chemical Heat Pumps

The rapid kinetics of Reaction 1, the ability to vary the hydrogen dissociation pressures by controlling the chemical composition of the A or B component, and

Table I. Representative Thermodynamic Data

Compound (α-phase)	Compound (β-phase) (limiting composi-tion)	$-\Delta H$ (kcal/ mol of H_2)	$-\Delta S$ (cal/deg mol of H_2)	Ref.	$(\Delta S_{CeCo_5H_3} - \Delta S_c)/$ R	Struc-ture β-phase[c]
$CeCo_5H_{0.6}$	$CeCo_5H_3$	9.3	32.6	16	0	β^{III}
$YCo_5H_{0.3}$[a]	YCo_5H_3	7.7	31.8[b]	17	−0.4	
$PrCo_5H_{0.3}$	$PrCo_5H_{3.5}$	9.2	30.1	16	−1.2	β^{II}
$LaCo_5H_{0.17}$	$LaCo_5H_4$	10.8	30.0	16	−1.3	β^{I}
$SmCo_5H_{0.3}$	$SmCo_5H_3$	7.8	29.0	16	−1.8	
$NdNi_5H_{0.5}$[a]	$NdNi_5H_{4.5}$	6.7	27.7	26	−2.5	
$LaNi_4CuH_{0.5}$[a]	$LaNi_4CuH_6$	8.1	27.1	26	−2.8	
$LaNi_5H_{0.5}$[a]	$LaNi_5H_{6.5}$	7.2	26.1	18	−3.3	

[a] Estimated from P–C diagrams.
[b] Calculated from equation: $\log P = 6.96 - 1679/T$.
[c] See Ref. 4.

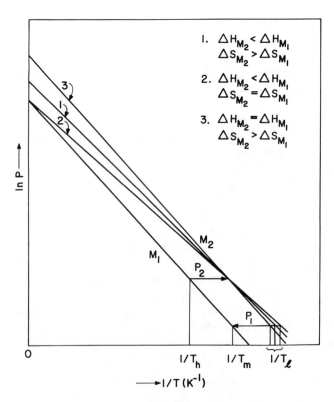

Figure 3.   Three examples of chemical heat pump thermodynamic relationships

the high hydrogen-carrying capacity of the AB$_5$ alloys suggest that these materials are useful in chemical heat pump applications (6).

The thermal energy of hydrogen absorption and desorption processes suggests the possibility of pumping heat from a low temperature to an intermediate temperature, by using a higher temperature source and appropriate pairs of metal–hydrogen systems. In such thermally driven heat pumps used for cooling or heating purposes, thermal energy does not have to be converted first to mechanical energy to drive, for example, a compression refrigerator. This avoids moving parts and vibration. More importantly, low grade heat sources, such as provided by solar collectors, could power these chemical heat pumps.

Suppose a thermally driven heat pump operates at temperatures T_h, T_m, and T_l. At the high temperature T_h, heat Q_h is supplied to the heat pump; at temperature T_m, heat Q_m is generated by the heat pump; and at the temperature T_l, heat Q_l is extracted from a low temperature source. For a Carnot cycle (reversible process) the following relations hold:

$$Q_m = Q_h + Q_l \tag{3}$$

$$Q_m/T_m = Q_h/T_h + Q_l/T_l \tag{4}$$

Substitution gives the expression for the efficiency of a Carnot process:

$$\eta_c = Q_m/Q_h = (1 - T_l/T_h)(1 - T_l/T_m)^{-1} \tag{5}$$

Consider a pair of different metal hydrides, M_1 and M_2, with hydrogen gas flowing freely between them. Suppose hydride M_2 operates between the temperatures T_l and T_m with corresponding pressures P_1 and P_2 (*see* Figure 3). Straight lines in Figure 3 are plots of ln P as $1/T$. Equilibrium dissociation pressures in the two-phase region are given by enthalpies ΔH_{M_1} and ΔH_{M_2} of Reaction 1, and the intercepts are given by entropies ΔS_{M_1} and ΔS_{M_2}.

To operate hydrides M_1 and M_2 in the heat pump mode, begin with M_1 saturated with hydrogen at T_m and P_1. Heat M_1 to T_h to raise the pressure to P_2 and to desorb the hydrogen. The released hydrogen is absorbed by M_2 at T_m and P_2, and the heat of absorption, ΔH_{M_2}, is rejected at T_m. The temperature of M_1 is lowered to T_m and M_2 to T_l. Hydrogen desorbing from M_2 at T_l absorbs heat from the environment while the heat of reaction, ΔH_{M_1}, is rejected at T_m. Hydrogen gas is the working fluid in this closed cycle which can be repeated indefinitely.

To choose metal hydride pairs efficiently for heat pump operation, the relationships among the thermodynamic quantities that govern Reaction 1 and the values for T_h, T_m, and T_l should be established. Three examples for the relationships between thermodynamic quantities of M_1 and M_2 are shown in Figure 3. These are the only possibilities that simultaneously satisfy the condition: $P(M_2) > P(M_1)$ for any given $1/T > 0$.

1. $\Delta H_{M_2} < \Delta H_{M_1}$; $\Delta S_{M_2} > \Delta S_{M_1}$

2. $\Delta H_{M_2} < \Delta H_{M_1}$; $\Delta S_{M_2} = \Delta S_{M_1}$

3. $\Delta H_{M_2} = \Delta H_{M_1}$; $\Delta S_{M_2} > \Delta S_{M_1}$

For the general case, $\Delta H_{M_2} \neq \Delta H_{M_1}$, $\Delta S_{M_2} \neq \Delta S_{M_1}$:

$$-R \ln P_2 = \frac{\Delta G_{M_2}}{T_m} = \frac{\Delta G_{M_1}}{T_h} \tag{6}$$

$$-R \ln P_2 = \frac{\Delta G_{M_2}}{T_1} = \frac{\Delta G_{M_1}}{T_m} \tag{7}$$

Straightforward substitution and rearrangement shows

$$\frac{\Delta H_{M_2}}{\Delta H_{M_1}} - \frac{T_m(\Delta S_{M_2} - \Delta S_{M_1})}{\Delta H_{M_1}} = \frac{T_m}{T_h} \tag{8}$$

$$\frac{\Delta H_{M_2}}{\Delta H_{M_1}} - \frac{T_1(\Delta S_{M_2} - \Delta S_{M_1})}{\Delta H_{M_1}} = \frac{T_1}{T_m} \tag{9}$$

For the special case, $\Delta S_{M_2} = \Delta S_{M_1}$:

$$\frac{\Delta H_{M_2}}{\Delta H_{M_1}} = \frac{T_m}{T_h} = \frac{T_1}{T_m} \tag{10}$$

from which it follows that

$$T_m^2 = T_h T_1 \tag{11}$$

For the special case, $\Delta H_{M_2} = \Delta H_{M_1} = \Delta H$:

$$\frac{\Delta S_{M_2} - \Delta S_{M_1}}{\Delta H} = \frac{1}{T_m} - \frac{1}{T_h} \tag{12} \text{ (from (8))}$$

$$\frac{\Delta S_{M_2} - \Delta S_{M_1}}{\Delta H} = \frac{1}{T_1} - \frac{1}{T_m} \tag{13} \text{ (from (9))}$$

from which it follows that

$$\frac{T_m}{T_h} + \frac{T_m}{T_1} = 2 \tag{14}$$

Equations 8 and 9 are used to calculate T_h and T_1 for a range of T_m values if ΔH and ΔS values for a pair of metal hydrides are given. Such calculations were made for the CaNi$_5$–hydride pair ($\Delta H = -7.55$ kcal, $\Delta S = -23.9$ cal/deg) and the LaNi$_5$–hydride pair ($\Delta H = -7.20$ kcal, $\Delta S = -26.1$ cal/deg). Results are listed in Table II. Calculations of T_1 (using the T_m and T_h values) for the special cases $\Delta S_{M_2} = \Delta S_{M_1}$ and $\Delta H_{M_2} = \Delta H_{M_1}$ are listed in Table II. As shown in Figure 3 for fixed T_m and T_h, the lowest refrigeration temperature (T_1) is reached when $\Delta S_{M_2} = \Delta S_{M_1}$ while the least effective heat pump action occurs when $\Delta H_{M_2} = \Delta H_{M_1}$. The differences in achievable refrigeration temperatures are $\sim$10% of ($T_m - T_1$) and therefore help to determine overall cycle efficiency.

Table II. Examples of Chemical Heat Pump Operating Temperatures

Hydride Pair Involving LaNi$_5$ and CaNi$_5$			T_m and T_h Values of Columns 1 and 2	
			$\Delta S_{M_2} = \Delta S_{M_1}$	$\Delta H_{M_2} = \Delta H_{M_1}$
$T_m(°C)$	$T_h(°C)$	$T_l(°C)$	$T_l(°C)$	$T_l(°C)$
20	64.5	−15.6	−18.6	−14.1
30	77.2	−7.5	−10.8	−6.0
40	90.0	+0.5	−3.1	+2.1
50	102.8	+8.5	+4.6	+10.2
60	115.8	+16.4	+12.2	+18.2

Rule of Reversed Stability

A scheme for predicting ternary hydride stability was proposed by workers at the Phillips Research Laboratories (*9, 15, 16, 17*). According to this scheme, the heat of formation of a ternary hydride AB$_n$H$_{2m}$ is given by:

$$\Delta H(AB_n H_{2m}) = \Delta H(AH_m) + \Delta H(H_n H_m) - \Delta H(AB_n) \qquad (15)$$

Most heats of formation of binary hydrides were determined experimentally, and they are used in Equation 15 with calculated $\Delta H(AB_n)$ values. Miedema's (*18*) approach for calculating $\Delta H(AB_n)$ is summarized by Stewart, Lakner, and Uribe (*8*):

"Assumes that the driving force for reactions between metals is a function of two factors: a negative one, arising from the difference in chemical potential, $\Delta \phi^*$, of electrons associated with each metal atom, and a positive one that is the difference in the electron density, Δn_{ws}, at the boundaries of Wigner–Seitz type cells surrounding each atom. Values of ϕ^* for the metals are approximated by the electronic work functions; n_{ws} is estimated from compressibility data. The atomic concentrations in the alloy must be included in the calculation."

The equation for $\Delta H(AB_n)$ calculations (*17*) is

$$\Delta H(AB_n) = N_o f(C_A^s, C_B^s) g P \left[-e(\Delta \phi_o^*)^2 + \frac{Q_o}{P}(\Delta n_{ws})^{2/3} - \frac{R}{P} \right] \qquad (16)$$

where C_A^s and C_B^s = surface concentrations of each component; P, Q_o, and R = constants that vary between systems of alloys, i.e., transition metal-transition metal alloys, transition-metal-p-metal alloys, etc.; e = electronic charge; N_o = Avogadro's number; and g is a function of the metal parameters n_{ws}, $V_m^{2/3}$, and the atomic concentrations (*17*). Values of ϕ^* and n_{ws} (*18*) and the constants P, Q_o, and R are listed for each of the systems studied.

The rule of reversed stability is illustrated by Equation 15. When the stability of the intermetallic compound AB$_n$ is great, the stability of the ternary hydride is low; therefore, the hydrogen dissociation pressure of the hydride is high (*8*). The rule of reversed stability aids in calculating approximate values of ΔH for hydride formation. However, recent attempts to perform such calculations (8) by using estimated entropy values of binary hydrides have shown that reliable estimates of ternary hydride equilibrium pressures cannot be made. The semi-empirically calculated $\Delta H(AB_n H_{2m})$ values represent the enthalpies for the phase changes in the plateau regions and are related to the equilibrium hydrogen pressures. The calculations of Steward, Lakner, and Uribe (8) show that plateau pressures are lower than the experimental values by factors ranging from 10^6 to 10^{14}. To obtain reliable estimates of ternary hydride equilibrium pressures, the rule of reversed stability should be modified or a new method should be devised.

Configurational Entropies and Stabilities of Intermetallic Hydrides

The rule of reversed stability regards the enthalpy as a direct measure of the relative stability of metal–hydrogen phases since it was assumed that the entropy changes of Reaction 1 are relatively constant (9). To calculate the free energies of Reaction 1, Steward et al. estimate entropy changes and assume that they depend only on the M:H ratio (8). Entropy changes in Reaction 1 are almost certainly dominated by the entropy of gaseous hydrogen (31 cal/deg mol H_2) since the contribution of hydrogen atoms to the lattice vibrational entropy of the solid AB_5 hydride is probably quite small near room temperature. The main effect is that an optical phonon branch is added at fairly high frequencies. The change in lattice vibrational entropy because of the lanthanide contraction would be small since the volume change in the two extreme cases of $LaCo_5$ and YCo_5 is only 6%, which is about one-half of that for the elements La and Y.

However, the experimental data (see Table I) show that entropy changes (ΔS) in a series of reactions of hydrogen with AB_5 compounds differ by up to 6.5 eu/mol H_2 at $300°K$. This gives differences of ~ 2 kcal/mol H_2 in the free energies of Reaction 1 and more than one order of magnitude change in the hydrogen dissociation pressure (20). Differences in entropy changes for a homologous series of Reaction 1 are based on changes in the configurational entropies of the AB_5H_n ternary hydrides; therefore, the configurational entropies are important in determining H_2 dissociation pressures, and future attempts to explain the theoretical predictions of ternary hydride dissociation pressures should account for the relationship between the hydride structures and their thermodynamic stabilities.

Hydrogen atoms in the $RECo_5H_n$ ($n \leq 4$) hydrides occupy sixfold tetrahedral sites with two RE and two Co, and threefold octahedral sites with two RE and four Co (4). For higher stoichiometry hydrides (AB_5H_n, $n > 4$), an additional two sets of twelvefold tetrahedral sites can be occupied (21). Configurational entropies ($S^{conf} = k \ln w$) were calculated on the assumption that the tetrahedral and octahedral sites are distinguishable and are equally available to hydrogen at temperatures near $300°K$ and that occupation of a particular site does not block other available sites. The number of possible hydrogen arrangements in a g-atom of metal was obtained by multiplying two combinatorial formulas

$$w = \frac{(XN)!}{(\theta XN)![(1 - \theta)XN]!} \tag{17}$$

where $X = 3$ for the octahedral sites and $X = 6$ for the tetrahedral sites in the RECo$_5$H$_n$ ($n \leq 4$) compounds; N is Avogadro's number and θ is the occupied fraction of the XN available sites (22). Several models fit the data for the compounds AB$_5$H$_n$ ($n > 4$). For one model, we used $X = 3$ for the octahedral sites and $X = 30$ for the tetrahedral sites. Equation 17 simplifies via Stirling's approximation to

$$\ln w = -XN[\theta \ln \theta + (1 - \theta) \ln (1 - \theta)] \tag{18}$$

Differences in $\Delta S^{conf}_{\alpha \to \beta}$ for different AB$_5$H$_n$ compounds compared with $\Delta S^{conf}_{\alpha \to \beta}$ for CeCo$_5$H$_3$ are listed in Table III. The values of these numbers (*see* Table III), calculated using the fractional site occupations for the β phase, can be compared with the experimentally determined entropy differences listed in Table I. The calculated configurational entropy differences (*see* Table III) agree satisfactorily with the experimental data (*see* Table I) currently available for seven AB$_5$H$_n$ compounds. Structures of some AB$_5$H$_n$ compounds deduced from neutron diffraction data (4) are listed in Table I. For compounds whose structures have not been determined, the occupation numbers listed in Table III are in best agreement with the thermodynamic data.

Although the maximum enthalpy difference for the AB$_5$H$_n$ group is 4.1 kcal/mol H$_2$, which is larger than the maximum entropy difference of 2.0 kcal/mol H$_2$ (at 300°K), the latter quantity helps determine the dissociation pressure ratio of two AB$_5$H$_n$ compounds at a given temperature. For example,

Table III. Calculated Configurational Entropies

Alloy	S^{conf}_α (cal/deg)	S^{conf}_β (cal/deg)	$S^{conf}_\beta - S^{conf}_\alpha = \Delta S^{conf}$ (mol of H$_2$)	$\Delta S^{conf}_{CeCo_5H_3} - \Delta S^{conf}_C$ (R)	β-phase occupancy
					3 oct.—6 tet.
CeCo$_5$	3.9	7.6	3.1	0.0	1/3^{ord}—1/3
YCo$_5$	2.4	7.6	3.9	−0.4	1/3^{ord}—1/3
PrCo$_5$	2.4	11.7	5.8	−1.3	1/2—1/3
LaCo$_5$	1.5	11.4	5.2	−1.1	2/3—1/3
SmCo$_5$	2.4	11.4	6.7	−1.8	1/3—1/3
					3 oct.—30 tet.
NdNi$_5$	3.4	19.4	8.0	−2.5	1/2^{ord}—1/10
LaNi$_4$Cu	3.4	27.2	8.7	−2.8	2/3—2/15
LaNi$_5$	3.4	32.5	9.7	−3.3	1/6—1/5

based on the enthalpy difference alone, the H_2 dissociation pressure of YCo_5H_3 at $300°K$ (~20 atm) should be comparable with that of $LaNi_5 H_6$ (~2 atm). The H_2 dissociation pressures differ by about one order of magnitude. The lower configurational entropy of the former compound is responsible for this behavior.

Although the model of three octahedral and six tetrahedral sites seems satisfactory for hydrides (AB_5H_n) where $n \leq 4$, several models yield comparable results for the three hydrides where $n > 4$. To select the best model, additional neutron diffraction data for hydrides with $n > 4$ would be useful.

Hydrogen Dissociation Pressure—Cell Volume Correlation

Considerations of enthalpy, as illustrated by the rule of reversed stability, and of configurational entropy provided insight into the factors governing the stabilities of the AB_5 hydrides. Theoretical understanding to predict dissociation pressures should be developed on heat pump application, for example. Although

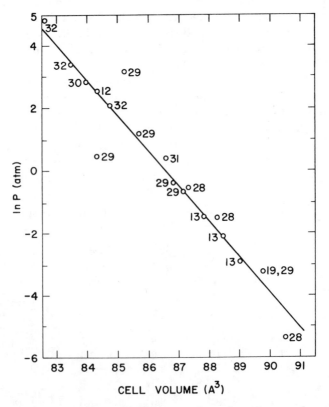

Figure 4. Alloy cell volume vs. ln P for AB$_5$ type hydrides (reference numbers given on figure)

Table IV.   Plateau Pressure and Alloy Cell Volume Data

Compound	P_{eq} (at 20°C)(atm)	Reference	Cell Volume (Å³)
LaCo₅	0.04	19, 29	89.74
CeCo₅	1.55	29	84.30
PrCo₅	0.51	29	87.13
NdCo₅	0.68	29	86.79
SmCo₅	3.30	29	85.67
GdCo₅	24	29	85.19
YCo₅	17.1[a]	30	83.96
LaNi₅	1.5[a]	31	86.54
PrNi₅	8[b]	32	84.73
SmNi₅	30[b]	32	83.44
GdNi₅	120[b]	32	82.58
LaCo₄Ni	0.055[c]	13	89.01
LaCo₃Ni₂	0.12[c]	13	88.44
LaCo₂Ni₃	0.22[c]	13	87.80
NdNi₅	12.7[a]	12	84.32
LaNi₄.₈Al₀.₂	0.57[a]	28	87.28
LaNi₄.₆Al₀.₄	0.22[a]	28	88.24
LaNi₄Al	0.005	28	90.51

[a] Calculated from given data or equation.
[b] Given at 23°C only.
[c] Estimated from data at 40°C.

an adequate theory does not exist, attempts were made to correlate dissociation pressures empirically with other observations.

Geometrical factors are important in determining transition metal hydride stability (*24*).   Recently, other authors cited the importance of crystal structure and geometrical factors to the affinity for hydrogen and the stability of metal alloy–hydrogen systems (*23, 24, 25, 26*).   Lundin et al. have shown a correlation of interstitial hole volume vs. the logarithm of hydrogen dissociation pressure (*26*).   Prediction of the proper alloy composition a priori would be helpful in searching for alloys with a particular plateau pressure at a given temperature. These considerations led to a direct correlation between crystal cell volumes and equilibrium plateau pressures (*12*).   Figure 4 shows a linear plot of the ln P_{plateau} vs. cell volume for room temperature data (*see* Table IV).

The theoretical explanation of this remarkable correlation is unclear.   Note however, that $\Delta n_{\text{ws}}^{2/3}$ in Equation 16 is proportional to the molar volumes ($V_m^{-4/3}$) of the alloys' elemental constituents.

The LaNi₅₋ₓAlₓ System: Versatile New Ternary Alloys for Metal Hydride Applications

Many AB₅ hydrides have $\Delta H \sim 8$ kcal so that at 0–100°C (of interest for chemical heat pump action), the hydrogen dissociation pressures change roughly by one order of magnitude.   This implies that a metal hydride pair could function in the heat pump mode even if their enthalpies were exactly the same, provided that their entropies of formation differed by ~5 eu. Discussion of configurational entropies showed that entropy differences of this magnitude occur among the AB_5 hydrides. Therefore, entropy considerations are important in selection of candidates for heat pump pairs, particularly since cycle efficiency criteria also depend on entropy.

For successful use of metal hydrides as chemical heat pumps, alloy systems should allow for changes in the free energies of formation continuously over a wide range. Given a particular hydride, it would be possible to specify ΔH and ΔS of a second hydride to reach a given T_h and T_l. Recent work in our laboratory has shown that $LaNi_{5-x}Al_x$ alloys provide a close approach to such a system (27).

X-ray diffraction measurements were made on some $LaNi_{5-x}Al_x$ alloys. Results show the a_0 and c_0 lattice parameters and the crystal cell volumes as a function of the aluminum concentration (see Figure 5). Substitution of one Al atom into the five c and g sites results in a sharp increase in c_0 and a smaller increase in a_0. Intensity calculations show that aluminum prefers to enter the $3g$ sites rather than the $2c$ sites. This finding is consistent with an earlier investigation of the YCo_5-YNi_5 system (28). Further aluminum substitution causes c_0 to level off but a_0 increases at a faster rate. The initial increase in c_0 must result from a lengthening of the bond between layers, while the subsequent leveling off in c_0 results from the tendency for the large La atom to collapse into the hexagonal ring of $Ni(Al)$ atoms above and below it.

Using the cell volume–decomposition pressure correlation, one predicts that aluminum substitutions in $LaNi_5$ should lower the hydrogen decomposition pressures by one order of magnitude for every two-Å^3 increase in cell volume (12).

Aluminum substitution for nickel in $LaNi_5$ lowers decomposition pressures without impairing the kinetics or the hydrogen-carrying capacity. They allow for continuous spanning of a wide range of decomposition pressures. In 0–20% Al, the plateau pressures of the $LaNi_5-LaNi_4Al$ hydride system are reduced by ~300 (27).

Alloys were prepared from metals of 99.9% purity by arc melting on a water-cooled copper hearth under an argon atmosphere. The alloys were homogenized at 800°C. Diffraction patterns of cast material were equally as sharp as those of homogenized alloys. X-ray diffraction patterns were taken with filtered FeK_α radiation. Computer programs verified x-ray pattern indexes.

Alloy samples, weighed to ±0.0001 g, were placed in a stainless steel type 316 reactor fitted with 1μ porous stainless steel filter disc. The reactor was connected to a stainless steel (type 316), high pressure manifold connected to a 0–2000 psia Heise pressure gage (accuracy ±0.1%), a vacuum pump, and a high pressure hydrogen gas cylinder. Hydrogen was Matheson's prepurified grade (99.95% min). Hoke valves rated to 3000 psi were used. The reactor was immersed in a water bath whose temperature was maintained to ±0.2°C by a Thermistemp temperature controller (for temperatures <80°C) and by a ni-

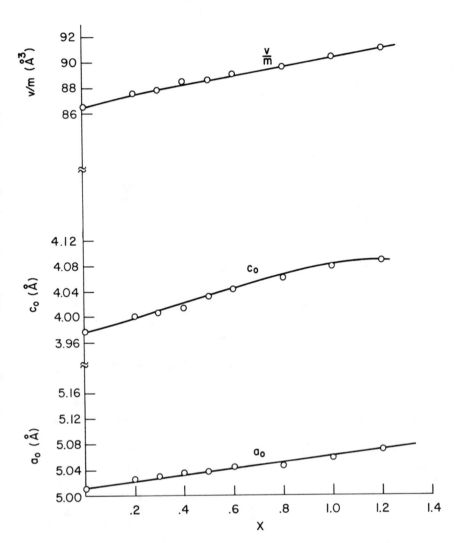

Figure 5.   Parameters $a_o, c_o,$ and v/M vs. x in $LaNi_{5-x}Al_x$ alloys

chrome-wound cylindrical furnace (for temperatures $\geq 80°C$). Samples were activated by evacuating the reactor containing the alloy and exposing the sample to high pressure hydrogen (300–800 psia) for 1–2 hr.

Pressure–composition data were obtained for $LaNi_{4.8}Al_{0.2}$, $LaNi_{4.6}Al_{0.4}$, $LaNi_4Al$, and $LaNi_{3.5}Al_{1.5}$. Preliminary data at several temperatures allow us to plot ln P vs. $1/T$ for the hydrides of these four compounds. From the slope and intercept of the van't Hoff plots (*see* Figure 6), ΔH and ΔS values for the reaction $LaNi_{5-x}Al_x + nH_2 \rightarrow LaNi_{5-x}Al_xH_{2n}$ are estimated and given in Table V.

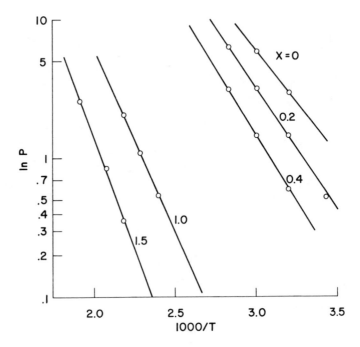

Figure 6.   1/T vs. ln P for LaNi$_{5-x}$Al$_x$ hydrides

Table V.   Thermodynamic Data for the LaNi$_{5-x}$Al$_x$–H$_2$ System[a]

Compound	ΔH(kcal/mol H$_2$)	ΔS(cal/deg mol H$_2$)	T (°C) for P = 2.0 atm
LaNi$_5$	−7.2 ± 0.1	−26.1 ± 0.4	∼25
LaNi$_{4.8}$Al$_{0.2}$	−8.3 ± 0.1	−27.3 ± 0.4	∼50
LaNi$_{4.6}$Al$_{0.4}$	−9.1 ± 0.2	−28.1 ± 0.7	∼70
LaNi$_4$Al	−12.7 ± 0.3	−29.2 ± 0.7	∼180
LaNi$_{3.5}$Al$_{1.5}$	−14.5 ± 0.6	−29.6 ± 1.4	∼240

[a] The increasing hydride stability can be illustrated by isobaric data as in the last column of Table V. The decomposition pressures of the LaNi$_{5-x}$Al$_x$ hydrides at 25°C are included in Figure 4 and follow the correlation curve in a regular manner.

Conclusion

Metal hydrides that have a linear dependence of ln P vs $1/T$ can be used in chemical heat pump systems. General equations were developed to relate the thermodynamic variables of the hydrides to the chemical heat pump operating temperatures. A series of hydrides with variable thermodynamic properties were prepared from alloys of composition LaNi$_{5-x}$Al$_x$ (x = 0–1.5).

Literature Cited

1. van Vucht, J. H. N., Kuijpers, F. A., Bruning, H. C. A. M., *Philips Res. Rep.* (1970), **25**, 133.
2. Kuijpers, F. A., van Mal, H. H., *J. Less-Common Met.* (1971), **23**, 395.
3. Zijlstra, H., *Chem. Technol.* (1972), **2**, 280.
4. Kuijpers, F. A., Loopstra, B. O., *J. Phys. Chem. Solids* (1974), **35**, 301.
5. Newkirk, H. W., "Hydrogen Storage by Binary and Ternary Intermetallics for Energy Applications," UCRL-52110, 1976.
6. Gruen, D. M., McBeth, R. L., Mendelsohn, M., Nixon, J. M., Schreiner, F., Sheft, I., *Proc. Intersoc. Energy Convers. Eng. Conf., 11th, 1976*, 681.
7. Buschow, K. H. J., van Mal, H. H., *J. Less-Common Met.* (1972), **29**, 203.
8. Steward, S. A., Lakner, J. F., Uribe, F., UCRL-52039 and 77455 (1976).
9. van Mal, H. H., Buschow, K. H. J., Miedema, A. R., *J. Less-Common Met.* (1974), **35**, 65.
10. Gruen, D. M., Mendelsohn, M., Dwight, A. E., *Proc. Rare Earth Res. Conf., 13th, 1977*.
11. Sandrock, G. D., *Proc. Intersoc. Energy Convers. Eng. Conf., 12th, 1977*.
12. Gruen, D. M., Mendelsohn, M., Sheft, I., *Proc. Symp. Electrode Mater. Processes Energy Conversion Storage*, The Electrochemical Society, 1977.
13. van Mal, H. H., Buschow, K. H. J., Kuijpers, F. A., *J. Less-Common Met.* (1973), **32**, 289.
14. van Mal, H. H., Thesis, Technische Hogeschool, Delft (1976).
15. Miedema, A. R., *J. Less-Common Met.* (1973), **32**, 117.
16. Buschow, K. H. J., van Mal, H. H., Miedema, A. R., *J. Less-Common Met.* (1975), **42**, 163.
17. Miedema, A. R., *J. Less-Common Met.* (1976), **46**, 67.
18. Miedema, A. R., Boom, R., de Boer, F. R., *J. Less-Common Met.* (1975), **41**, 283.
19. Kuijpers, F. A., *J. Less-Common Met.* (1971), **27**, 27.
20. Gruen, D. M., Mendelsohn, M. H., *J. Less-Common Met.* (1977) **55**, 149.
21. Bowman, A. L., Anderson, J. L., Nereson, N. G., *Proc. Rare Earth Res. Conf., 10th, Arizona, 1973*.
22. Gibb, T. R. P., Jr., *J. Phys. Chem.* (1964), **68**, 1096.
23. Buschow, K. H. J., van Mal, H. H., *J. Less-Common Met.* (1972), **29**, 203.
24. Beck, R. L., "Investigation of Hydriding Characteristics of Intermetallic Compounds," Denver Research Institute, DRI-2059 (1962).
25. Bechman, C. A., Goady, A., Takeshita, T., Wallace, W. E., Craig, R. S., *Inorg. Chem.* (1976), **15**, 2184.
26. Lundin, C. E., Lynch, F. E., Magee, C. B., *Proc. Intersoc. Energy Convers. Eng. Conf., 11th, 1976*, p. 961.
27. Mendelsohn, M., Gruen, D. M., Dwight, A. E., *Nature* (1977) **269**, 45.
28. Dwight, A. E., *J. Less-Common Met.* (1975), **43**, 117.
29. Kuijpers, F. A., Thesis, Technische Hogeschool, Delft (1973).
30. Takeshita, T., Wallace, W. E., Craig, R. S., *Inorg. Chem.* (1974), **13**, 2282.
31. Lundin, C. E., Lynch, F. E., Denver Research Institute, First Annual Report, No. AFOSR, F44620-74-C0020, (DDC #ADA006423) (1975).
32. Anderson, J. L. et al., Los Alamos Report, LA-5320-MS (1973).

RECEIVED July 18, 1977. Work performed under the auspices of U.S. Energy Research Development Administration.

24

The Titanium–Molybdenum–Hydrogen System: Isotope Effects, Thermodynamics, and Phase Changes

JAMES F. LYNCH, J. J. REILLY, and JOHN TANAKA[1]

Brookhaven National Laboratory, Upton, NY 11973

Pressure–composition–temperature and thermodynamic relationships of of the titanium–molybdenum–hydrogen (deuterium) system are reported. β-TiMo exhibits Sieverts' Law behavior only in the very dilute region, with deviations toward decreased solubility thereafter. Data indicate that the presence of Mo in the β-Ti lattice inhibits hydrogen solubility. This trend may stem from two factors: for Mo contents >50 atom %, an electronic factor dominates whereas at lower Mo contents, behavior is controlled by the decrease in lattice parameter with increasing Mo content. Evidence suggests that Mo atoms block adjacent interstitial sites for hydrogen occupation. Thermodynamic data for deuterium absorption indicate that for temperatures below 297°C an inverse isotope effect is exhibited, in that the deuteride is more stable than the hydride. There is evidence for similar behavior in the tritide.

Efficient separation of the isotopes of hydrogen is of interest in nuclear reactor technology. Separation is usually achieved by chemical exchange or by cryogenic distillation. We recently began investigations wherein stability differences exhibited by metal protides, deuterides, and tritides are exploited to achieve isotope separation. Most metal–hydrogen systems exhibit the so-called normal isotope effect near room temperature where the protide is more stable than the deuteride which is in turn more stable than the tritide. However, protium–tritium exchange experiments (1) established an inverse isotope effect at or near room temperature for a large number of binary titanium alloys. Of these alloys, the titanium–molybdenum system exhibits one of the largest separation factors at room temperature. This result first sparked our

---

[1] Present address: Chemistry Department, University of Connecticut, Storrs, CT 06268.

interest in the titanium–molybdenum–hydrogen system.   Our interest was intensified when we found that systematic examinations of the hydriding properties of binary titanium alloys are lacking.   Thus, this work was prompted by a desire to characterize a metal–hydrogen system that exhibits an anomalous isotope effect and to provide data in an area where results are conspicuously absent.

Literature Background

It is appropriate to review briefly the existing data pertaining to titanium–hydrogen and binary titanium alloy–hydrogen systems.   The titanium–hydrogen system was reviewed comprehensively by Mueller et al. (2) in 1968; the reader is referred there for a more detailed description.   The phase diagram presented by McQuillan (3) generally is accepted.   The system exhibits an eutectic.   The terminal phases are the solution of hydrogen in the low temperature modification of titanium (α-Ti, hcp) that is designated the α-phase and the nonstoichiometric dihydride that is designated the γ-phase.   The solution of hydrogen in the high temperature modification of titanium (β-Ti, bcc) is designated the β-phase.   The transformation of α-Ti to β-Ti occurs at about 870°C in the pure metal; however, the solution of hydrogen in β-Ti stabilizes the high temperature form to temperatures as low as 300°C.   The α-phase converts directly to the γ-phase with the addition of hydrogen at temperatures below the eutectic (~300°C).   Below 67°C, the fcc γ-phase transforms to a tetragonal structure (4) that is designated the δ-phase, via a contraction along the [001] direction (5).   At room temperature, the onset of the δ-phase occurs at the composition $TiH_{1.83}$ (6).

Structural aspects of the γ-phase have been examined by x-ray and neutron diffraction for both the hydride (5, 6, 7) and the deuteride (5, 8).   Both systems are fcc in the γ-phase.   Nagel and Perkins (6) report that at room temperature the γ-phase exists for compositions ranging from $TiH_{1.65}$ to $TiH_{1.85}$; over this composition range the lattice parameter increases from 4.405 to 4.416 Å.   Higher hydrogen contents exhibit the tetragonally distorted δ-phase.   Goon and Malgiolio (7) suggested that this distortion reflects the change of some hydrogen positions from tetrahedral to octahedral interstices in the fcc lattice; however, this behavior has not been confirmed.   Miron et al. (9) recently examined the β-phase deuteride at the eutectic composition $TiD_{0.66}$ via neutron diffraction.   Their result establishes the β-phase structure as bcc with deuterium atoms situated in tetrahedral interstices.   The hydride likely exhibits similar behavior.   Miron's data indicate a nonrandom distribution of deuterium in the β-phase.   It is not clear, however, whether this result reflects an ordered deuterium structure or merely random concentration inhomogeneities.

Pressure–composition–temperature (P–C–T) relationships were reported for hydrogen and deuterium in α-titanium.   Both systems obey Sieverts' Law in the dilute region (10); i.e., the square root of equilibrium pressure is linearly proportional to the solute content.   McQuillan (3) extended the hydrogen P–C–T data into the γ-phase.   For a maximum equilibrium hydrogen pressure of 500

Torr, he reports a hydrogen-to-titanium atom ratio (H/Ti) of 1.7 (468°C). A subsequent study indicates a maximum H/Ti value is 1.92 (P = 50 atm, room temperature) (6). Morton and Stark (11) extended the deuterium P–C–T data into the γ-phase; they observe D/Ti = 1.8.

The equilibrium P–C–T relationships have been used to determine thermodynamic parameters of absorption at low solute concentration (H/Ti → 0). Results reported in various studies are in reasonable agreement; for example, values reported for the enthalpy of hydrogen absorption range from −10.8 kcal/H (7) to −12.1 kcal/H (10). Values of $\Delta \overline{H}_D$ range from −11.26 kcal/D (11) to −12.00 kcal/D (10). Similar thermodynamic data for the solution of hydrogen in β-Ti indicate values of $\Delta \overline{H}_H$ ranging from −12.9 kcal/H (12) to −13.9 kcal/H (7) at H/Ti → 0. A recent calorimetric examination of hydrogen and deuterium in titanium (13) reports thermodynamic data in agreement with the results obtained via P–C–T studies; limiting values of $\Delta \overline{H}_H$ = −10.6 kcal/H and $\Delta \overline{H}_D$ = −10.1 kcal/D were measured at 434°C. Kleppa reports partial molar enthalpies corresponding to the phase transitions $\alpha \to \beta$ ($\Delta \overline{H}_H^{\alpha \to \beta} \cong$ −12.5 kcal/H) and $\beta \to \gamma$ ($\Delta \overline{H}_H^{\beta \to \gamma} \cong$ −18.8 kcal/H). The calorimetric data indicate a small temperature dependence in the partial molar enthalpy of hydrogen absorption; the enthalpy becomes more exothermic with increasing temperature. The reason for this trend is not unambiguously clear; Kleppa suggests a quasifree particle behavior of hydrogen atoms in the titanium interstitial sites.

A discussion of titanium alloy–hydrogen systems should incorporate the recent developments regarding the reaction of hydrogen with titanium intermetallic compounds. Miedema et al. (14) investigated the properties of intermetallic hydrides in some detail. The intermetallic compound of iron and titanium, FeTi, has received considerable attention because of its potential as a hydrogen-storage medium (15, 16). However, the data regarding the titanium intermetallic–hydrogen systems are extensive and are outside the scope of this discussion. We restrict our comments to the hydrogen solution in random binary alloys of titanium.

Virtually all of the reported structural data on titanium alloy hydrides and deuterides indicate that the solute atoms occupy tetrahedral interstitial sites in the metal lattice. Neutron diffraction data obtained for deuterium in Ti/34 atom % Zr and in Ti/34 atom % Nb (17) indicate tetrahedral site occupancy in the bcc β-phase. Similarly, data reported for deuterium in Ti/19 atom % V and in Ti/67 atom % Nb (18) indicate tetrahedral site occupancy in the fcc γ-phase. Crystallographic examination of the γ-phase Ti–Nb–H system (19) reveals that increasing niobium content linearly increases the lattice parameter of the fcc γ-phase for Nb contents ranging from 0 to 70.2 atom %. Vanadium, on the other hand, exerts the opposite effect (6); at H/M = 1.85, the γ-phase lattice parameter decreases with increasing vanadium contents.

Nagel and Perkins (6) studied carefully the effect of added vanadium on the γ → δ transition in titanium hydrides via x-ray analysis. They observed a shift in (H/M)$_{crit}$ (the hydrogen-to-metal ratio at the γ → δ transition) to lower values for small vanadium additions; e.g., $(H/M)_{crit} = 1.85$ in pure Ti whereas $(H/M)_{crit} = 1.68$ for Ti/8 atom % V. Additionally, the γ-phase again becomes more stable at higher hydrogen contents. Thus, for example, in Ti/2 atom % V at room temperature, the δ-phase exists between $H/M = 1.75$ and 1.85 whereas the γ-phase was observed for $H/M < 1.75$ and $H/M > 1.85$. For samples of vanadium content above 26 atom %, no tetragonal δ-phase was observed. A similar trend was observed in the critical H/M values for the $\alpha \rightarrow \gamma$ transition. For pure titanium, the α-phase is converted completely to γ at $H/M = 1.7$ whereas for Ti/20 atom % V, the conversion is completed at $H/M = 1.6$. At higher vanadium contents this trend is reversed.

The effect of added alloy constituents on the maximum hydrogen solubility has been reported by numerous workers. The data indicate that many binary titanium alloys form a nearly stoichiometric dihydride over an appreciable range of alloy partner composition. For example, McQuillan and Upadhyaya (*20*) report the maximum value $H/M \cong 2$ for titanium–niobium alloys to 65 atom % Nb. Similar results have been reported for titanium–vanadium alloys (to 50 atom % V) (*6*) for titanium–zirconium alloys (over the entire range of zirconium compositions) (*21, 22*), and for titanium–molybdenum alloys (to 50 atom % Mo) (*23*). In contrast to these results, there is evidence that the presence of aluminum in titanium inhibits hydrogen solubility. Rudman (*24*) has examined the solubility of hydrogen in α-Ti$_3$Al. A limiting value of $H/M \cong 0.35$ is reported at 450°C and 1000 Torr equilibrium pressure. This result indicates that aluminum in the metal lattice blocks sites that are potentially available for hydrogen occupation. If this interpretation is correct, then it is expected that aluminum similarly will influence the entire range of titanium–aluminum compositions.

Studies of the effect of added alloy constituents on P-C-T and thermodynamic relationships largely have been restricted to the bcc β-phase because the solubility of most metals in α-Ti is extremely limited. McQuillan (*25*) examined the effect of small additions (to $\sim$5 atom %) of vanadium, chromium, nickel, and cobalt to β-Ti in the temperature range 750°–950°C. He reported ideal behavior to $H/M = 0.05$ (i.e., the P-C-T data follow Sieverts' Law). Saito and Someno (*26*) recently extended McQuillan's study and confirmed similar behavior for alloys with iron, manganese, copper, zirconium, niobium, molybdenum, and tantalum. In the Sieverts' Law region, all of these alloying elements except zirconium, niobium, and tantalum decrease the solubility of hydrogen. Analysis of the P-C-T data analysis reveals that the partial molar enthalpy of absorption at low hydrogen contents, $\Delta \overline{H}_H$ ($n \rightarrow 0$), increases smoothly (i.e., becomes less exothermic) with increasing alloy partner content although the magnitude of the effect is relatively small. For example, McQuillan (*25*) reports an increase in $\Delta \overline{H}_H$ ($n \rightarrow 0$) for titanium–cobalt alloys from -13.9 kcal/H in pure β-Ti to -12.8 kcal/H in Ti/5 atom % Co. Titanium alloys with vanadium exhibit an anomaly (*25*) in that $\Delta \overline{H}_H$ values ($n \rightarrow 0$) exhibit a maximum at $\sim$1 atom % V and then decrease at higher vanadium contents. There is open disagreement regarding the effect of niobium addition; McQuillan, et al. (*27*) report an increase in $\Delta \overline{H}_H$ ($n \to 0$) with increasing niobium content whereas Saito and Someno (26) report a decrease.

Variations in $\Delta \overline{H}_H$ ($n \to 0$) were correlated with variations in the lattice parameter of the parent alloy (28). This correlation presumably reflects the view that the insertion of solute hydrogen atoms becomes more unfavorable as the lattice parameter (and consequently the volume of the interstitial site) decreases. This correlation appears valid in titanium alloy systems for which data are available, except for titanium–vanadium. The lattice parameter of β-Ti/V decreases (29), but $\Delta \overline{H}_H$ ($n \to 0$) becomes more exothermic as the vanadium content is increased (25, 30). This correlation supports the trend in enthalpy with niobium content observed by Saito and Someno (26). The lattice parameter of β-Ti/Nb alloys increases with niobium content; thus, it is expected that $\Delta \overline{H}_H$ ($n \to 0$) becomes more exothermic.

It is pertinent to note the data regarding the solution of hydrogen in titanium–molybdenum alloys. Examination of the titanium–molybdenum phase diagram reveals that the addition of molybdenum markedly stabilizes the bcc β-phase of titanium. Molybdenum is soluble in β-Ti throughout the entire composition range. The β-phase alloys of titanium and molybdenum absorb hydrogen exothermically for molybdenum contents below approximately 60 atom %. The alloys form a nearly stoichiometric dihydride up to 50 atom % Mo; subsequent Mo additions rapidly reduce the solubility (23). The low hydrogen content P–C–T relationships follow with Sieverts' Law (26, 27) although the addition of molybdenum decreases hydrogen solubility in the Sieverts' Law region (i.e., for fixed H/M, the equilibrium hydrogen pressure increases with increasing molybdenum content). The partial molar enthalpy of absorption, $\Delta \overline{H}_H$ ($n \to 0$), becomes less exothermic with increasing molybdenum content (26).

For temperatures near ambient, most metal–hydrogen systems exhibit the so-called normal isotope effect, i.e., the stability order is protide > deuteride > tritide. There are, however, a few systems in which replacement of hydrogen by deuterium or tritium results in a more stable compound. A detailed review of the hydrogen isotope effect exhibited by metal–hydrogen systems is outside the scope of this discussion. It is pertinent, however, to note briefly those systems that are relevant to this work. Our discussion is limited to temperature at or near room temperature, and the inverse isotope effect is defined by the stability order tritide > deuteride > protide.

The group IV B elements titanium, zirconium, and hafnium exhibit the normal isotope effect. Most of the data for the titanium–hydrogen system have been obtained at elevated temperatures. However, extrapolation of the available data (11, 13, 31) to room temperature indicates a normal effect for hydrogen and deuterium. The group VB metals vanadium, niobium, and tantalum, on the other hand, exhibit inverse isotope effects; indeed, these are the only pure metals that exhibit the inverse effect near room temperature. Extensive data have been reported for these systems. The P–C–T data obtained by Wiswall and Reilly (32) for vanadium hydrogen and deuterium clearly show a greater stability for the deuteride.   Similarly, hydrogen–tritium exchange experiments indicate an inverse isotope effect.   Comparison of the vanadium–deuterium phase diagram (*33, 34, 35*) with that of vanadium–hydrogen (*36*) supports this result.   Steward (*37, 38, 39*) demonstrated an inverse effect for the niobium hydrogen and deuterium systems.   Likewise, tantalum exhibits an inverse effect as evidenced by P–C–T measurements (*40*) and phase diagram determinations (*41*).

Although the pure titanium–hydrogen system exhibits the normal isotope effect, many titanium alloys show the inverse effect.   The exchange of protium–tritium mixture with the hydrided phase of these alloys has demonstrated an inverse protium–tritium isotope effect in Ti–V, Ti–Mo, Ti–Cr, Ti–Mn, and the ternary alloy TiCrMn (*1*).   On the other hand, Ti–Co, Ti–Fe, and Ti–Ni systems exhibit the normal isotope effect.   Clearly much can be learned from a study of these systems.

Experimental

Sample Preparation.   The metals used in this research were obtained from the Materials Research Corp., Orangeburg, NY.   MARZ grade (purity > 99.9%) materials were used.   Samples were prepared by arc-melting the desired quantities of the elements under argon.   Each specimen was melted twice, manually crushed, and remelted several times since the initial melt often contained discrete lumps of molybdenum in the alloy button.   After a homogeneous melt was obtained as indicated by micrograph data, the sample was annealed.   To eliminate oxidation during annealing, each sample was placed in a small titanium can that was closed by arc-welding the lid under argon; the can containing the sample subsequently was annealed in an alumina tube under 1 atm He at 1250°C for 48 hr.   Although the samples were not quenched, metallographic and x-ray data indicated that samples prepared in this manner were homogeneous solid solutions.   Figure 1 displays representative metallographs of samples before and after annealing.   It is evident that the annealing treatment removes coring and inhomogeneity.   Alloy samples that were brittle were crushed in a steel mortar and passed through a 20-mesh sieve.   Ductile (high titanium) alloys were rolled and chopped into pieces before hydriding.   In any case, the hydriding reaction resulted in particle attrition, and the particle size of all samples after the experiment was always finer than 20 mesh.   Compositions of the samples prepared are indicated in Table I as determined by an electron microprobe.   Variation in the microprobe data arises from determinations at randomly selected sites on the polished alloy surface.   Composition of the alloy used for exchange was assumed to be the average of the microprobe determinations.

Pressure–Composition–Temperature Studies.   Prior to obtaining reproducible pressure–composition–temperature (P–C–T) data, it was necessary to activate thoroughly the samples.   Activation usually was achieved by outgassing at high temperature (~400°C), contacting with hydrogen at moderate pressure (~100 psi), cooling rapidly to room temperature, contacting with hydrogen at high pressure (~1000 psi), and cycling over −78° to 25°C.   This treatment was repeated when necessary.   Samples that were exposed to this procedure yielded reproducible maximum solubilities of hydrogen.

P–C–T relationships were obtained via two techniques.   A desorption technique was used for equilibrium hydrogen pressures above 0.5 atm.   The

 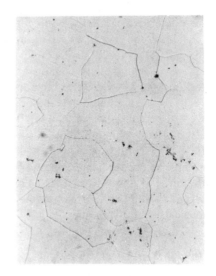

Ti$_2$Mo as Cast   TiMo Annealed 48 Hr @ 1200°C

10 HF   10 HNO$_3$ 30 Lactic   10 HF   10 HNO$_3$ 30 Lactic

Figure 1. Representative metallographs of Ti/Mo samples before and after annealing, etched and magnified ×125. Ti$_2$Mo is presented after melting, crushing, and remelting six times. TiMo is presented after similar casting, then annealing at 1200°C for 48 hr.

sample reactor, desorption apparatus, (42) and a detailed discussion of the experimental procedure (43) have been described in the literature. The reported temperatures were maintained constant to ±0.5°C. For temperatures below 80°C the sample was maintained at constant temperature with a water bath, while at high temperatures an electric furnace was used. The temperature was monitored with a chromel–alumel thermocouple.

For hydrogen equilibrium pressures below 100 Torr, an absorption technique was used. Thoroughly activated samples were outgassed at 500°C to a vacuum

Table I.   Alloy Characterization

Alloy Composition as Prepared	*Electron Microprobe*	*X-ray Alloy*
TiMo$_{1.96}$	TiMo$_{1.94}$	bcc a = 3.161 Å
Ti$_{0.99}$Mo	Ti$_{1.06}$Mo	bcc a = 3.180 Å
	Ti$_{1.11}$Mo	
Ti$_{2.06}$Mo	Ti$_{2.24}$Mo	bcc a = 3.207 Å
	Ti$_{2.28}$Mo	
Ti$_{3.79}$Mo	Ti$_{4.01}$Mo	bcc a = 3.241 Å
	Ti$_{4.19}$Mo	
	Ti$_{4.23}$Mo	
	Ti$_{4.34}$Mo	

of 1×10^{-5} Torr, then maintained at the desired temperature with an electric furnace. Ultra pure hydrogen (purity $>$ 99.999%) at a known initial pressure and temperature was admitted to the sample from a calibrated dosing volume. Absorption was followed via the change in pressure monitored by a pressure transducer (MKS, 0–100 Torr or 0–10 Torr full scale), and the signal was displayed on a strip chart recorder. At equilibrium (indicated by no further pressure change) the final pressure was recorded, and the process was repeated. Thus, it is possible to monitor continuously the hydrogen content of the sample.

For P-C-T relationships involving deuterium, a similar procedure was followed. However, studies were limited to the range $0 < D/M \lesssim 0.01$ where D/M represents the deuterium-to-metal atom ratio, and only a 10-Torr, full-scale transducer was used. Deuterium was supplied to the system from a metal deuteride that was prepared in this laboratory. An alloy of composition $TiFe_{0.98}$ $Mn_{0.02}$ (mass 200 g) was activated thoroughly, then outgassed completely at $500°C$. The alloy was exposed to deuterium (ultrapure) at 1000 psi for 4 hr, cooled to $-196°C$, and evacuated. The deuteride reservoir was closed, warmed to room temperature, and moved to the absorption manifold where it served as a high-purity deuterium source.

X-ray Studies. X-ray diffraction data were obtained via the Debye–Scherrer technique at room temperature using Zr-filtered Mo radiation. Data were obtained for hydrogen-free samples and their hydrides of various hydrogen contents. Most of the low-content, hydrogen alloy samples were prepared via absorption. Active specimens were exposed to the desired hydrogen quantity at room temperature. After about 4 hr, the sample was heated to $100°C$ and then cooled to room temperature to ensure equilibrium hydrogen distribution throughout the sample. Since we observed that these samples slowly lost hydrogen in air, each sample was poisoned with CO, removed, and examined immediately. The hydrogen content was confirmed by volumetric measurements on the hydrogen evolved by thermal decomposition of the hydride. With high hydrogen content, samples were equilibrated with high pressure hydrogen at room temperature for 24 hr. Then, hydrogen was withdrawn until the desired content was obtained, as evidenced by the P-C-T data. The sample was cooled to $-196°C$, and the hydrogen remaining in the gas phase was pumped off. Air was admitted to poison the sample that then was warmed, vented, and immediately x-rayed. Again, the hydrogen content was determined by volumetric measurements of the hydrogen evolved by thermal decomposition of the hydride.

Hydrogen–Tritium Exchange Studies. The alloy to be studied was placed between porous nickel frits in a tube 4 in. long with $3/8$-in. i.d. This sample tube was inserted as part of a closed loop of $1/4$-in. stainless steel tubing.

Gas circulation was carried out using a hydride pump as described previously (44). The loop and circulation system were fitted appropriately with valves so that it could be pressurized to 1000 psi with hydrogen or evacuated. It was connected also to a calibrated gas handling and sampling system that removed samples for counting. The tritium was counted with a proportional counting Bernstein–Ballentine tube (45). The initial tritium concentration was adjusted so that 50 Torr in the counting tube, followed by enough P-10 counting gas (90% argon–10% methane) to bring the pressure to 750 Torr, gave an initial count of 90,000–100,000 cpm.

Approximately 10 g alloy sample was hydrided by first activating the metal using rapid hydriding–dehydriding cycles and then by exposing the sample to 800–1000 psi hydrogen for 8–10 hr. Then the pressure over the sample and pump was reduced to 225 psi. A section of the loop containing a 50 mL tank (83 mL total volume) was evacuated and filled to 500 Torr with a protium–tritium mixture. The low-pressure tritium mixture in the 50-mL tank portion of the loop then was diluted further by pressurizing to 225 psi with normal hydrogen. The pump was started, and the tritium in the sample tank was thoroughly mixed with all the gas in the system except for that in the sample tube. The volume of the latter section was 29 mL compared with 248 mL for the remainder of the system. Once the tritium was mixed thoroughly in the larger volume, the valves to the sample volume were opened and the gas circulated. Periodic counts were made on the circulating gas until equilibrium was established (3–26 days, depending on the the sample). After equilibrium was established, the sample was isolated from the larger volume. The gas in the latter volume was removed, and its tritium content was determined. Then the sample tube was heated and the hydride decomposed. The tritium content of the evolved gas was determined similarly. Alpha, the ratio of the tritium counts in the solid phase to those in the gas phase, was determined from the ratio of the total activity of all the gas obtained from dehydriding the alloy hydride to the total activity in the gas phase in equilibrium with this solid.

Results and Discussion

Metallurgical Aspects of the Titanium–Molybdenum System. The portion of the titanium–molybdenum phase diagram relevant to this work is shown in Figure 2. It is apparent that above 900°C, titanium and molybdenum form a continuous series of solid solutions; at lower temperatures, however, a miscibility gap exists. In spite of this, we successfully prepared homogeneous solid solutions of titanium/molybdenum that were stable at room temperature; indeed, the x-ray diffraction and metallographic characteristics of these samples were unchanged over several months. However, many aspects of the experimental procedure used here involve exposing the specimens to high temperatures for varying lengths of time. For samples of molybdenum content less than 40 atom %, this can result in the solid solution decomposing. Therefore, we attempted to ascertain the effect of temperature on the samples examined here.

A sample with composition Ti_2Mo was prepared as described earlier. After obtaining an x-ray powder pattern, the sample was activated and was loaded with hydrogen to a content of $H/M \cong 0.30$, whereupon the hydrogen pressure fell to virtually zero. The sample was heated to 400°C, resulting in a pressure increase to $P \cong 10$ Torr. However, instead of remaining constant, the pressure immediately began to decrease again, albeit quite slowly. After 25 days, the pressure had fallen to $P \cong 4$ Torr. At this point, the sample was evacuated for 24 hr to 10^{-5} Torr then cooled to room temperature. The powder pattern again was determined. The results are displayed in Table II; it is evident that this treatment resulted in significant decomposition of the solid solution to a molybdenum-poor solution in α-titanium and to a molybdenum-rich solution in β-titanium. The pressure behavior indicates that the decomposition begins immediately upon exposure to elevated temperature, although the decomposition rate is quite slow. An x-ray of the specimen that was obtained three weeks after this experiment was identical with the decomposed result, indicating that the decomposition is irreversible at room temperature.

A similar Ti₂Mo sample was prepared and annealed, and the powder pattern was determined. The sample was outgassed at 450°C for 2 hr, and activated as described previously, whereupon a hydrogen-to-metal atom ratio of $H/M \cong$ 1.80 was obtained. The sample again was outgassed at 450°C for 2 hr, cooled to room temperature, and the powder pattern redetermined. Both diffraction

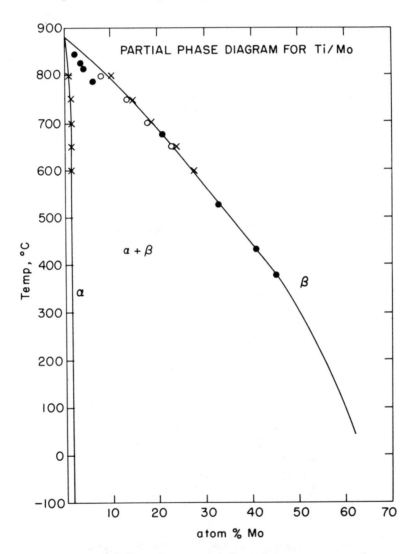

Figure 2. Partial phase diagram for the Ti/Mo system over the temperature range −100°–900°C; molybdenum contents from 0 to 70 atom %. (×,○) Ref. 54. (●) Ref. 55.

Table II. X-Ray Results Showing Decomposition of Homogeneous Ti₂Mo into α-Ti Plus Mo-Rich Solid Solution[a]

H-Free Ti₂No[b]	Ti₂Mo after 400°C Anneal[c]	α-Ti[d]
	2.5533	2.557
2.2684	2.2716	
	1.7560	1.726
1.5993	1.6070	
	1.4844	1.475
	1.3488	1.332
1.3081	1.3091	
	1.2699	1.276
1.1355	1.1347	
1.0153	1.0159	
	0.9517	0.9458
0.9242	0.9276	
0.8565	0.8569	
0.7572	0.7569	
0.7164	0.7194	
0.6847	0.6865	
0.6557	0.6564	
0.6304	0.6298	

[a] Data shown are d-values obtained via molybdenum-radiation in a Debye–Scherrer camera.
[b] After arc melting, annealing 48 hr at 1200°C.
[c] After annealing 25 days at 400°C in hydrogen.
[d] X-ray powder file data. ASTM Special Technical Publication 48-J 1960. File #5-0682.

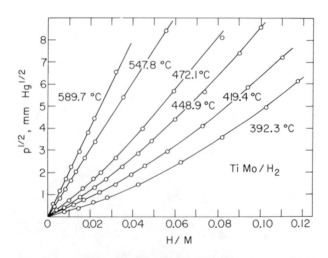

Figure 3.   Hydrogen absorption isotherms exhibited by TiMo over the temperature range 589.7°–392.3°C. Data are presented as $P^{1/2}$ (mm Hg)$^{1/2}$ vs. hydrogen-to-metal atom ratio.

patterns were identical, indicating that a relatively short, high-temperature treatment does not lead to significant decomposition of the alloy. However, in view of the earlier indication that decomposition begins immediately upon exposure to high temperature, we ensured that exposure of the samples to high temperature was minimized. The fact that the *P–C–T* data proved to be reproducible indicates that the effect of β-Ti/Mo decomposition on these results is minimal.

P–C–T Determinations; Low Pressure Studies. Absorption isotherms obtained for the reaction of hydrogen with TiMo are shown in Figure 3 for 590°–392°C. These temperatures are above the decomposition temperature of β-TiMo (*see* Figure 2); consequently, decomposition of the solid solution plays no role here. These data follow Sieverts' Law only in the very dilute region—to hydrogen-to-metal ratios (H/M) of about 0.02. Thereafter, deviations in the direction of decreased solubility are observed. Data in the region of Sieverts' Law can be used to determine the relative partial molar enthalpy and entropy at infinite dilution (*47*). From Sieverts' Law (Equation 1), where K_s is a tem-

$$P^{1/2} = K_s n \tag{1}$$

$$K_s = 1/n_s \exp(\Delta \overline{H}_H^0/RT) \exp(-\Delta \overline{S}_H^0/R) \tag{2}$$

perature-dependent constant, and $n \equiv$ H/M, we get Equation 2, where n_s is the geometrically limiting value of H/M. Thus, in the region of Sieverts' Law, plots of $\ln K_s$ vs. $1/T$ are linear; values of $\Delta \overline{H}_H^0$ are determined from the slope, and, if n_s is known, $\Delta \overline{S}_H^0$ values are determined from the intercept. These results, shown in Table III, are determined via least-square analysis of $\ln K_s$ vs. $1/T$; when determining $\Delta \overline{S}_H^0$, we assumed that $n_s = 1$. These data are compared with similar results for hydrogen in β-titanium (*48*). The enthalpy determined for TiMo is about 2.5 kcal/H less exothermic than that reported for β-titanium. This result is not surprising since adding molybdenum decreases the lattice parameter of β-titanium (*see* x-ray studies). We previously noted the $\Delta \overline{H}_H^0$ trend toward less exothermic values with decreasing lattice parameter that is exhibited by many metal–hydrogen systems.

On the basis of dissociative solution, i.e., hydrogen is absorbed as H atoms rather than H_2 molecules, the relative partial molar free energy of absorption is related to the equilibrium constant via the equilibrium hydrogen pressure (Equation 3). Values of the relative partial molar enthalpy as a function of hy-

$$\Delta \overline{G}_H = RT \ln P^{1/2} = \Delta \overline{H}_H - T \Delta \overline{S}_H \tag{3}$$

drogen content thus are obtained via plots of $\ln P^{1/2}$ vs. $1/T$ at fixed values of H/M. This relationship, presented in Figure 4, shows that the enthalpy increases (i.e., becomes less exothermic) with increasing hydrogen content. The relative partial molar enthalpy is independent of temperature over the range examined

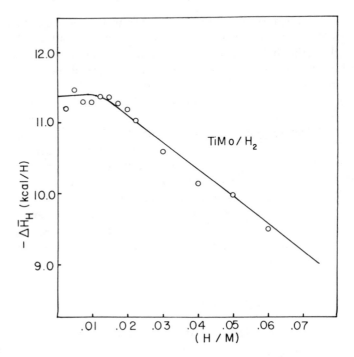

Figure 4. Relative partial molar enthalpy of hydrogen absorption exhibited by TiMo as a function of hydrogen content. Data are presented as $-\Delta\bar{H}_H$ (kcal/g-atom hydrogen) vs. hydrogen-to-metal atom ratio.

here. An extrapolation of $\Delta\bar{H}_H$ to $n = 0$ yields a value $\Delta\bar{H}_H$ ($n \to 0$) slightly more exothermic than that obtained via plots of $\ln K_s$ vs. $1/T$; -11.20 kcal/H vs. -10.37 kcal/H.

It is of interest to compare the relative partial molar entropy at infinite dilution exhibited by $TiMo/H_2$ with that exhibited by β-Ti/H_2. Gallagher and Oates (48) reported $\Delta\bar{S}_H^{xs}$ values for absorption of hydrogen by β-Ti; $\Delta\bar{S}_H^{xs}$, the excess entropy, is obtained via Equation 4, where X_H refers to the atom frac-

$$RT \ln P^{1/2}/X_H = \Delta\bar{H}_H - T(\Delta\bar{S}_H^{xs} + R \ln n_s) \qquad (4)$$

tion of dissolved hydrogen. Gallagher and Oates assumed that hydrogen occupies tetrahedral interstitial sites in bcc β-Ti. Since there are six tetrahedral sites per metal atom, the value of n_s is fixed at six. Since our value of $\Delta\bar{S}_H^0$ in Table III is obtained for $n_s = 1$, a meaningful comparison is made if the $\Delta\bar{S}_H^{xs}$ value of Gallagher and Oates is adjusted by $R \ln 6$; i.e.,

$$\Delta\bar{S}_H^0 \,(\beta\text{-Ti}) = \Delta\bar{S}_H^{xs} + R \ln 6 = -16.21 + 3.56 = -12.65 \text{ eu/H.}$$

Thus, for similar values of n_s, the entropy at infinite dilution for TiMo/H_2 is about 3.4 eu/H more negative than that for β-Ti/H_2.   An explanation for this difference might be that the molybdenum atoms in the metal lattice block potentially available interstitial sites for hydrogen occupation, resulting in nonrandom occupation of sites at low hydrogen content.   Rudman (24) proposed such a role for aluminum in titanium.   We presently are gathering more data for alloys of varying molybdenum composition to test this hypothesis.

Metal/hydrogen systems have been treated successfully by the assumption that, at low hydrogen concentrations, deviations from Sieverts' Law result from an elastic interaction energy among the absorbed hydrogen atoms.   The interaction is attractive in nature since including a hydrogen atom into an interstitial site results in a local expansion of the metal lattice.   It is well known that lattice dilation exerts an attractive interaction with interstitial solute atoms.   Under this

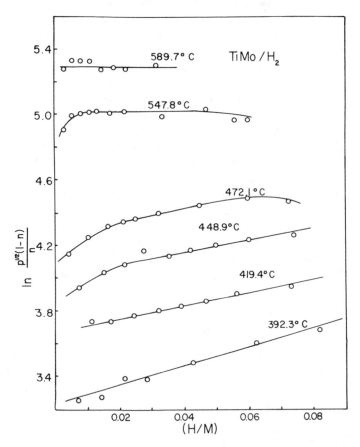

Figure 5.   Plots of $\ln P^{1/2(1-n)/n}$ *vs. hydrogen-to-metal atom ratio obtained for TiMo/H_2 over the temperature range 589.7°–392.3°C*

$$\ln \frac{P^{1/2}(n_s - n)}{n} = \frac{\Delta \overline{H}_H^0}{RT} - \frac{\Delta \overline{S}_H^0}{R} + \frac{2W_{HH}n}{RT} \qquad (5)$$

assumption, the isothermal behavior in the dilute hydrogen content region can be described by Equation 5, where n and n_s are the H/M and geometrically limiting H/M values, respectively, and W_{HH} is the interaction energy per mole of hydrogen atoms. Thus, plots of $\ln P^{1/2}(n_s - n)/n$ vs. n yield straight lines with slope $= 2 W_{HH}/RT$. This relationship is displayed in Figure 5 over the temperature range examined here, where we have assumed that n_s (the geometrically limiting H-to-M atom ratio) is unity. The data are insufficient to confirm the linearity predicted by Equation 5. Of more immediate concern, however, is the fact that the interaction energy obtained from the plots in Figure 5 is repulsive in the range of hydrogen contents examined here (i.e., the sign of W_{HH} is positive). This result reflects the behavior of $\Delta \overline{H}_H$—as the hydrogen content is increased, $\Delta \overline{H}_H$ values become less exothermic in the region of hydrogen contents studied here. However, it challenges the assumption on which Equation 5 is based. There can be no doubt that absorption of hydrogen atoms into interstitial positions results in a lattice expansion since x-ray data for this system exhibits an increase in lattice parameter with increasing hydrogen content. Thus, it is difficult to understand a repulsive interaction among the absorbed hydrogen atoms. A possible explanation is to postulate an additional term in Equation 5. This is unsatisfactory, however, since the nature of this term is speculative at best. Further, the onset of the $\beta \rightleftarrows \gamma$ plateau requires that the attractive interaction energy dominate at compositions approaching the plateau. $TiMo/H_2$ is not unique in its deviations from Sieverts' Law. Similar behavior was observed recently in $LaNi_5/H_2$ (49) and in $TiCr_2/H_2$ (50). It is evident that an understanding of this behavior requires further study.

Deuterium absorption isotherms were determined similarly for TiMo. The temperature range examined was the same as that for hydrogen absorption, but the range of deuterium contents was significantly smaller, extending only to D/M $\cong 0.01$. The data are presented in Figure 6. It is evident that Sieverts' Law is

Table III. Thermodynamic Parameters of Hydrogen (Deuterium) Absorption by TiMo

System	$\Delta \overline{H}_H^0$ (kcal/H)	$\Delta \overline{S}_H^{xs}$ (eu/H)	$\Delta \overline{H}_D^0$ (kcal/D)	$\Delta \overline{S}_D^{xs}$ (eu/D)
TiMo	-10.37^a	-16.05	-11.632^a	-18.26
	-11.20^b	—	—	—
β-Ti	-13.90^c	—	—	—
	-12.92^d	-16.21^e	—	—

[a] Result obtained via temperature dependence of Sievert's constants; n_s, the geometrically limiting value of H/M, is assumed equal to 1.
[b] Result obtained via extrapolation of $\Delta \overline{H}_H$ (n) to $n = 0$.
[c] Ref. 25.
[d] Ref. 12.
[e] Ref. 48; n_s is assumed equal to 6.

followed in this region. The relative partial molar thermodynamic parameters were obtained by using Equation 2. These results are included in Table III, where we again assumed $n_s = 1$ in calculating $\Delta \bar{S}_D^0$. The enthalpy of deuterium absorption is about 1.2 kcal/g-atom more exothermic than that for hydrogen. The corresponding entropy is about 2.2 eu/g-atom more negative. The variation of $\Delta \bar{H}_D$ with D/M is not determined because of the limited range of deuterium contents examined here.

Over the temperature range examined here, the protium–deuterium isotope effect is normal; i.e., the protide is more stable than the deuteride. However, comparing the thermodynamic data in Table III indicates that for temperatures

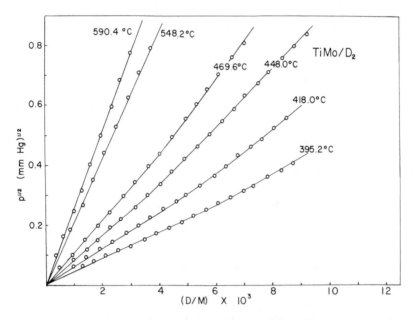

Figure 6. Deuterium absorption isotherms exhibited by TiMo over the temperature range 590.4°–395.2°C. Data are presented as $P^{1/2}(mm\ Hg)^{1/2}$ vs. deuterium-to-metal atom ratio $\times 10^3$.

below 297°C, the trend reverses and the deuteride becomes the more stable isotope. Similar behavior is exhibited by the protium–tritium isotope effect at high solute contents; the exchange experiments described subsequently indicate that near room temperature the tritide is more stable than the hydride. However, extrapolation of values of ln α vs. $1/T$ to $\alpha = 1$ indicate that above 152°C, the trend is reversed, and the protide becomes more stable.

P–C–T **Determinations; High Pressure Studies.** Typical desorption isotherms obtained for the Ti–Mb–H system at 40°C are shown in Figure 7 for samples ranging in molybdenum content from 33 to 66 atom %. The plateau pressure corresponding to the $\beta \rightleftarrows \gamma$ phase equilibrium increases with molyb-

Table IV. Partial Molar Enthalpy of Dihydride Formation

System	$-\Delta\overline{H}_H$ (kcal/H)
β-Ti[a]	18.6—19.4
Ti_2Mo	6.75—9.67
$Ti_{1.24}Mo$	3.80—4.90
TiMo	2.16—3.80

[a] From Ref. *13*. Calorimetric result obtained at 441°C.

denum content to the extent that the alloy $TiMo_{1.96}$ does not form a dihydride even at 600 atm and 0°C. Throughout this discussion, the hydrogen solution in the metal is designated the β-phase. We chose this nomenclature to be similar with hydrogen in β-titanium and to emphasize that the hydrogen solution in metal does not exhibit the hcp structure observed in α-titanium. It is evident from Figure 7 that the $\beta \rightleftarrows \gamma$ plateau pressures are not truly flat, but rather exhibit a pronounced slope. A possible explanation for this behavior may stem from the fact that concentration gradients in the alloy can result in sloping plateaus. It is likely that cooling these samples from near their melting point through the miscibility gap can result in local concentration variations even though the sample shows no evidence of phase inhomogeneity; indeed, the electron microprobe data (*see* Table I) indicates that this is the case.

The thermodynamic parameters of dihydride formation were obtained from plots of $\ln P(\beta \rightleftarrows \gamma)$ vs. $1/T$, where $P(\beta \rightleftarrows \gamma)$ refers to the plateau pressure. The results are given in Table IV. Because of the sloping nature of the plateaus, the thermodynamic data are presented over a small range rather than at a fixed value. Nonetheless, the stability of the dihydride decreases with increasing molybdenum content.

Recent augmented plane wave calculations (*51*) have demonstrated the role of the electron band structure of the alloy in hydrogen absorption. Results indicate that including hydrogen in the metal lattice results in electron band structure perturbation, such that some initially empty states are pulled below the Fermi energy and become available for occupation by hydrogen electrons. In this regard, the variation in maximum hydrogen solubility with alloy partner content should be considered. Such data is plotted in Figure 8 for the titanium–molybdenum system and the vanadium–chromium system (*52*) and was obtained at 40°C with a hydrogen charging pressure of 60 atm. At low alloy contents, the solubility is not affected greatly; however, the relationship undergoes

Table V. X-Ray Lattice

Sample	H-Free	β-Max (bcc)
Ti_2Mo	3.209 ± .005	3.318 ± .004
$Ti_{1.44}Mo$	3.184 ± .005	3.299 ± .003
$Ti_{1.24}Mo$	3.172 ± .012	3.296 ± .004
TiMo	3.179 ± .002	3.262 ± .006

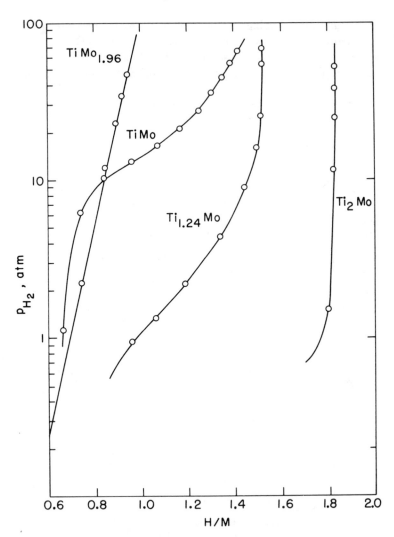

Figure 7. Hydrogen desorption isotherms exhibited by the Ti/Mo system at 40°C. Data are presented as P (atm) vs. hydrogen-to-metal atom ratio for samples of molybdenum content ranging from 33 to 66 atom % Mo.

Parameter Data (Å)

γ-Min (fcc)	(H/M) β-Max
4.437 ± .014	0.71—0.77
4.397 ± .007	0.65—0.69
4.402 ± .007	(0.69)
4.407 ± .010	0.49—0.57

an abrupt change so that at higher alloy content, the solubility is affected strongly. It is significant that in the two systems, the abrupt change in H/M occurs at different alloy compositions but at a similar electron-to-atom ratio (e/a $\sim$ 5.1 in the hydrogen free alloy). This behavior is compatible with Switendick's result (*51*).

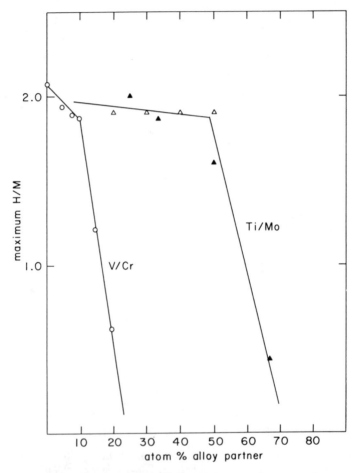

Figure 8.   Maximum hydrogen solubility as a function of atom % alloy partner exhibited by V/Cr and Ti/Mo alloys.   Data obtained at 40°C for a hydrogen charging pressure of 60 atm. (O) V/Cr, Ref. 52; (▲) Ti/Mo, Ref. 56; (△) Ti/Mo, this work.

Alloying molybdenum with titanium (similarly, chromium with vanadium) results in an increased electron density in the metal.   At e/a values $\lesssim$ 5.1, empty electron states are still available near the Fermi level; however, at higher e/a, the solute hydrogen atoms evidently no longer can pull empty states below the Fermi level, and hydrogen solubility is sharply inhibited.

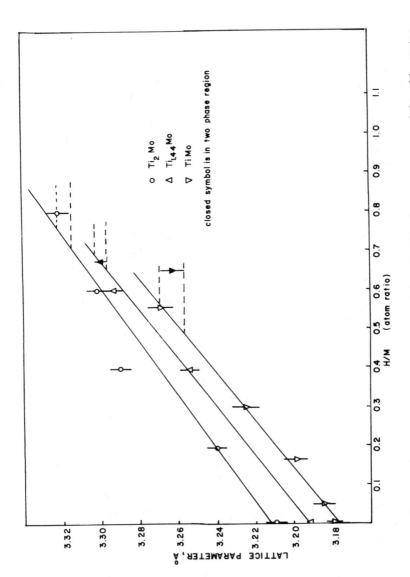

Figure 9. Variation in lattice parameter as a function of hydrogen content exhibited by Ti/Mo alloys. Data are presented as lattice parameter (Å) vs. hydrogen-to-metal atom ratio.

Table VI.   Tritium Isotope Effects

Material Equilibrated	Max H/M	Temperature
$TiMo_2H_{1.10}$ as melted	0.44	$-20°$
$TiMo_{1.94}H_{0.9}$ annealed	0.31	$-20°$
$TiMo_{1.94}H_{0.9}$ annealed	0.31	$40°$
$Ti_{1.09}MoH_{3.29}$ annealed	1.52	$-20°$
$Ti_{1.09}MoH_{2.99}$ as melted	1.43	$40°$
$Ti_{1.09}MoH_{2.53}$ annealed	1.52	$40°$
$Ti_{2.26}MoH_{5.90}$ annealed	1.81	$-20°$
$Ti_{2.26}MoH_{4.77}$ as melted	1.46	$40°$
$Ti_{2.26}MoH_{5.9}$ annealed	1.81	$40°$
$Ti_{2.26}MoH_{5.6}$ annealed	1.81	$103.2°$
$Ti_{2.26}MoH_{5.1}$ annealed	1.81	$157.5°$
$Ti_{4.19}MoH_{8.2}$ annealed	1.64	$110°$
$Ti_{4.19}MoH_{8.1}$ annealed	1.64	$173.6°$
$Ti_{4.19}MoH_{7.7}$ annealed	1.64	$195.8°$

X-Ray Studies. The x-ray data that were determined at room temperature for the hydrogen-free alloys and the hydrides are presented in Table V and Figure 9. The data reveal that the parent alloys are bcc, exhibiting a decrease in lattice parameter with increasing molybdenum content; such a trend was noted in the literature (29). Hydrogen solution in the metal (here designated the β-phase) results in a bcc lattice expansion without a change in crystal structure; Figure 9 indicates that within experimental error, the lattice expansion is linear in hydrogen content. Powder patterns were determined for samples in the two-phase region which allowed simultaneous determination of the parameters of the β-phase and γ-phase (β-max and γ-min in Table V). The effect of the molybdenum content on these two phases is similar to that observed in the parent alloys; i.e., the lattice parameter decreases with increasing molybdenum content.

The x-ray data presented in Figure 9 permit an estimation of the hydrogen composition at the $\beta \rightleftarrows \gamma$ phase boundary via determination of the lattice parameter of the bcc β-phase obtained on samples examined in the two-phase region. These results are included in Table V. The accuracy of these data are sufficient to conclude that the composition of β-max (i.e., the maximum hydrogen composition exhibited by the single β-phase) decreases with molybdenum content. This result agrees qualitatively with the data presented in Figure 7.

Hydrogen–Tritium Exchange Studies. Isotope exchange experiments were conducted as described previously for the γ-phase hydrides (except $TiMo_2$, that does not form the γ-phase). The results of the isotope separation determinations are shown in Table VI, where the parameter α is defined as:

$$\alpha = \frac{T/H \text{ in solid}}{T/H \text{ in gas}}$$

in Titanium–Molybdenum Alloys

cpm Solid/cpm Gas	Activity Balance
1.87	96.8
1.85	101
1.24	101
2.40	106
1.61	100
1.59	104
1.98	96.6
1.61	100.8
1.64	104.8
1.31	105
1.19	103
1.55	89
1.15	80
1.0	46.6

The exchange experiments clearly demonstrate that an inverse isotope effect is exhibited by the γ-phase hydrides since the observed values of α exceed unity. The limited data that is available suggests a linear correlation between $1/T$ and $\ln \alpha$. A similar correlation was reported by Sicking (53) in the palladium–hydrogen–tritium system. There is, however, no clear trend in the temperature effect with respect to alloy composition. The alloy TiMo shows the largest temperature effect, and the alloy Ti_2Mo shows the least. The temperature effects for $TiMo_2$ and Ti_4Mo fall in between those for TiMo and Ti_2Mo.

When the isotope effect is examined with respect to alloy composition, no correlation is observed. Although the isotope effects cannot be compared directly for all four alloys at the same temperature, three can be compared at either $-20°$ or $40°C$. The isotope effect exhibits a maximum with TiMo at $-20°C$ whereas TiMo and Ti_2Mo exhibit virtually the same isotope effect at $40°C$. Clearly there is no correlation of α with alloy composition.

A question arises as to the sources of error in determining α. Although many duplicate runs were not made because of the time needed for each determination, the agreement shown in the duplicate runs indicate that reproducibility is good. The good tritium material balances also lend confidence to the results.

Extrapolation of the observed separation factors indicates that the reverse isotope effect persists to about $155°C$ in TiMo. Although the observed isotope effect is determined for the γ-phase hydrides, it is likely that absorption data at low solute contents will exhibit the same trend.

Summary

Well-characterized β-Ti/Mo alloys varying in molybdenum content from 33 to 66 atom % were used to determine the thermodynamic parameters of protium and deuterium absorption, the partial molar enthalpy of diprotide formation, the relative stabilities of protides and tritides, and the structural changes during hydrogen absorption. A summary of the important results reveals:

(1)   a marked inverse hydrogen–tritium isotope effect exhibited by the γ-phase hydrides near room temperature, with TiMo exhibiting the most pronounced effect;

(2)   a decrease in dihydride stability with increasing molybdenum content;

(3)   an increase in values of $\Delta \overline{H}_H$ with hydrogen content coupled with an apparent repulsive hydrogen–hydrogen interaction, exhibited by $TiMo/H_2$ in the region of dilute hydrogen solution;

(4)   evidence in $\Delta \overline{S}_H^{xs}$ behavior that molybdenum presence in the β-Ti lattice blocks interstitial sites that potentially are available for hydrogen occupation at low hydrogen contents.

Acknowledgment

The authors gratefully acknowledge R. H. Wiswall for his criticisms and suggestions during the course of this work.   The expert assistance of A. Holtz and A. Cendrowski in preparing and characterizing the alloys also is gratefully acknowledged.

Literature Cited

1. Tanaka, J., Wiswall, R. H., Jr., Reilly, J. J., unpublished data.
2. Mueller, W. M., "Metal Hydrides," W. M. Mueller, J. P. Blackledge, G. G. Libowitz, Eds., Chapter 8, Academic, New York, 1968.
3. McQuillan, A. D., *Proc. R. Soc.* (1951) **A204**, 309.
4. Azarkh, Z. M., Garrilov, P. J., *Sov. Phys. Crystallogr.* (1970) **15**, 231.
5. Yakel, H. L., Jr., *Acta Crystallogr.* (1958) **11**, 46.
6. Nagel, H., Perkins, R. S., *Z. Metallkd.* (1975) **66(6)**, 362.
7. Goon, E. J., Malgiolio, J., AEC Report NYO-7547, April 1, 1958.
8. Sidhu, S. S., Heaton, L., Zauberis, D. D., *Acta Crystallogr.* (1956) **9**, 607.
9. Miron, N. F., Shcherbak, V. I., Bykov, V. N., Levdik, V. A., *Kristallografiya* (1974) **19(4)**, 754.
10. Giorgi, T. A., Ricca, F., *Nuovo Cimento, Suppl.* (1967) **V**, 472.
11. Morton, J. R., Stark, D. S., *Trans. Faraday Soc.* (1960) **56**, 351.
12. Takeuchi, S., Honma, T., Ikeda, S., *Nippon Kinzoku Gakkaishi* (1966) **30 (12)**, 1173.
13. Dantzer, P.,Kleppa, O. J., Melnichak, M. E., *J. Chem. Phys.* (1976) **64**, 139.
14. Miedema, A. R., Buschow, K. H. J., Van Mal, H. H., *J. Less Common Met.* (1976) **49**, 463.
15. Reilly, J. J., Johnson, J. R., *World Hydrogen Energy Conf., 1st, Florida 1976.*
16. Reilly, J. J., Wiswall, R. H., Jr., *Inorg. Chem.* (1974) **13**, 218.
17. Miron, N. F., Shcherbak, V., Bykov, V. N., Ledvik, V. A., Report 1972, FE1–341 22pp., CA 82 163254.
18. Chertkov, A. A., Somenkov, V. A., Shil'shtein, S. Sh., Kalanov, M., Mikheeva, V. I., *Strukt. Faz. Fazovye Prevrasch. Diagr. Sostoyaniya Metall. Sist.* (1974) 18, CA 83 069441.
19. Lileev, Yu. Ya., Chertokov, A. A., Malyuchkov, O. T., *Izv. Vyssh. Uchebn. Zaved. Chern. Metall.* (1970) **13(3)**, 123.
20. Upadhyaya, G. S., McQuillan, A. D., *Trans. Metall. Soc. AIME* (1962) **224**, 1290.

21. Pessall, N., McQuillan, A. D., *Trans. AIME* (1962) **224**, 536.
22. Labaton, V. Y., Garner, E. V., Whitehead, E., *J. Inorg. Nucl. Chem.* (1962) **24**, 1197.
23. Grushina, V. V., Rodin, A. M., *Zh. Fiz. Khim.* (1963) **37**, 288.
24. Rudman, P. S., Reilly, J. J., Wiswall, R. H., *Berg. Bunsenges. Phys. Chem.* (1971) **81**, 76.
25. McQuillan, A. D., *J. Inst. Met.* (1951) **80**, 363.
26. Saito, H., Someno, M., *Nippon Kinzoku Gakkaishi* (1972) **36**(8), 791.
27. Jones, D. W., Pessall, N., McQuillan, A. D., *Philos. Mag.* (1961) **6**, 455.
28. Artman, D. D., Lynch, J. F., Flanagan, T. B., *J. Less Common Met.* (1976) **45**, 215.
29. Pearson, W. B., "A Handbook of Lattice Spacings and Structures of Metals and Alloys," p. 875, Pergamon, New York, 1958.
30. Schurmann, T., Kootz, T., Preisendanz, H., Schuler, P., Kander, G., *Z. Metallkd.* (1974 **65**, 249.
31. Jones, P. M. S., Ellis, P., Aslett, T., *Nature* (1969) **223**, 829.
32. Wiswall, R. H., Jr., Reilly, J. J., *Inorg. Chem.* (1972) **11**, 1691.
33. Hardcastle, K. I. Gibb, T. R. P., Jr., *J. Phys. Chem.* (1972) **76**(6), 927.
34. Westlake, D. G., Mueller, M. H., Knott, H. W., *J. Appl. Crystallogr.* (1972) **6**, 206.
35. Westlake, D. G., Mueller, M. H., *Ber. Kernforschungsanlage Juelich* (1972) **6**(1), 66.
36. Maeland, A. J., *J. Phys. Chem.* (1964) **68**, 2197.
37. Steward, S. A., *J. Chem. Phys.* (1975) **63**(2), 975.
38. Westlake, D. G., *J. Chem. Phys.* (1976) **65**, 5030.
39. Steward, S. A., *J. Chem. Phys.* (1976) **65**, 5031.
40. Pryde, J. A., Tsong, I. S. T., *Trans. Faraday Soc.* (1971) **67**, 297.
41. Slotfeldt-Ellingsen, D., Pedersen, B., *Phys. Status Solidi A* (1974) **25**(1), 115.
42. Reilly, J. J., Wiswall, R. H., *Inorg. Chem.* (1967) **6**, 2220.
43. Reilly, J. J., Wiswall, R. H., *Inorg. Chem.* (1970) **9**, 1678.
44. Reilly, J. J., Holtz, A., Wiswall, R. H., Jr., *Rev. Sci. Instrum.* (1971) **42**, 1485.
45. Bernstein, W., Ballentine, R., *Rev. Sci. Instrum.* (1950) **21**, 158.
46. Christman, D. R., *Nucleonics* (1961) **19**, 51.
47. Flanagan, T. B., Oates, W. A., *Ber. Bunsenges. Phys. Chem.* (1972) **76**, 706.
48. Gallagher, P. T., Oates, W. A., *Trans. Metall. Soc. AIME* (1969) **245**, 179.
49. Tanaka, S., Flanagan, T. B., *J. Less Common Met.* (1977) **51**, 79.
50. Lynch, J. F., Johnson, J. R., Reilly, J. J., to be published.
51. Switendick, A. C., *J. Less Common Met.* (1976) **49**, 283.
52. Lynch, J. F., Reilly, J. J., Millot, F., unpublished data.
53. Sicking, G., *Z. Phys. Chem. (Frankfurt am Main)* (1974) **93**, 53.
54. Hansen, M., "Constitution of Binary Alloys," 2nd ed., McGraw–Hill, New York, 1958.
55. DeFontaine, D. et al., *Acta Metall.* (1971) **19**, 1153.
56. Grushina, V. V., Rodin, A. M., *Russ. J. Phys. Chem. Engl. Transl.* (1963) **37**, 288.

RECEIVED July 19, 1977. This work was supported by the Division of Basic Energy Sciences, U.S. Energy Research and Development Administration, Washington, D. C.

25

Results of Reactions Designed to Produce Ternary Hydrides of Some Rarer Platinum Metals with Europium or Ytterbium

RALPH O. MOYER, JR., ROBERT LINDSAY, and DAVID N. MARKS

Trinity College, Hartford, CT 06106

Bimetallic ternary hydrides in which at least one of the metal constituents is an alkali or an alkaline earth element and the second a transition element are reviewed first. The results of the preparation, structure, and properties of materials formed by reactions between EuH_2 or YbH_2 and ruthenium or iridium are described and interpreted. Eu_2RuH_6 or Yb_2RuH_6 can be formed by reacting EuH_2 or YbH_2 with ruthenium at $800°C$ in a hydrogen atmosphere. These ternary hydrides are isostructural with their alkaline earth–ruthenium analogs Sr_2RuH_6 and Ca_2RuH_6, with the metal atoms arranged in a fluorite-type lattice. The magnetic susceptibilities are consistent with $+2$ oxidation states for europium in Eu_2RuH_6 and for ytterbium in Yb_2RuH_6. The electrical conductivities of Eu_2RuH_6 and Yb_2RuH_6 show a temperature dependence typical of semiconductors. Ir data for Eu_2RuH_6 and Yb_2RuH_6 suggest absorption bands which could be attributed to ruthenium–hydrogen stretching and bending modes. Recent work concerning the preparation of the hydride systems europium–iridium–hydrogen and ytterbium–iridium–hydrogen are discussed.

Presently, there is considerable interest in using hydrogen as a fuel and in using some metal hydrides as convenient, safe, and economical storage devices. In view of the present concern for alternative energy supplies, these notions, although not exactly new, are nonetheless more appealing than when first proposed. G. F. Jaubert (1) obtained a French patent in 1902 for the application of metal hydrides, and particularly calcium hydride, for the production of hydrogen. Five years later he received a British patent describing an apparatus mounted on a

0-8412-0390-3/78/33-167-366/$05.00/0

© American Chemical Society

carriage for an almost-continuous production of hydrogen from calcium hydride (2, 3). The binary hydride was considered portable hydrogen for filling dirigible hulls.   Bimetallic hydride systems, i.e., ternary hydrides, of the general stoichiometry MM'H$_x$, where M' is a d-block transition element and M is a rare earth, a d-block transition element, or an actinide element, have been studied extensively.   These systems, with the exception of those where M is europium or ytterbium, will not be discussed.   Recent reports by H. W. Newkirk (4) and Van Mal (5) give reviews of these ternary hydrides along with summaries of their properties.   In addition, the large class of complex metal ternary hydrides of the representative elements, such as LiAlH$_4$ and NaBH$_4$, some of which are used extensively as reducing agents, will not be considered.

This chapter commences with a review of a limited number of ternary hydride systems that have two common features.   First, at least one of the two metal constituents is an alkali or alkaline earth element which independently forms a binary hydride with a metal hydrogen bond that is characterized as saline or ionic.   The second metal, for the most part, is near the end of the d-electron series and with the exception of palladium, is not known to form binary hydrides that are stable at room temperature.   This review stems from our own more specific interest in preparing and characterizing ternary hydrides where one of the metals is europium or ytterbium and the other is a rarer platinum metal.   The similarity between the crystal chemistry of these di-valent rare earths and Ca^{2+} and Sr^{2+} is well known so that in our systems, europium and ytterbium in their di-valent oxidation states are viewed as pseudoalkaline earth elements.

Li–M–H Systems Where M = Eu, Sr, or Ba

Messer and his co-workers (6, 7) reported the formation of LiEuH$_3$, LiSrH$_3$, and LiBaH$_3$ by heating a binary mixture of the respective elements in 1 atm H$_2$ at $700°-750°$C for 20–30 min and then at $500°-600°$C for 2–3 hr before cooling. X-ray powder diffraction patterns were indexed on the basis of a primitive cubic cell with $a = 3.796$ Å for LiEuH$_3$, $a = 3.833$ Å for LiSrH$_3$, and $a = 4.023$ Å for LiBaH$_3$.   Each ternary hydride adopted the perovskite structure with $\rho = 3.378$ g/cm^3 for LiEuH$_3$, $\rho = 2.877$ g/cm^3 for LiSrH$_3$, and $\rho = 3.756$ g/cm^3 for LiBaH$_3$.

Greedan (8) reported the preparation of LiEuH$_3$ and LiSrH$_3$ by heating the respective binary hydrides at $700°$C for $\frac{1}{2}$ hr in approximately 1 atm H$_2$ according to the following equation, where M = Sr or Eu:

$$LiH + MH_2 \rightarrow LiMH_3$$

He also reported the preparation of large single crystals (13 mm in diameter and 13–25 mm in length) of LiEuH$_3$ and LiSrH$_3$ by reacting the respective binary hydrides and growing the crystals from the melt by a modified Bridgeman–Stockbarger technique.   Single crystals of LiSrH$_3$ were orange brown; those of

$LiEuH_3$ were deep red. Room temperature electrical resistivities for $LiEuH_3$ and $LiSrH_3$ were approximately 10^7 ohm-cm, showing little change after annealing.

Potassium–Magnesium–Hydrogen System

Ashby and his group (9) reported the formation of $KMgH_3$ by the hydrogenolysis of $KMg(sec$-$C_4H_9)_2H$ in benzene according to the following equation:

$$KMg(sec\text{-}C_4H_9)_2H + 2H_2 \rightarrow KMgH_3 + 2C_4H_{10}$$

A $0.5M$ benzene solution of $KMg(sec$-$C_4H_9)_2H$ was hydrogenated at 3000 psi of hydrogen at $25°C$ for 4 hr. Because $KMg(sec$-$C_4H_9)_2H$ is soluble in benzene, the technique avoided any necessity for hydrogenolysis in more basic solvents such as ether, which might have competed with the hydride ion for a coordinating position on the magnesium.

$KMgH_3$ was a yellow solid that reacted violently in air. Like the lithium–alkaline earth hydrides, $KMgH_3$ possessed the perovskite structure and was isostructural with $KMgF_3$. Two broad absorption bands were observed for $KMgH_3$ at 1150 cm^{-1} and 680 cm^{-1} in the ir spectrum. Because similar positions were observed for MgH_2, it was concluded that equivalent environments, with respect to hydrogen, existed between six-coordinate magnesium in the rutile MgH_2 and the six-coordinate magnesium in the perovskite $KMgH_3$.

Li–M–H Systems Where M = Group VIIIB Metal

Two ternary hydrides of the lithium–rhodium–hydrogen system, Li_4RhH_4 and Li_4RhH_5, are known. J. D. Farr (10) reported the formation of Li_4RhH_4 by reacting freshly ground LiH and rhodium at $600°C$ in an argon atmosphere. Li_4RhH_4 also was formed by heating Li_4RhH_5 to $400°C$ in a vacuum according to the following equation:

$$Li_4RhH_5 \rightarrow Li_4RhH_4 + \tfrac{1}{2}H_2$$

Li_4RhH_5 was prepared by heating lithium and rhodium or LiH and rhodium in hydrogen at $600°C$ and slowly cooled at $440°C$ in a period of 1 hr while maintaining a hydrogen atmosphere of approximately 550 Torr. Li_4RhH_4 and Li_4RhH_5 were reactive to water and alcohol, resulting in hydrogen evolution and reformation of rhodium.

Single crystal x-ray diffraction techniques were used to elucidate the structures of Li_4RhH_4 and Li_4RhH_5 (11). The unit cell of Li_4RhH_4 was found to be tetragonal with $a = 6.338$ Å and $c = 4.113$ Å, with a probable space group of $I4/m$. The density was 2.707 g/cm^3, consistent with two formula units per unit cell. Li_4RhH_5 was orthorhombic with $a = 3.880$ Å, $b = 9.020$ Å, and $c = 8.895$ Å; the space group was $Cmcm$. The number of formula units per unit cell was four, giving a calculated density of 2.895 g/cm³.

The magnetic susceptibility of Li_4RhH_4 and Li_4RhH_5 was measured at $51°$, $75°$, and $297°$K in a field of 4–10 kG (*11*). The molar susceptibility for Li_4RhH_4 was 1.2×10^{-6} cgs and for Li_4RhH_5 was 2.1×10^{-6} cgs; both compounds showed temperature independent paramagnetic behavior. Room temperature electrical resistivities were 0.5 ohm-cm for Li_4RhH_4 and 1.5 ohm-cm for Li_4RhH_5 (*11*).

Graefe and Robeson (*12*) presented evidence for the preparation of impure Li_3IrH_6, lithium palladium hydride, and lithium platinum hydride. Li_3IrH_6 was formed by reacting LiH and iridium at 500 Torr of H_2 for 8 hr at $539°$C. Li_3IrH_6, a light yellow solid, reacted readily with the atmosphere.

Potassium–Rhenium–Hydrogen and Potassium–Technetium–Hydrogen Systems

One of the most interesting accounts in ternary hydride chemistry surrounds K_2ReH_9, potassium nonahydridorhenate(VII), formed by the reduction of potassium perrhenate with potassium or lithium. Prior to 1960 the product was believed to be potassium rhenide. Floss and Grosse in 1960 (*13*) suggested that the product of the reduction of perrhenate with potassium in aqueous ethylenediamine was a complex ternary hydride of rhenium with the formula $K(ReH_4)\cdot 2$-$4H_2O$. At about the same time, Ginsberg and co-workers (*14*) independently announced in a note that the product prepared in a fashion similar to that reported by Floss and Grosse was a hydride with rhenium in a positive oxidation state. Later Ginsberg et al. (*15*) assigned the formula K_2ReH_8 to potassium rhenium hydride. It was not until 1964 that the structure of potassium rhenium hydride was resolved unequivocally by neutron diffraction analysis and the composition was shown to be K_2ReH_9 (*16*). On the basis of single crystal x-ray diffraction, K_2ReH_9 was hexagonal with unit cell dimensions of $a = 9.607$ Å and $c = 5.508$ Å and three formula weights per unit cell (*17*). The probable space group was D_{3h}^3-$P\bar{6}2m$. The neutron diffraction studies supported a nine-coordinate rhenium atom with six of the nine hydrogens at the corners of a trigonal prism, with rhenium at its center and the remaining three hydrogens beyond the centers of the rectangular faces (*16*). The hydrogen elemental analysis by a thermal decomposition technique supported a H:Re = 8.7 and by a combustion technique, a H:Re = 8.9 (*16*). The magnetic susceptibility for potassium rhenium hydride (*18*) was found to be temperature independent from $80°$ to $295°$K, with a x_m of $-64 \pm 10 \times 10^{-6}$ cgs/mol, uncorrected for the core diamagnetism. After these corrections were applied, a small positive value was calculated, i.e., $x_m = +25 \times 10^{-6}$ cgs/mol, entirely consistent with rhenium in the $+7$ oxidation state. The ir spectrum of K_2ReH_9 showed two intense bands, one in the metal–hydrogen stretching region at 1846 cm^{-1} and the second at 735 cm^{-1} in the metal–hydrogen bending region (*15*).

Potassium technetium hydride was formed by the reduction of ammonium pertechnate with potassium in a mixed solvent of anhydrous ethylenediamine-absolute ethanol containing 2% potassium ethoxide (19). K_2TcH_9 was similar to K_2ReH_9 in its chemical reactivity, its structure, and ir spectrum. According to the x-ray powder diffraction evidence, K_2TcH_9 was isostructural with K_2ReH_9. The ir spectrum of K_2TcH_9, scanned from 4000 to 400 cm^{-1}, was precisely the same as the pattern observed for K_2ReH_9 except that for K_2TcH_9, all the bands were shifted approximately 50 cm^{-1} lower. This shift indicated the transition metal–hydrogen bond was weaker for K_2TcH_9 than for K_2ReH_9 and was used to explain the greater chemical reactivity of K_2TcH_9. Like K_2ReH_9, K_2TcH_9 reacted with water, giving off hydrogen and leaving behind a precipitate, presumably the transition metal. K_2TcH_9 was soluble in strong aqueous alkali with only minor decomposition.

M–Re–H Systems Where M = Na, K, or $(C_2H_5)_4N$

The sodium salt, the quaternary ammonium salt, and a mixed sodium–potassium salt of nonahydridorhenate(VII), as well as a new route to the formation of K_2ReH_9, were reported (20). Na_2ReH_9 was formed by reducing an ethanolic solution of sodium perrhenate with sodium according to the following equation:

$$NaReO_4 \xrightarrow[C_2H_5OH]{\text{excess Na}} Na_2ReH_9 + NaOC_2H_5 + NaOH$$

The ternary hydride was purified by reprecipitation from aqueous methanolic sodium hydroxide. Na_2ReH_9 was soluble in water and methanol, slightly soluble in ethanol, and insoluble in 2-propanol, acetonitrile, ether, and THF. Decomposition started at approximately 245°C when heated in a vacuum, with the evolution of hydrogen and sodium as the temperature increased. Na_2ReH_9 was a precursor for the preparation of the tetraethylammonium salt, the potassium salt, and the mixed potassium sodium salt of nonahydridorhenate(VII).

$[(C_2H_5)_4N]_2ReH_9$ was formed by the metathesis reaction between Na_2ReH_9 and $[(C_2H_5)_4N]_2SO_4$, and the product was purified by reprecipitation with acetonitrile. $[(C_2H_5)_4N]_2ReH_9$ was soluble in water, ethanol, 2-propanol, and acetonitrile but insoluble in ether and THF. Thermal decomposition in a vacuum occurred at 115°–120°C, giving off hydrogen and ethane.

The interesting product formed by the treatment of a solution of Na_2ReH_9 with excess KOH, followed by precipitation with methanol, was not K_2ReH_9 but rather $NaKReH_9$. The x-ray powder diffraction pattern showed clearly that the mixed cation ternary hydride was not a mixture of Na_2ReH_9 and K_2ReH_9. The formation of K_2ReH_9 from Na_2ReH_9 took a more circuitous route. Na_2ReH_9 first was converted to $BaReH_9$, followed by the formation of K_2ReH_9 by the metathesis reaction between $BaReH_9$ and K_2SO_4.

Magnesium-Nickel-Hydrogen System

Reilly and Wiswall (*21*) reported the synthesis of Mg_2NiH_4 by reacting the Mg_2Ni intermetallic compound with hydrogen at 300 psi and 325°C, according to the following chemical equation:

$$Mg_2Ni(s) + 2H_2 \rightarrow Mg_2NiH_4(s)$$

The ternary hydride was formed also at a reduced pressure and temperature of 200 psi and 200°C, respectively, after several cycles of hydriding and decomposition. Mg_2NiH_4, a rust-colored solid with a nonmetallic luster, reacted sluggishly with water but more vigorously with nitric acid solution, giving off hydrogen. Mg_2NiH_4 appeared to be unreactive to air upon short exposure.

The x-ray diffraction powder pattern for Mg_2NiH_4 was indexed on the basis of a tetragonal crystal system with unit cell dimensions of $a = 6.464$ Å and $c = 7.033$ Å. The measured density was 2.57 g/cm^3, compatible with four formula units of Mg_2NiH_4 per unit cell.

Calcium-Silver-Hydrogen System

A new ternary hydride of $CaAg_2H$ was formed by either hydriding the intermetallic $CaAg_2$ alloy at 575°–600°C for 16–18 hr or reacting CaH_2 and silver in a 1:2 molar ratio at 600°–650°C in a hydrogen atmosphere for 16 hr (*22*). The x-ray powder diffraction pattern for $CaAg_2H$ was indexed orthorhombic with unit cell dimensions of $a = 5.45$ Å, $b = 5.19$ Å, and $c = 9.86$ Å. The density for $CaAg_2H$ was 6.10 g/cm^3 and was compatible with four formula units per unit cell. A comparison of the magnetic susceptibility of the $CaAg_2$ intermetallic and $CaAg_2H$ showed a shift from a net positive or paramagnetic susceptibility in the former to a net negative or diamagnetic susceptibility in the latter. It has been speculated that electrons from the hydrogen fill up holes in the conduction band in the $CaAg_2$ that are responsible for the spin paramagnetism and thus suppress the susceptibility (*22*).

Ca-M-H and Sr-M-H Systems Where M = Group VIIIB Metal

$SrPd_2H$, Sr_2PdH_4, and $Ca_3Pd_2H_4$ were reported by Stanitski and Tanaka (*23*). $SrPd_2H$ and Sr_2PdH_4 were formed by reacting stoichiometric mixtures of SrH_2 and palladium for 10 hr at 760°C and 600 Torr H_2 for $SrPd_2H$ and at 805°C and 625 Torr H_2 for Sr_2PdH_4. Both ternary hydrides were black, crystalline, nonvolatile solids and were reactive to water or acidic solutions evolving hydrogen. $SrPd_2H$ also was formed by hydrogenation of the $SrPd_2$ intermetallic alloy. X-ray powder diffraction analysis showed that the metal atoms in $SrPd_2H$ were arranged in a cubic $MgCu_2$, Laves-phase array with the unit cell length of 7.97 Å. $Ca_3Pd_2H_4$ was formed by a solid state reaction between CaH_2 and palladium at 850°C in H_2 at 625 Torr. The x-ray powder pattern for $Ca_3Pd_2H_4$

was indexed body-centered cubic with a = 5.22 Å. $SrPd_2H$, Sr_2PDH_4, and $Ca_3Pd_2H_4$ were either diamagnetic or weakly paramagnetic.

The ternary hydrides Ca_2IrH_5, Sr_2IrH_5, Ca_2RhH_5, Sr_2RhH_5, Ca_2RuH_6, and Sr_2RuH_6 were formed by reacting the alkaline earth binary hydride with the platinum group metal (24). Here, the conditions were a temperature of 800°C in approximately 1 atm H_2 for about 12 hr. Like the calcium–palladium and strontium–palladium systems, these hydrides were nonvolatile and reacted with water or acidic solutions, with the evolution of hydrogen. Structural studies were carried out by x-ray diffraction and neutron diffraction analysis. The x-ray diffraction powder pattern for each of the six was indexed on the basis of a face-centered cubic (24), and the respective cell dimensions are given in Table I. Structural studies showed that the metal atoms were arranged in a fluorite-type lattice consistent with the space group $Fm3m$, wherein the alkaline earth atoms were occupying the fourfold sites and the rarer platinum metals were occupying the eightfold sites.

Deuterium sites for Sr_2IrD_5 and Sr_2RuD_6 were located from neutron diffraction studies, and the positions that gave the best fit (24) are summarized in Table II. With these coordinates, bond distances also were determined and are shown in Table III. The transition elements were viewed as six coordinate with respect to deuterium, where the atomic ratios were D:Ru = 6 for Sr_2RuD_6 and D:Ir = 5 plus one random vacancy for Sr_2IrD_5.

Table I. Ternary Hydride Physical Properties

Hydride	Color	Cell Dimension[a]	Density, Exp.	g/cc Calc.[b]
Ca_2IrH_5	Black	7.29	4.80	4.85
Sr_2IrH_5	Black	7.62	5.46	5.56
Ca_2RhH_5	Black	7.24	3.40	3.31
Sr_2RhH_5	Black	7.60	4.20	4.24
Ca_2RuH_6	Light Green	7.24	3.32	3.31
Sr_2RuH_6	Light Green	7.60	4.25	4.24

[a] Cubic crystal system.
[b] Calculation based on four formula units per unit cell.

Table II. Structural Parameters for Sr_2IrD_5 and Sr_2RuD_6 (Space Group $Fm3m$)

Atom	Position[a]	Multiplicity
Sr	± (¼, ¼, ¼)	8
Ir, Ru	(0, 0, 0)	4
D	± (¼, 0, 0)	
	± (0, ¼, 0)	24
	± (0, 0, ¼)	

[a] Plus face centering, i.e., (0, ½, ½) and permutations should be added to each atomic position.

Table III. Bond Distances (in Å) for the Sr_2MD_x Structure

	Ir	Ru
M–D	1.70	1.69
Sr–D	2.70	2.69
Sr–M	3.30	3.29

Table IV. Electrical Resistivity at Room Temperature

Metal	$\rho(ohm\text{-}cm)$[a]	Ternary Hydride	$\rho(ohm\text{-}cm)$
Ir	6.1×10^{-6}	Ca_2IrH_5	$>10^7$
Rh	4.7×10^{-6}	Ca_2RhH_5	~10
Ru	10×10^{-6}	Ca_2RuH_6	$>10^6$
Ca	4.6×10^{-6}	Sr_2IrH_5	$>10^6$
Sr	25×10^{-6}	Sr_2RhH_5	~10
		Sr_2RuH_6	$>10^7$

[a] Ref. 41.

All of the above ternary hydrides with iridium, rhodium, and ruthenium as the second metal showed weak temperature-independent paramagnetism or diamagnetism between 77° and 300°K (24). The electrical resistivities of the ternaries showed a sharp increase from those of the individual metallic constituents, as is presented in Table IV. These trends toward appreciably higher resistivities in the ternary hydride suggest that hydrogenation has modified the electron population of the conduction bands of the metals.

Eu–M–H and Yb–M–H Systems Where M = Ru

The results on the alkaline earth–group VIIIB ternary hydride systems suggested that systems where the alkaline earth is replaced by europium or ytterbium would be worthy of study for several reasons. First of all, the structural similarities between EuH_2, YbH_2, and CaH_2, SrH_2, and BaH_2 (all orthorhombic) and between $LiEuH_3$ and $LiSrH_3$ (both perovskite) suggested that the rare earth metals might substitute at alkaline-earth metal sites. Secondly, the distinctive magnetic character of the two rare earth metals, europium and ytterbium, with well-defined atomic moments arising from relatively well shielded $4f$ electrons seemed to offer a way to study valence states from data obtained by magnetic susceptibility measurements. In particular, the tendencies of these rare earth metals to form both di-valent and tri-valent compounds raised the question as to how hydrogenation would affect these preferences. Finally, there was interest in further investigating the shift in electrical conductivity that already had been identified in the alkaline earth–rarer platinum metal–hydrogen systems and which might offer evidence concerning the kind of electron exchange participated in by the hydrogens. The remainder of this article will be concerned with the background, experimental techniques, and interpretation of results obtained on investigations carried out so far in the laboratory of the senior author of this chapter.

There has been some previous work both on binary hydrides of europium and ytterbium and on binary alloys of europium and ytterbium with platinum group metals, which should be mentioned as a prelude to any consideration of the europium or ytterbium ternary hydride systems. The hydrides of europium and ytterbium are distinctive in comparison with the other lanthanide-series hydrides in that it has not been possible to prepare any hydrides higher than the dihydride at 1 atm H_2 (25). Furthermore, the EuH_2 and YbH_2 have the ortho-rhombic structure like the alkaline earth dihydrides, in contrast to the cubic fluorite structure that is favored by the other lanthanide dihydrides. EuH_2 is ferromagnetic with a Curie temperature of about $25°K$; above this temperature, its susceptibility can be fitted to a Weiss–Curie law which yields an effective Bohr magneton number quite close to that expected for Eu^{2+} (26). YbH_2 has been reported as being weakly paramagnetic with an order of magnitude which is felt to reflect the theoretically zero-atomic moment for Yb^{2+} (27). There is evidence of an increase in magnetic susceptibility for $YbH_{2.55}$ (28). The electrical resistivity of YbH_2 (29) has been shown to be of the order of magnitude of 10^7 ohm-cm, which is some 12 orders of magnitude greater than pure ytterbium metal (30). There are no values reported for the electrical resistivity of EuH_2, although it is presumably an insulator (26).

The question as to the electronic configuration of the hydrogen in EuH_2 and YbH_2 is still open. The magnetic studies by Wallace et al. (26, 27) and the conductivity studies on YbH_2 by Heckman and Hills (29) have been interpreted as favoring an anionic or H^- model for the hydrogen. Mössbauer studies by Mustachi on both of these hydrides also yielded results that support the anionic model (31). However, there is a problem with the inference when trying to explain the relatively high Curie temperature ($25°K$) in EuH_2. According to theories of cooperative magnetic behavior, the exchange effects necessary to produce ferromagnetism require the conduction electrons of the metals; any removal of these electrons in hydride formation might be expected to suppress the ferromagnetism. Explaining the ferromagnetism while retaining the anionic model requires the more speculative arguments of either direct exchange between europium ions or an indirect routing of exchange via the H^- ions (26, 31). There are other lanthanides where the experimental evidence appears to favor a protonic model. Schreiber and Cotts (32) and Kopp and Schreiber (33) concluded from NMR data on lanthanum–hydrogen and cerium–hydrogen that these systems were described better by the protonic model.

The properties of some rare-earth binary alloys with platinum group metals are also important in view of the role they can play in the chain of preparing ternary hydrides. Many of the alloys of the series R–M, where R is a rare earth element and M is a Group VIIIB metal, have been investigated structurally and magnetically. The alloys with iridium all have cubic structures, whereas those with ruthenium and osmium are hexagonal as long as the rare earth is heavier than neodymium. The cubic structures are of the C15-type Laves phase while the hexagonal structures have the C14-type Laves phase (34). The alloys that were studied magnetically (35) were ferromagnetic, with Curie temperatures varying systematically from close to zero to 90°K and then back to zero as the rare earth metal progressed from cerium to lutetium. The alloys specifically included: $CeIr_2$, $PrIr_2$, $PrOs_2$, $PrRu_2$, $NdIr_2$, $NdOs_2$, $NdRu_2$, $SmIr_2$, $SmOs_2$, $EuIr_2$, $GdIr_2$, $GdOs_2$, $GdRu_2$, $TbIr_2$, $TbOs_2$, $DyIr_2$, $DyOs_2$, $HoIr_2$, $HoOs_2$, $ErIr_2$, $ErOs_2$, $TmIr_2$, $YbIr_2$, and $LuRu_2$. The ferromagnetism was attributed to the exchange interaction between conduction electrons and the spin of the 4f-shell electrons. $EuIr_2$ was reported as having a large (77°K) Curie temperature and an extremely small saturation moment. However, later work suggested that the ferromagnetism may have been caused by the presence of EuO as an impurity (36). Some more recent results on $EuIr_2$ will be discussed later in connection with attempts to hydrogenate it.

Experimental

Synthesis. Europium and ytterbium metals were purified by vacuum distillation at 800°C. Tank hydrogen was purified by passing the gas through a heated palladium tube filter, Model H-1-DH, purchased from Matthey Bishop, Inc.

The usual synthesis route first was to prepare europium or ytterbium hydride by heating the metals in approximately 1 atm H_2 at 500°C. Ternary hydrides were then prepared by heating a compressed pellet containing a homogenous mixture of powders of europium or ytterbium hydride and the rarer platinum metal at 800°C for approximately 18 hr and in approximately 1 atm H_2. Metals were handled in a glove bag in a protective atmosphere of argon; the hydrides were handled in a nitrogen atmosphere.

The $EuIr_2$ alloy was prepared by direct combination of the metals. Europium chips, $1/8$ in. on edge were placed in a $1/4$ in.-diameter cavity of a die. Iridium metal powder, −325 mesh and 99.9% pure, was sprinkled over the rare earth element, and the mixture was compressed at 5000 psi. The compressed pellet was placed in a molybdenum boat, which then was transferred to a quartz sleeve, followed by insertion into a quartz reaction tube. The tube was attached to a glass vacuum line and evacuated. Argon was added to approximately 1 atm; and the compressed pellet was heated to 900°C and held at the temperature for 14 hr. The product was air quenched to room temperature. The product was crushed in an agate ball mill, compressed into a pellet again, and reheated in the same manner as before.

Hydrogen Analysis. The thermal decomposition technique was used to determine the hydrogen elemental composition. The sample was heated in vacuo to 925°C and was maintained at that temperature until all evolved gas was transferred by way of a Toepler pump to a calibrated gas buret.

X-Ray Diffraction Analysis. X-ray powder diffraction data for indexing were obtained with a General Electric XRD-6 diffractometer. Samples were sealed in glass capillaries (0.5 mm o.d.) and exposed to nickel-filtered $CuK\alpha$ radiation. KCl or iridium was used as an internal standard. X-ray powder diffraction intensity data were obtained with a General Electric XRD-5 unit. Here, the samples were mixed with a protective coating of petroleum jelly, smeared on a glass slide, and exposed to nickel-filtered $CuK\alpha$ radiation. Intensity data were obtained by measuring the area under the peaks on the chart paper with a planimeter.

Magnetic Susceptibility Measurements. Magnetic susceptibilities were measured by the Faraday method and were accurate to $\pm2\%$. Samples, usually as powders, were placed in a cylindrical Teflon boat having an internal volume of 0.30 cc. Measurements were made at 2963, 4357, 5724, 7077, and 8243 G, and at temperatures $77°-295°K$. It was possible to correct for ferromagnetic impurities from any field dependent effects.

Electrical Resistivity Measurements. Electrical resistances were measured by the voltage–current method. The powder sample was compressed into a pellet with a diameter of $\frac{1}{4}$ in. at 5000 psi. The pellet was placed between two brass electrodes, and the electrical resistance of the sample was determined from the measurements of the current through the sample and the potential across the sample.

Differential Thermal Analysis. High temperature differential thermal analyses were obtained with a Dupont Model 1200 instrument. Samples were heated from room temperature to $950°C$ at a rate of $20°C/min$ in a slow stream of hydrogen. Molybdenum cups were used to hold the sample and alumina reference. The instrument was calibrated with sodium chloride (mp $800°C$).

Ir Analysis. Ir spectra of Eu_2RuH_6 and Yb_2RuH_6 were made with a Beckman Ir-12 spectrophotometer. The samples were scanned as KBr discs from 4000 to 200 cm^{-1}.

Results and Discussion

Europium–Ruthenium–Hydrogen and Ytterbium–Ruthenium–Hydrogen Systems. Eu_2RuH_6 (37) and Yb_2RuH_6 (38) were formed by heating the respective binary rare earth dihydride and ruthenium in a molar ratio of 2:1. The pellet containing the homogeneous mixture expanded substantially during the reaction, accompanied by an absorption of approximately one mole of hydrogen per mole of ruthenium. Eu_2RuH_6 was a brick red crystalline solid; Yb_2RuH_6 was black. Both reacted to acidic solutions by giving off hydrogen and precipitating ruthenium, and both were unreactive to the atmosphere for brief periods of time. Structural analyses by x-ray diffraction techniques showed that as far as the arrangement of the metals were concerned, Eu_2RuH_6 was isostructural with Sr_2RuH_6 with $a = 7.566$ Å, and Yb_2RuH_6 was isostructural with Ca_2RuH_6 with $a = 7.248$ Å.

The magnetic susceptibility studies supported a $+2$ oxidation state for europium in Eu_2RuH_6 and for ytterbium in Yb_2RuH_6. Figure 1 is a plot of the reciprocal of x_A, the corrected magnetic susceptibility per g-atom of europium vs. T, the absolute temperature of the Eu_2RuH_6 sample. A straight line fits the experimental points from $85°$ to $296°K$. The Weiss–Curie law $x_A = C_A/(T - \theta)$ was fitted by $x_A = 7.58/T - 43$. The constant $C_A = 7.58$ corresponded to $\mu_{eff} = 7.82\ \mu_B$. This value agreed with Eu^{2+} having a theoretical magnetic moment of $7.94\ \mu_B$. The ruthenium was believed to be in the $+2$ oxidation state and was not contributing paramagnetically to the bulk susceptibility.

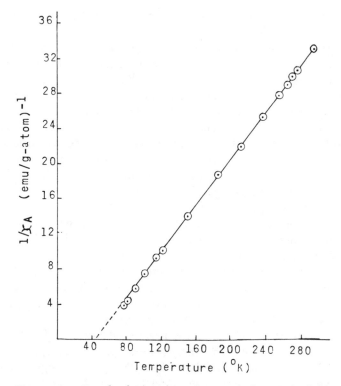

Figure 1.   Graph of $1/\chi_A$ (emu/g-atom)$^{-1}$ vs. T (°K) for Eu_2RuH_6

The value of the Weiss constant θ was 43 ± 1°K and suggested that Eu_2RuH_6 becomes ferromagnetic at some temperature below 77°K.   The magnetic susceptibility for Yb_2RuH_6 appeared to be consistent with di-valent ytterbium, which theoretically has zero atomic moment.   The small paramagnetism observed experimentally was assigned to Yb_2O_3 impurity.

The magnitude of the electrical resistivity as well as its behavior with temperature is shown in Table V.   The data in Table V were fitted to the equation $\rho = \rho_0 e^{\Delta E/2kT}$, where the activation energies ΔE were 0.15 eV for Eu_2RuH_6 and 0.18 eV for Yb_2RuH_6.   The hydrogenation process appears to have produced

Table V.   Electrical Resistivity

Eu_2RuH_6		Yb_2RuH_6	
T (°K)	ρ(ohm-cm)	T (°K)	ρ(ohm-cm)
294	2.5×10^5	293	3.06×10^4
209	1.2×10^6	200	2.27×10^5
167	4.1×10^6	145	1.09×10^6
77	4.0×10^9	77	6.52×10^8

ternary hydrides with semiconducting properties from the constituent metal-lic-type electrical conductors.

A profile of the differential thermal analysis of Eu_2RuH_6 is shown in Figure 2. Two endotherms were observed upon heating the sample: an intense dip at $903°C$ and weaker dip at $794°C$. Upon cooling, two exotherms were found at $817°$ and $788°C$. The $903°C$ endotherm was almost certainly caused by the thermal decomposition of Eu_2RuH_6. The endotherm at $794°C$ was assigned to the decomposition of EuH_2, and the exotherm at $788°C$ was believed to be caused by the formation of EuH_2. The exotherm at $817°C$ was assigned cautiously to the formation of a hydride by a route involving a reaction between the intermetallic compound of europium or ruthenium or by the chemical combination of europium, ruthenium, and hydrogen.

Preliminary interpretation of the ir spectra for Eu_2RuH_6 and Yb_2RuH_6 suggested that there were absorptions that could be assigned to the ruthenium–hydrogen stretching and bending vibrations. The results of the ir data are found in Table VI.

Chatt and Hayter (39) reported the results of ir spectra of several six-coordinate hydrido complexes of ruthenium in the +2 oxidation state. They reported that hydrido complexes that have one hydrogen show a band between 1750 and 1980 cm^{-1} attributable to ruthenium–hydrogen stretch. A shift in the stretching frequency of more than 350 cm^{-1} lower resulted when a second hydrogen was placed in the position trans to the first. For example, the absorption band observed at 1976 cm^{-1} for trans-$[RuHI\{o\text{-}C_6H_4(PEt_2)_2\}_2]$ was assigned to the ruthenium–hydrogen stretching frequency, and the band at 1617 cm^{-1} was assigned to the ruthenium–hydrogen stretching frequency for trans-$[RuH_2\{o\text{-}C_6H_4(PEt_2)_2\}_2]$. The strong ir absorption bands in the 1377 to 1500 cm^{-1} region for Eu_2RuH_6 and Yb_2RuH_6 therefore were assigned to ruthenium–hydrogen

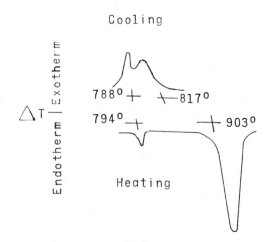

Figure 2.   DTA plot for Eu_2RuH_6

Table VI. Ir Absorption Data for Eu_2RuH_6 and Yb_2RuH_6 [a]

Eu_2RuH_6	Yb_2RuH_6
520 (m)	528 (m)
570 (m)	573 (m)
712 (w, sharp)	
875 (w, sharp)	886 (w)
1377 (s)1480 (s)	1550 (s)

[a] In cm^{-1}; w = weak, m = medium, s = strong.

stretching bands (ν_{Ru-H}), and the absorption bands located between 600–500 cm^{-1} were assigned to ruthenium–hydrogen bending vibration (δ_{Ru-H}).

Europium–Iridium–Hydrogen and Ytterbium–Iridium–Hydrogen Systems. When EuH_2 or YbH_2 was heated with iridium at 800°C in approximately 1 atm H_2, the intermetallic compounds $EuIr_2$ or $YbIr_2$, containing very little hydrogen, were produced. These products will be referred to in the discussion as $EuIr_2H_x$ and $YbIr_2H_x$. In the case of the europium–iridium system, x was determined to be between 0.1 and 0.2. In contrast to the europium– and ytterbium–ruthenium ternary hydrides, the reaction of EuH_2 or YbH_2 with iridium failed to show the pellet expanding when heated, and there was considerable hydrogen evolution. The x-ray diffraction patterns for both the $EuIr_2H_x$ and $YbIr_2H_x$ showed a fcc structure with unit cell dimensions of $a = 7.571$ Å and $a = 7.456$ Å, respectively. Structure factor analysis of the above products indicated that the metal atoms in both compounds were arranged in a cubic C15-type Cu_2Mg Laves phase. These findings are to be compared with the well-known Laves phase of $EuIr_2$ and $YbIr_2$ (40) with respective unit cell dimensions of 7.566 Å and 7.477 Å. As a check $EuIr_2$ was prepared by heating the respective elements at 900°C in an argon atmosphere. Results of the x-ray diffraction patterns showed fcc structure with $a = 7.570$ Å, but there was strong evidence from the patterns that the actual composition should more properly be described as a Laves phase deficient in iridium, close to $EuIr_{1.5}$. Furthermore, a finely divided powder sample of the $EuIr_2$ alloy in a hydrogen atmosphere was heated slowly to 500°C over a period of 30 hr, held at 500°C for approximately 18 hr, and cooled to approximately 100°C over a period of 8 hr. No hydrogen absorption, as noted by manometric measurements, was observed. It has been observed recently that when EuH_2 was combined with iridium in a 2:1 molar ratio and heated to 650°–700°C in approximately 1 atm H_2, a new ternary hydride was formed accompanied by an absorption of hydrogen. The new ternary hydride phase was indexed fcc and predicted to be isostructural with Sr_2IrH_5. The upper limit of the reaction temperature is critical. The iridium-deficient Laves phase mentioned above will form, accompanied by the evolution of hydrogen, should the reaction temperature approach 800°C.

The magnetic and electrical properties of $EuIr_2$ and $EuIr_2H_x$ were investigated to see if any differences in these properties could be detected between the two materials. Over the temperature range from 77° to 295°K, the molar susceptibilities x_M of both materials appeared to be paramagnetic, with little difference between them. However, the magnitudes were such that the presence of even a few percent of EuO would be sufficient to render any quantitative interpretation of these data subject to considerable uncertainty, a point which has been made previously by Bozorth et al. (35, 36) in discussing their results on $EuIr_2$. The resistivity measurements indicated that both preparations were metallic conductors of the same order of magnitude.

Literature Cited

1. Jaubert, G. F., French Pat. 327,878 (1902).
2. Jaubert, G. F., British Pat. 25,215 (1907).
3. Farber, E., "Chymia," Vol. 8, H. M. Leicester, Ed., p. 165, University of Pennsylvania, Philadelphia, 1962.
4. Newkirk, H. W., "A Literature Study of Metallic Ternary and Quaternary Hydrides," UCRL-51244, Rev. 1, Lawrence Livermore Laboratory, University of California, Livermore, 1975.
5. Van Mal, H. H., "Stability of Ternary Hydrides and Some Applications," *Philips Res. Rep. Suppl.* (1976) 1.
6. Messer, C. E., Eastman, J. C., Mers, R. G., Maeland, A. J., *Inorg. Chem.* (1964) 3, 776.
7. Messer, C. E., Hardcastle, K., *Inorg. Chem.* (1964) 3, 1327.
8. Greedan, J. E., *J. Cryst. Growth* (1970) 6, 119.
9. Ashby, E. C., Kovar, R., Arnott, R., *J. Am. Chem. Soc.* (1970) 92, 2182.
10. Farr, J. D., *J. Inorg. Nucl. Chem.* (1960) 14, 202.
11. Lundberg, L. B., Cromer, D. T., Magee, C. B., *Inorg. Chem.* (1972) 11, 400.
12. Graefe, A. F., Robeson, R. K., *J. Inorg. Nucl. Chem.* (1967) 29, 2917.
13. Floss, J. G., Grosse, A. V., *J. Inorg. Nucl. Chem.* (1960) 16, 37.
14. Ginsberg, A. P., Miller, J. M., Cavanaugh, J. R., Dailey, B. P., *Nature* (1960) 185, 528.
15. Ginsberg, A. P., Miller, J. M., Koubek, E., *J. Am. Chem. Soc.* (1961) 83, 4909.
16. Abrahams, S. C., Ginsberg, A. P., Knox, K., *Inorg. Chem.* (1964) 3, 558.
17. Knox, K., Ginsberg, A. P., *Inorg. Chem.* (1964) 3, 555.
18. Ibid (1962) 1, 945.
19. Ginsberg, A. P., *Inorg. Chem.* (1964) 3, 567.
20. Ginsberg, A. P., Sprinkle, C. R., *Inorg. Chem.* (1969) 8, 2212.
21. Reilly, J. J., Wiswall, R. H., *Inorg. Chem.* (1968) 7, 2254.
22. Mendelsohn, M. H., Tanaka, J., Lindsay, R., Moyer, R. O., Jr., *Inorg. Chem.* (1975) 14, 2910.
23. Stanitski, C., Tanaka, J., *J. Solid State Chem.* (1972) 4, 331.
24. Moyer, R. O., Jr., Stanitski, C., Tanaka, J., Kay, M., Kleinberg, R., *J. Solid State Chem.* (1971) 3, 541.
25. Mueller, W. M., "Metal Hydrides," W. M. Mueller, J. P. Blackledge, G. C. Libowitz, Eds., p. 384ff, Academic, New York, 1968.
26. Zanowick, R. L., Wallace, W. E., *Phys. Rev.* (1962) 126, 537.
27. Wallace, W. E., Kuhota, Y., Zanowick, R. L., "Non Stoichiometric Compounds," ADV. CHEM. SER. (1963) 39, 122ff.
28. Hardcastle, K. I., Warf, J. C., *Inorg. Chem.* (1966) 5, 1728.
29. Heckman, R. C., Hills, C. R., *Bull. Am. Phys. Soc.* (1965) 10, 126.
30. Kayser, F. X., Soderquist, S. D., *Scr. Metall.* (1969) 3, 259.
31. Mustachi, A., *J. Phys. Chem. Solids* (1975) 35, 1447.
32. Schreiber, D. S., Cotts, R. M., *Phys. Rev.* (1963) 131, 1118.
33. Kopp, J. P., Schreiber, D. S., *J. Appl. Phys.* (1967) 38, 1373.

34. Compton, V. B., Matthias, B. T., *Acta Crystallogr.* (1959) **12,** 651.
35. Bozorth, R. M., Matthias, B. T., Suhl, H., Corenzuit, E., Davis, D. D., *Phys. Rev.* (1959) **115,** 1595.
36. Matthias, B. T., Bozorth, R. M., Van Vleck, J. H., *Phys. Rev. Lett.* (1961) **1,** 160.
37. Thompson, J. S., Moyer, R. O., Jr., Lindsay, R., *Inorg. Chem.* (1975) **14,** 1866.
38. Lindsay, R., Moyer, R. O., Jr., Thompson, J. S., Kuhn, D. A., *Inorg. Chem.* (1976) **15,** 3050.
39. Chatt, J., Hayter, G., *J. Chem. Soc.* (1961) 2605.
40. Elliott, R. P., *Proc. Rare Earth Res., 4th, New York, 1965,* L. Eyring, Ed., p. 224, Gordon and Breach.
41. E. W. Washburn, Ed., "International Critical Tables of Numerical Data, Physics, Chemistry, and Technology," p. 104, 1st ed., Vol. 1, McGraw–Hill, New York, 1926.

RECEIVED August 1, 1977.

26

Applicability of Surface Compositional Analysis Techniques for the Study of the Kinetics of Hydride Formation

S. A. STEWARD and R. M. ALIRE

Lawrence Livermore Laboratory, University of California, Livermore, CA 94550

Passivation of active metals to hydrogen reaction has been recognized as an important problem in basic metal–hydrogen studies, especially in their technological application to various situations. Few investigations have addressed these difficulties. The advent of modern surface analytical techniques such as photoelectron spectroscopy, Auger electron spectroscopy, and ion spectrometry offer a tremendous opportunity to attack the passivation question. Each of these techniques is discussed with regard to their capabilities and application to hydride kinetics.

Passivation of bulk active metals to hydrogen absorption has been a subject of concern and frustration to interested scientists since the beginning of research in the field. An understanding of the phenomena that create surface passivation in metals and nonmetals would be not only intrinsically satisfying but also immensely important to the many areas of applied hydride technology. Metals, alloys, and intermetallic compounds are being considered as storage materials in hydrogen-energy conceptual studies and as a safe method for storing tritium, the radioactive isotope of hydrogen that will be a fuel in fusion reactors. Scavenging of tritium in fusion-reactor fuel cycles is also important, and passivation caused by chemically reactive impurities could be a severe problem. Neutron generators for medical and materials research often consist of metal tritide targets such as titanium or scandium tritide, in which surface impurities prevent the formation or resist the degradation of the target. Finally, hydrogen isotope permeation through nonreactive metals or alloys such as stainless steel presents a containment problem in many applications. Because surface compounds are thought to inhibit that permeation, the effect of surfaces on hydrogen diffusion into metals must be better understood. The generally accepted cause of deactivation has been the formation of a stable oxide or other compound of

0-8412-0390-3/78/33-167-382/$05.00/0
© American Chemical Society

the parent metal on its surface, thereby severely reducing or effectively pre-cluding hydrogen diffusion into the metal bulk. Solutions to this recurrent problem have been: (1) reactivating metal surfaces by heating in vacuo so that the impurity elements (oxygen, nitrogen, and carbon) diffuse into the bulk, thus removing the surface barrier; (2) repeating the hydrogen exposure to pulverize the material through embrittlement, which produces a large fresh surface area; or (3) cleaning the specimen by sputtering or heating in an ultra-high vacuum and then thinly coating it with palladium to protect the surface but allowing rapid hydrogen diffusion to occur. Each of these methods has disadvantages from an experimental viewpoint, and they might not be suitable in a technological ap-plication.

Techniques have been developed only in the last decade for elemental ex-amination of solid surfaces. The word surface should be defined before a dis-cussion of the available tools is meaningful. Strictly, the surface is the outermost monolayer of atoms, ions, or molecules on a solid. Although this monolayer is the most important with regard to external chemical or mechanical attack, ma-terials exposed to typical atmospheres have much thicker coverings, ranging from fractions of a nanometer to a few micrometers in depth.

Because electron-spectroscopic and ion-scattering methods yield information about the first seven atom layers, applying these techniques to metal–hydrogen kinetic problems requires incorporating depth-profiling capability. That is, an argon-sputtering gun must be incorporated into the analysis system to remove undesired surface material up to several nanometers deep.

Available Surface Techniques

During the last decade more than 25 electron-spectroscopic and ion-scat-tering techniques have been developed, and more than 56 surface techniques were tabulated in 1973 (*1*). It is impossible to describe in a few pages all of these varied methods.

In Table I, well-known techniques are listed by excitation source and mea-sured emission. Although mentioned there, x-ray fluorescence (XRF), electron microprobes (EMP), and scanning electron microscopes (SEM) will be excluded here because they are not sensitive to the surface, which is generally considered to be 2.5–5.0 nm deep.

Two texts are good sources of background information; one is edited by A. W. Czanderna (*2*) and the other by P. E. Kane and G. B. Larabee (*3*). If the reader has a deeper interest in the subject, the primary journals in the field are *Surface Science* and the *Journal of Vacuum Science and Technology*. Adequate descriptions of the equipment are available in the introductory reviews.

The category of electron spectroscopy includes those instruments that measure intensity of ejected secondary electrons as a function of their energy. These electrons have a very short, mean free path in solids and therefore are in-herently sensitive to the first few atomic layers on the surface.

Table I. Better-Known Spectroscopic Techniques According to the Excitation Source and Measured Emission[a]

Source / Emission	hν	e⁻	Ion
hν	XRF	EMP	
e⁻	XPS (ESCA) UPS AES	SEM AES EM	SEM
Ion			IMMA SIMS ISS (RS)

[a] The acronyms are: (XRF) x-ray fluorescence, (XPS) x-ray photoelectron spectroscopy, (UPS) uv photoelectron spectroscopy, (AES) Auger electron spectroscopy, (EMP) electron microprobe, (SEM) scanning electron microscopy, (EM) electron microscopy, (IMMA) ion microprobe mass analyzer, (SIMS) secondary ion mass spectrometry, (ISS) ion scattering spectrometry, and (RS) Rutherford scattering.

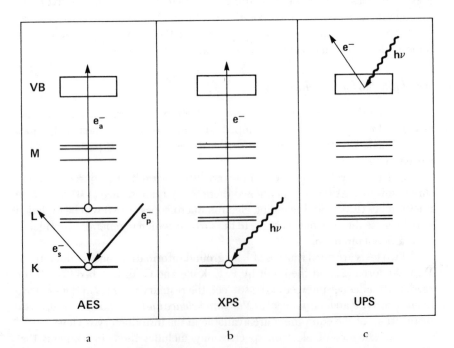

Figure 1.   Electronic processes in: (a) Auger electron spectroscopy (AES), (b) x-ray photoelectron spectroscopy (XPS), and (c) uv photoelectron spectroscopy (UPS)

Electron Spectroscopy. The energy of these ejected secondary electrons and the accompanying chemical information depends upon the excitation source selected and its energy. To probe the innermost electrons of an atom, bombardment energies of several kilovolts (1–10 keV) are given to electrons in Auger-electron spectroscopy (AES) and to photons in x-ray photoelectron spectroscopy (XPS). In uv photoelectron spectroscopy (UPS), the photon energy is only a few electron volts; the helium resonance at 21.2 eV is the most energetic source commonly used. Because of this low excitation energy, UPS obtains its information from the less tightly bound valence electrons near the Fermi level. Thus, it is very sensitive to the chemical state of the species in question.

The following briefly describes important electron spectroscopic techniques and is an introduction for persons who are relatively unfamiliar with these tools of modern surface science but who are interested in their application to metallic or organometallic hydrides.

AUGER ELECTRON SPECTROSCOPY. Even though Pierre Auger (Oh-zhay) discovered the electronic effect named after him in 1925 (4), it was not until the late sixties that Harris (5) demonstrated that electronic differentiation could give acceptable sensitivity to Auger spectra, and other persons (6) demonstrated the feasibility of using more readily available, low-energy electron diffraction (LEED) optics for electron energy analysis. Soon afterward, Palmberg and co-workers (7) produced the even more sensitive cylindrical mirror analyzer.

The Auger electron emission process is illustrated in Figure 1a. The excitation source in a typical spectrometer is a 1–100 μm diameter electron beam with an energy of 1–5 keV and a current of 1–80 μA. A minute fraction of these primary electrons eject inner core electrons from atoms in the surface region of the sample. An electron from a higher energy level drops down to this void, releasing energy as x-ray fluorescence or ejecting a higher-level electron, the Auger electron. In lighter elements, atomic number ≤ 33 (arsenic), the Auger process predominates whereas x-ray fluorescence prevails as the atomic mass increases. Consequently, the technique is particularly suitable for detecting reactive low Z elements such as carbon, oxygen, and nitrogen. An energy analyzer detects the kinetic energy of the ejected electrons and their intensity. This signal is amplified electronically and differentiated, resulting in an Auger spectral plot. Because there are relatively few Auger electrons produced compared with those from other sources, this signal must be differentiated to extract a sizable peak from the large electronic background. Since most Auger processes result in electrons with kinetic energies between 20 and 2000 eV, commercial instruments are designed to operate in this range.

An advantage of the scanning Auger microprobe (SAM) is that it can scan a surface and provide a map of each element in question. As such, these results are similar to an electron microprobe except the SAM is much more surface sensitive.

C. Chang wrote a well-known review article (8) in 1971 that covers AES fundamentals. Also, there is a chapter by A. Joshi et al. in Czanderna's book and

Chang's contribution in Kane's book. All three are recommended as introductions to the method. A comprehensive review of AES and XPS was written recently by Carlson (9), but it is quite sophisticated and does not emphasize solid surfaces. The novice surface scientists should reserve this important text for later reading.

ELECTRON ENERGY LOSS SPECTROSCOPY (ELS). Optical surface phonons caused by long wavelength charge density fluctuations in solids can be detected by measuring inelastically scattered electrons using a spectrometer with resolution greater than 30 meV. High-resolution electron, energy-loss spectra (HRELS) (10) can provide orientational information of gas adsorbates on surfaces. About a 5–10 meV resolution is needed for these studies. The number and location of the loss peaks are related to the polarity and position of the adsorbate. The energy range is in the ir and is, consequently, a surface vibrational-type spectroscopy. Peaks correspond to the various vibrational modes of the adsorbates themselves as well as the effects caused by the interaction with the surface; e.g., dissociative or molecular adsorption. Combining HRELS with low-energy electron diffraction provides a powerful tool to analyze microscopic and long-range structure on a surface.

PHOTOELECTRON SPECTROSCOPY. As a subdivision of electron spectroscopy, photoelectron or photoemission spectroscopy (PES) includes those instruments that use a photon source to eject electrons from surface atoms. The techniques of x-ray photoelectron spectroscopy (XPS) and uv photoelectron spectroscopy (UPS) are the principles in this group. Auger electrons are emitted also because of x-ray bombardment, but this combination is used infrequently.

Variations in fixed-angle PES are increasing in prominence. Low-photon incidence angles increase the surface sensitivity of PES. Although the photon beam penetrates to the same depth, the component normal to the surface is much less at small incidence angles.

Photoelectron emission intensity is a function of the incident photon energy and kinetic energy of the emitted electrons for a specific collection geometry and polarization (11). A map of the surface thus can be obtained, yielding information on d-band structure, adsorbate orbital symmetries, and geometric structure. To obtain the necessary data, emitted electrons are collected at small, known solid angles and energy windows. Since synchrotron radiation is required because of its intensity and continuous nature, the present availability and usefulness of the modification is limited. This technique is designated by the term angle-resolved (ARPES, ARUPS, ARXPS).

X-RAY PHOTOELECTRON SPECTROSCOPY. During the 1960s, Kai Siegbahn and co-workers (12) developed a surface-sensitive, high-resolution photoelectron spectroscopy that used x-rays as the irradiation source. He designated it as Electron Spectroscopy for Chemical Analysis or ESCA. The generic denotation for x-ray photoelectron spectroscopy of XPS is more accurate than ESCA and is preferred for clarity and for indicating the relationship of UPS.

Figure 1b graphically illustrates the atomic XPS process. A monochromatic x-ray source is focused on a specimen. Secondary electrons are ejected and their kinetic energy and intensity are monitored by an electron-energy analyzer. These data are processed, forming spectra. Because binding energies are of greater interest than kinetic energy, most spectra include such a scale that is simplistically the x-ray source energy minus the electron kinetic energy. Strictly, the work function of the spectrometer and the electron recoil energy should also be subtracted, but they are comparatively small. Many spectrometers perform this arithmetic internally and plot the spectra directly in terms of binding energy.

X-ray sources on most instruments use either K_α peaks of magnesium (1254 eV) or aluminum (1487 eV). The K_α peaks of light elements have a smaller full width at half maximum (FWHM), 0.7–0.9 eV, than those from the more energetic heavy elements, such as copper and molybdenum. That is, they are almost monochromatic and yield more narrow photoelectron lines, which are a measure of the spectrometer resolution. These two sources are suitable particularly because the x-rays produced have sufficient energy to excite electrons below 1000 eV, the region where most of the useful photoelectron peaks occur.

Three outstanding features of XPS exist. First, because the excitation source is photonic, there is no measurable damage to the surface through sputtering or impurity deposition, such as carbon, caused by the electron beam in AES. Second, the mean free path of electrons in metals and their oxides is less than 2.5 nm and in organic materials (13) is less than 10 nm. Thus, the technique can be sensitive to the outer monolayers, depending upon the skill of the operator and the quality of his equipment. Third, the technique can detect chemical shifts, i.e., changes in peak location, that result from the relative changes in oxidation state of an atom. Through calibration we can determine the chemical state of surface species. The chemical shift is a change in electron binding energy resulting from the modified shielding effect of the atomic electron cloud as electrons are added or removed through chemical interaction. For further reading, the chapter by Riggs and Parker in Czanderna's book (2) and the chapter by Hercules in Kane's book (3) are recommended. David Hercules published similar material in 1970 (14).

UV PHOTOELECTRON SPECTROSCOPY. As mentioned earlier, uv photoelectron spectroscopy (UPS) is similar to XPS but differs in the photon source used. Since the emission lines of hydrogen and the rare gases are of suitable energy, discharge lamps containing these gases are ideal for the task. They are also desirable because the radiation is very monochromatic and intense. Resonance radiations of He(I) (21.2 eV) and He(II) (40.8 eV) and used more commonly since they are the most energetic of gaseous emissions. The advantage of UPS is that it probes valence electronic orbitals having low binding energy, which are the ones most influenced by chemical bonding; so potentially it can yield more chemical information than XPS can. Because of the high flux of uv radiation and the low escape depth of the emitted electrons, the technique results in high signal strength and sensitivity.   It is also less destructive than XPS because of its lower excitation energy.   For comparing the outer techniques, Figure 1c shows the UPS excitation and emission process.

The bulk of UPS literature has appeared since 1970, and it has concentrated in two areas: (1) interpreting spectra of organic vapors and (2) studing transition metal electronic band structure changes caused by sorption of simple gases on the metal surfaces.   Interpreting uv photoelectron spectra is very difficult and until a data base of spectral measurements is accumulated, it will be used infrequently in surface chemical studies.

Currently there are no reviews of the UPS literature nor suitable reviews of available instruments.   The only similar article is a somewhat sophisticated review by Eastman (15) that theoretically interprets photoemission spectra of metals and instrumentation that was available in 1971.

Ion Spectrometry.   Ion beams offer another way to probe solid surfaces. The size of an ion compared with an electron has both an advantage and a disadvantage.   Because of its large size, an ion such as helium or argon cannot penetrate as deeply into a sample as an electron or photon of comparable energy, and the same is true for ejection.   As a consequence, techniques using ions as excitation or emission sources should be sensitive to the surface monolayer alone. However, because ions have a mass comparable with the surface atoms, kinetic energy exchange can occur, causing significant sputtering.

These two points separate the two common ion spectrometries.   Ion scattering spectrometry (ISS) describes the energy measurement of elastically scattered primary ions from a surface.   Secondary ion mass spectrometry (SIMS) examines the sputtered fragments from the surface, so a mass spectrometer is substituted for the energy analyzer in ISS.   A general disadvantage of ion spectrometry is that sputtering always will occur and is not independent of the method as it is in the electron spectroscopies.   Excellent introductions to both techniques are included in Czanderna's book (2), with a chapter by Buck on ISS and one by McHugh on SIMS.

ION SCATTERING SPECTROMETRY.   Although scattering of low-energy ions from surfaces first was studied more than 20 years ago, it was not until 1957 that Smith (16) used inert gas ions on molybdenum and nickel, thereby demonstrating the potential of the method for surface analytical investigations.

The primary advantages of ISS are the location of adatoms by shadowing, the study of matrix effects in the yield spectra, the identification of isotopes of low Z elements, and the inherent ability of probing only the surface monolayer. These potential benefits are offset, however, by the sputtering effect of the ions, even at the low energies involved (0.5-3 keV).

The ISS technique is quite simple.   A monoenergetic, well-collimated beam of a selected noble gas ion (^{3}He, ^{4}He, ^{20}Ne, or ^{40}Ar) strikes the target surface.   The energy and the number of scattered primary ions is measured at a constant angle of reflection.   If the scattering angle is fixed at 90°, the ratio of the scattered ion energy, E_1, to incident ion energy, E_0 is given by the simple expression:

$$\frac{E_1}{E_0} = \frac{M_2 - M_1}{M_2 + M_1}, \tag{1}$$

where M_2 is the surface atom mass and M_1 is the incident ion mass. Then, ion yield spectra are obtained for a given signal intensity as a function of the reduced energy E_1/E_0. Thus, the analysis does not require removing a surface as in SIMS, but sputtering will occur because of the ion bombardment.

More than one probe gas usually is used to study a given specimen because, as shown by Equation 1, the probe atomic mass cannot be less than the sample atomic mass, and the peak resolution decreases as the difference between these masses increases. Semi-quantitative data is received if the instrument is calibrated with samples of known surface composition. Qualitative data is available readily from using Equation 1.

SECONDARY ION MASS SPECTROMETRY. The preliminary development of SIMS began much earlier than the other surface techniques, starting with negative ion formation studies by Arnot and co-workers (*17*) in the late 1930s. It was the late 1950s, however, before SIMS was considered seriously as a surface analytical tool. Surface compositional analysis by SIMS uses the same type of bombarding ion beam as in ISS. Incident ion energies are usually higher (>5 keV) than in ISS, giving a higher secondary ion yield per primary ion. However, the sputtering rate still can be controlled by varying the ion current density that strikes the target. A fraction of the sputtered surface atoms are ejected as positive or negative ions. Their intensity vs. mass is monitored, producing spectra for both positive and negative ions.

There are four basic components of a SIMS instrument: a primary ion source with beam conditioning, a system energy filter, a mass-to-charge analyzer, and an ion detector. In many systems, the ion gun used for routine sputter cleaning is modified and functions also as the SIMS primary ion source. The analyzer and detection system often is a quadrupole residual gas analyzer. The angles of the mass analyzer segment to the target and to the ion source are not critical for analysis as they are in ISS since sputtered species will come out at many angles.

Because the technique is essentially a mass spectrometer, the detection sensitivity can be quite high, approaching 1 ppb for many elements. This advantage, the capability of analyzing all elements and readily distinguishing between isotopes, and the built-in depth-profiling ability make SIMS a very powerful qualitative surface probe.

Comparison of the Major Surface Techniques

No single surface-analysis tool can solve every problem facing the surface scientist. For this reason, the inclusion of two or more techniques in one system is useful. If the researcher expects to attack certain problems, he can select a combination of techniques such that the strengths of one will overlap the weaknesses of the others. With a surface laboratory that provides analytical services for various research or applied efforts, more versatile techniques, such as AES or XPS, might be selected. Adding a new technique to an existing system or to one being planned usually does not increase greatly the investment when compared with the enlarged capability. For example, adding a helium discharge lamp to an XPS system with a suitable analyzer easily allows for collecting uv photoelectron spectra; adding a beam conditioner to an ion-sputter gun and an energy filter on a quadrapole mass analyzer gives one SIMS capability. Finally, including two or more instruments in the same vacuum chamber alleviates problems of air exposure when transferring a sample to another system and reduces the uncertainties so that different samples studied at facilities by other experimentalists can be compared.

To illustrate the appearance of AES, XPS, and UPS spectra, we have reproduced in Figure 2 (18) spectra of nickel and nonstoichiometric nickel oxide that were taken with the three techniques. It also illustrates different and complementary information available from a multi-technique analysis on one sample.

Both oxygen and nickel are clearly present in the Auger and XPS spectra. UPS shows a broad valence electron distribution because of the bonding interaction of the nickel and oxygen. Examining the XPS and Auger spectra of clean nickel shows the enhancement of the nickel peaks as the oxygen disappears. Also, the nickel $2p_{3/2}$ peak energy decreases as the nickel goes to its zero oxidation state. The UPS shows that bonding electrons concentrate near the Fermi level in nickel.

A comparable spectrum of polycrystalline nickel that was taken with an ISS by Goff and Smith (19) (Figure 3) shows the presence of carbon and oxygen on the surface, resulting from adsorbed hydrocarbons, CO, or possibly nonstoichiometric NiO_x, as indicated by the other methods.

Generalizations can be made concerning common characteristics of the surface techniques we have discussed. All are sensitive to the complete set of elements from lithium to uranium. SIMS is the only technique that can detect hydrogen and helium, except for very low angle scattering by ISS that has been done only by Brongersma. As stated earlier, the ion spectrometries have a built-in sputtering effect, so they are destructive to some degree. AES, XPS, and UPS are nondestructive although some small effects, negligible for most purposes, do occur. Any time that sputtering is used, sample consumption is unavoidable. All five techniques discussed above can measure surface concentrations, but in each case, comparative standards are required to establish relative sensitivities of each technique to the elements of interest. Sputtering rates must be determined for quantitative depth profiling and for using SIMS as a quantitative tool.

Direct chemical information is obtainable only from XPS and UPS, with XPS by far the easier to interpret. Similar information can be determined indirectly by an experienced investigator using AES by noting small peak shifts

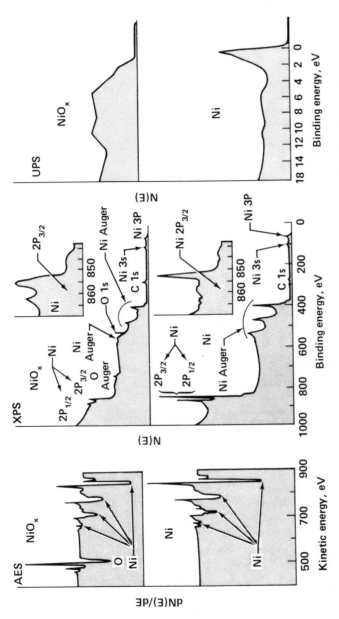

Figure 2. AES, XPS, and UPS spectra of a nickel metal and an oxidized nickel surface

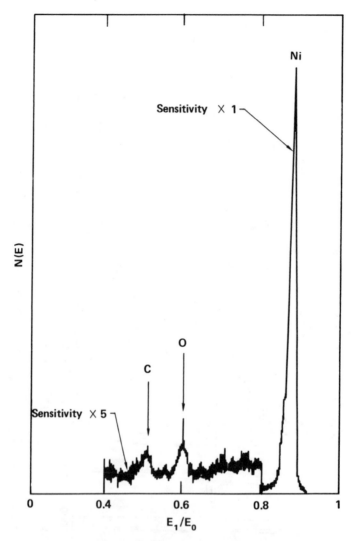

Figure 3. An ISS spectrum of He⁺ scattered from polycrystalline nickel

(<1 eV) and changes in peak shape. For some systems, the energy dependence of the ISS yield spectra change for surface atoms in different chemical environments; recent work in this area shows that ISS can produce considerable chemical information. Ionic fragments obtained in SIMS give some clues to chemical combinations. The other factors are listed more easily according to the technique:

   AES—It is very sensitive to the lighter elements. The electron beam can enhance impurity deposition at the surface spot being studied, e.g., carbonization.

Quantification is difficult and detection is limited to about 0.1% of the surface concentration. As with XPS, the higher energy of the secondary electrons makes the technique less surface sensitive. However, it is used widely and is a good all-around technique.

XPS—For chemical bonding information this technique is unsurpassed, but it is also limited to 0.1% sensitivity. Much fewer surface contamination problems exist with the x-ray source here than that in AES, but this technique is less surface sensitive (1–10 nm) than AES.

UPS—Photoelectron spectra at these low energies are more surface sensitive than XPS, and consequently, the signals are more intense at an equivalent excitation intensity. Interpreting the spectra is still difficult.

ISS—Low energy ion scattering is very sensitive to the surface monolayer; it can resolve low Z isotopes and has a detection limit of 10 ppm to 1%, depending on the instrument. Depth profiling is a long operation unless the probing beam also is supplemented with a separate ion gun.

SIMS—The detection limits are by far the best of the five techniques and is the only one that offers isotopic resolution for all elements. Quantification is difficult because of the wide variation in sputtering yields produced by different ion beams.

Application of Surface Analysis to the Kinetics of Hydride Formation

Understanding the influence of surface impurities on the formation of solid metal hydrides is one of the most important problems facing the hydride community. Participants at an ERDA-sponsored Hydride Research Meeting on April 29 and 30, 1976, concluded that surface studies were the first priority in technical application and second in scientific interest (20).

Even though authorities in the field of hydride research have emphasized the importance of surface studies, literature on the subject almost is nonexistent. To the best of our knowledge, only a few papers address the subject directly. In 1974, Zuechner examined tantalum surfaces with SIMS (21). Tantalum metallic, oxide, and carbide peaks were compared with as-received oxide-coated foil, sputter-cleaned oxide-coated foil, and cleaned palladium-coated foil. Palladium effectiveness as a surface protector had been shown earlier (22), and the later study investigated the protective mechanism. One each of the tantalum-positive and tantalum-negative secondary ion spectra taken by Zuechner are shown in Figure 4. Several impurities are present in both. Depth profiles were recorded for tantalum, oxide, and carbide ions. Although sputtering rates were not determined, carbide and oxide species disappeared as the surface was removed. The impurity coating was estimated to be 100 atomic layers deep.

Van Deventer and co-workers (23) recently investigated the thickness of vanadium oxide on as-received metal and heated specimens. Oxide thicknesses of up to 10 μm were observed, depending on the treatment. Hydrogen pretreatment seemed to increase the hydrogen permeability through the specimens. Similar effects of hydrogen pretreatment frequently have been seen.

Results of tritium diffusion into oxide-covered niobium (24) indicate that the release rates are affected by the oxide thickness. The release rate through

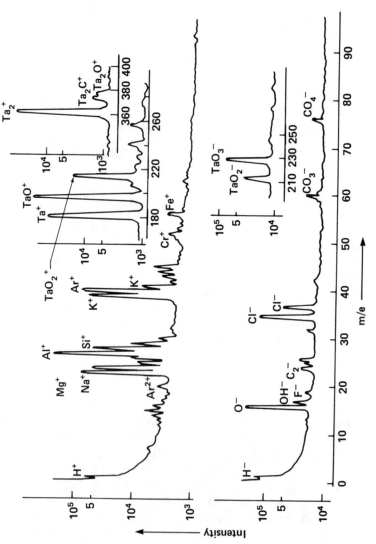

Figure 4. Positive and negative secondary ion mass spectra (SIMS) of tantalum

the films was lower than that for the bulk metal, an expected result. The AES used to determine the film thickness was imprecise since the sputtering rates were not known and specimen handling was not well defined. The reaction rate of deuterium with thin films of titanium has been reported by Malinowski (25). Exposing freshly evaporated titanium to carbon monoxide reduced the deuterium pumping speed relative to an unexposed surface. There is a voluminous amount of literature containing studies of gas adsorption on metal surfaces and surface compounds. Even though they do not address the present problem directly, useful information can be derived by reviewing the significant papers.

Solid metal hydrides specifically have been reviewed here, but XPS and UPS can serve as tools to study vapors or volatile liquids. Much of the original work with these two methods involved organic molecules; only later were solid surfaces studied. Therefore, they should always be considered as helpful analytical instruments for examining the bonding chemistry of organometallic compounds. This symposium covered mainly organometallic hydrides, and they are prime candidates for photoelectron spectroscopy study.

There are two important criteria that must be observed when designing experiments in this area. First, surface analyses should be performed on the same samples that are used in diffusion or permeability experiments. More importantly, these experiments should be carried out in situ to avoid the slightest chance of air contamination, even with argon transfer or other external movement. A processing chamber should be connected to the work chamber of the analysis system. Such an approach would allow for treating a sample with varying amounts of different gases; e.g., oxygen and carbon monoxide, as a function of pressure and temperature. Surface analyses before and after such treatment could be compared with hydrogen diffusion measurements in the same processing chamber.

Such experiments should reveal the extent and mechanism of passivation in various metal systems. These phenomena, when understood, can help to prevent the problem of nonreactivity of many metals and alloys in applied systems. We have begun experiments of this type.

The selection order of the techniques that would be most useful to hydride studies varies according to the particular experiment planned. One possible order of preference would be the following.

(1) AES—This technique is the most commonly available and will determine the elemental composition of the surface.

(2) SIMS—The presence of hydrogen and its isotopes can be determined and some chemical information can be derived from the kinds of ionic fragments sputtered from the surface.

(3) XPS—Although this technique possibly could give only hydrogen interaction data with light elements, its great potential is its ability to provide chemical bonding information about the metal surface.

(4) ISS and UPS—Preference of one of the these techniques over another is difficult to determine. ISS could yield surface structural comparisons with

the bulk, and UPS, because of its valence electron sensitivity, might detect hydrogen interactions with a metal surface.

Conclusion

We briefly reviewed important modern surface techiques and indicated their application to metal–hydrogen studies. Experimental investigations in this area should be significant for solving both fundamental and applied problems.

Acknowledgment

The authors thank A. W. Czanderna for his thoughtful criticism of the manuscript. We also wish to thank the authors at Physical Electronics Industries, Inc., and the publishers of I.R. and of the American Institute of Physics for their kind permissions to reproduce an illustration.

Literature Cited

1. Dylla, H. F., "Comparison of Techniques for Surface Measurements," *Electron. Quart. Progr. Rept.* (1973) *110*, M.I.T. Research Lab.
2. A. W. Czanderna, Ed., "Methods of Surface Analysis," Elsevier, Amsterdam, 1975.
3. P. F. Kane, G. B. Larabee, Eds., "Characterization of Solid Surfaces," Plenum, New York, 1974.
4. Auger, P., *J. Phys. Radium* (1925) **6**, 205.
5. Harris, L. A., *J. Appl. Phys.* (1968) **39**, 1419.
6. Weber, R. E., Peria, W. T., *J. Appl. Phys.* (1967) **38**, 4355.
7. Palmberg, P. W., Bohn, G. K., Tracy, J. C., *Appl. Phys. Lett.* (1969) **15**, 254.
8. Chang, C. C., *Surf. Sci.* (1971) **53**, 25.
9. Carlson, T. A., "Photoelectron and Auger Spectroscopy," Plenum, New York, 1975.
10. Froitzheim, H., "Electron Energy Loss Spectroscopy," *Topics in Current Physics, Vol. 4: Electron Spectroscopy for Chemical Analysis,* H. Ibach, Ed., Springer, New York, 1977.
11. Lapeyre, G. J., Smith, R. J., Anderson, J., *J. Vac. Sci. Technol.,* (1977) **14(1)**, 384.
12. Siegbahn, K. et al., "ESCA, Atomic, Molecular, and Solid State Structure Studied by Means of Electron Spectroscopy," Almquist and Wiksells, Eds., Uppsala, Sweden, 1967.
13. A. W. Czanderna, Ed., "Methods of Surface Analysis," p. 104, Elsevier, Amsterdam, 1975.
14. Hercules, D. M., *Anal. Chem.* (1970) **42**, 20A.
15. Eastman, D. E., "Techniques of Metals Research VI," E. Passaglia, Ed., Chapter 6, Interscience, New York, 1971.
16. Smith, D. P., *J. Appl. Phys.,* (1967) **38**, 340.
17. Arnot, F. L., Milligan, J. C., *Proc. R. Soc. London, Ser. A* (1936) **156**, 538.
18. Riach, G. B., Riggs, W. M., *Ind. Res.* (1976) **18 (10)**, 74.
19. Goff, R. F., Smith, D. P., *J. Vac. Sci. Technol.* (1970) **7**, 72.
20. "Hydride Research, Summary of Research and Discussions," U.S. ERDA Div. Phys. Res., April, 1976.
21. Zuechner, H., Boes, N., *Z. Phys. Chem., N.F.* (1974) **93**, 65.
22. Boes, N., Zuechner, H., *Phys. Status Solidi A* (1973) **17**, K111.

23. Van Deventer, E. H., Renner, T. A., Pelto, R. H., Maroni, V. A., *J. Nucl. Mater.* (1977) **64**, 241.
24. Chandra, D., Elleman, T. S., Verghese K., *J. Nucl. Mater.* (1976) **59**, 263.
25. Malinowski, M. E., *J. Nucl. Mater.* (1976) **63**, 386.

RECEIVED July 29, 1977.

AUTHOR INDEX

Names and numbers in bold are authors and first pages, respectively, of complete chapters. Numbers in italics show the page on which the complete reference is listed. Numbers in parentheses are reference numbers which indicate that an author's work is referred to, followed by pages of that referral.

A

Abel, E. W.   *59* (48) 52
Abragam, A.   *262* (23) 251,256,257
Abrahams, S. C.   *58* (6) 37; *72* (11) 62,63; *90* (18) 74,75,87,88; *91* (28) 76,87; *213* (10) 202; *380* (16) 369
Achiwa, K.   *134* (4) 122
Adams, D. M.   *246* (1) 233
Adams, R. D.   *59* (12) 37,58
Adamson, A. W.   *200* (54) 197
Adkins, C. M.   *311* (18) 308
Ahmad, N.   *200* (25) 183,288,294,295; *301* (50) 297
Albano, V. G.   *9* (3) 1, (11) 3, (22) 4; *10* (29) 6, (30) 6; *91* (30) 77,87, (31) 77,86; *110* (6) 94, (16) 94,95; *213* (13) 203
Alefeld, G.   *300* (23) 287,295, (24) 287,294,295, (25) 287,288,294,295; *301* (50) 297
Alire, R. M.   382
Alt, H. G.   *159* (27) 155
Amano, M.   *300* (13) 285
Ambrose, H. A.   *301* (37) 288
Ames, G. R.   *135* (53) 132
Anantharaman, T. R.   *72* (39) 71
Anderson, D. W. W.   *169* (14) 164
Anderson, J.   *396* (11) 386
Anderson, J. L.   *341* (21) 334, (32) 336,337
Anderson, J. S.   *300* (5) 285
Andrews, M.   *25* (22) 20, (23) 20; *59* (32) 40; *72* (22) 63; *91* (59) 88; *246* (19) 235,238, 240,243
Andrews, M. A.   215; *230* (7) 215, 219, 220, 221,227, (13) 219,221-224, (14) 219,220; *246* (36) 236; *262* (37) 256,258,260
Anisimov, K. N.   *159* (15) 151
Anker, W. M.   *9* (22) 4; *110* (16) 94,95
Aresta, M.   *199* (10) 182
Arnot, F. L.   *396* (17) 389
Arnott, R.   *380* (9) 368
Artman, D. D.   *365* (28) 346
Aslanov, L.   *91* (43) 81
Aslett, T.   *365* (31) 346
Ashby, E. C.   *380* (9) 368
Aston, J. G.   *301* (38) 288
Auger, P.   *396* (4) 385

Axelrod, M.   *135* (41) 129
Ayron, P. A.   *25* (20) 18
Azarkh, Z. M.   *364* (4) 343

B

Bailey, D.   *262* (39) 256
Bailey, M. F.   *25* (18) 18
Bailey, P. M.   *120* (26) 120
Baird, P. H.   *247* (37) 236
Baker, R. W.   *58* (7) 37, (8) 37,56
Ballentine, R.   *365* (45) 349
Balzani, V.   *199* (9) 181
Baranowski, B.   *300* (10) 285,287, (20) 287, (21) 287, (22) 287; *301* (28) 287
Barefield, E. K.   *148* (2) 137
Barker, G. K.   *120* (6) 111
Barnaal, D. E.   *262* (25) 251
Barnard, C. F. J.   *200* (39) 190,197
Barnes, G. H.   *120* (16) 117
Barnes, R. G.   *262* (13) 249,252
Basch, H.   *214* (22) 207
Bateman, L. R.   *10* (34) 9; *110* (12) 94
Bau, R.   9 (4) 1, (10) 3; *10* (25) 5,8; *25* (10) 15,18, (11) 15,18, (13) 12, (15) 17,20, (22) 20, (23) 20, (24) 20; *59* (19) 38, (20) 38, (21) 38,39, (24) 38, (25) 38, (26) 38, (32) 40; *60* (52) 54, (63) 58; **61;** *71* (5) 62; *72* (17) 63, (18) 63, (22) 63, (23) 63, (24) 63, (25) 63, (26) 62, (27) 62; **73;** *91* (34) 80,88, (44) 81,87,88, (45) 83,87, (46) 84,87,88; (51) 85,88, (52) 87, (59) 88; *92* (61) 88, (62) 88, (69) 82, (70) 85,86; *110* (29) 105; *213* (11) 203, (12) 203; *230* (8) 216, (23) 225; *231* (30) 227; *246* (18) 235,238, (19) 235, 238,240,243, (22) 235,243; *247* (37) 236, (46) 238; *262* (35) 256, 260; *263* (43) 256
Baumann, H.   *110* (14) 94
Bechman, C. A.   *326* (9) 313; (11) 313,318; *341* (25) 337
Beck, R. L.   *310* (12) 304; *341* (24) 337
Beck, W.   *230* (4) 215
Bell, A. P.   *148* (10) 137,139
Bell, B.   *90* (2) 73, (11) 73
Bell, L. G.   *148* (17) 139

399

SUBJECT INDEX

413

The text of this book is set in 10 point Highland with two points of leading. The chapter numerals are set in 26 point Times Roman; the chapter titles are set in 18 point Baskerville Bold.

The book is printed offset on Text White Opaque, 50-pound. The cover is Joanna Book Binding blue linen.

Jacket design by Linda Mattingly. Editing and production by Saundra Goss.

The book was composed by the Mack Printing Co., Easton, PA., printed and bound by The Maple Press Co., York, PA.